Trends in Applications of Mathematics to Mechanics

Proceedings of the 7th Symposium,
Held in Wassenaar, The Netherlands,
December 7-11, 1987

Editors: J. F. Besseling and W. Eckhaus

With 62 Figures

Springer-Verlag Berlin Heidelberg New York
London Paris Tokyo

Professor Dr. Johannes F. Besseling
Technological University, Mekelweg 2, NL-2628 CD Delft, The Netherlands

Professor Dr. Wiktor Eckhaus
Rijksuniversiteit Utrecht, Mathematisch Instituut, Budapestlaan 6,
NL-3584 CD Utrecht, The Netherlands

ISBN-13: 978-3-642-73935-4 e-ISBN-13: 978-3-642-73933-0
DOI: 10.1007/978-3-642-73933-0

Softcover reprint of the hardcover 1st edition 1988

2155/3150 - 543210 – Printed on acid-free paper

Trends in Applications of Mathematics to Mechanics

Preface

The 7th Symposium on Trends in Applications of Mathematics to Mechanics was the continuation of a series of meetings held every two years under the auspices of the International Society for Interaction of Mechanics and Mathematics. The scientific committee consisted of G. Iooss, W.T. Koiter and the organizers. The symposium was attended by 53 participants from 15 countries and a total of 35 lectures were delivered. The presentations were divided into the following four sections on specific themes:

I. Perturbation methods and applications
II. Instability, bifurcations and transition to chaos
III. Multibody dynamics and control
IV. Mechanics and mathematics of non-classical materials.

Furthermore, there was a fifth "free" section, meant to introduce recent new interactions of mathematics and mechanics. We wish to thank all participants for their contributions to the lectures and the discussions. We also gratefully acknowledge the financial support from the following sources: Afdeling Mechanica (Division of Mechanics of the Royal Institute of Engineers in the Netherlands), Olivetti Netherlands, Shell International, Wiskundig Genootschap (Dutch Union of Mathematicians).

Delft, Utrecht
May 1988

J.F. Besseling
W. Eckhaus

The International Society for
Interactions of Mechanics and Mathematics
mourns the death of the Honorary Member

W.K. NOWACKI

and the Ordinary Members

R.D. MINDLIN

YA.B. ZEL'DOVICH

Contents

Part III Multibody Dynamics and Control

Part IV Mechanics and Mathematics of Non-Classical Materials

Part V New Interactions of Mathematics and Mechanics

Part I

Perturbation Methods and Applications

Asymptotic Analysis of Stabilizing Fronts

A. van Harten

Data Processing Division, Rijkswaterstaat and
Mathematical Institute, University of Utrecht, The Netherlands

Singular perturbation problems of dynamic free boundary type give rise to interesting phenomena, such as moving fronts. Here we analyse the evolution of fronts in non-linear diffusion problems in situations where in the long run the front stabilizes at its stationary position. This is done by constructing asymptotic approximations of the solution using the technique of matched asymptotic expansions. Four cases are compared, where the main difference lies in the behaviour of the non-linearity at the front and the order of the jump occuring at the front. In some cases the dynamics of the moving front is governed by an internal driving mechanism induced by the layer at the front. In other cases the outer solution produces the main effect of the dynamics in an external way.

0. Introduction

Some recent applications of singular perturbation theory are found in the field of free boundary problems. For example it is used in the theory of premixed flames with high activation energy, cf. [1], [2]. Another example is the propagation of shocks with small viscosity, cf. [3]. In these examples solutions behave like a travelling front for all time.
Here we consider a number of non-linear diffusion problems where the stationary solution has a free boundary and the time evolution from a generic class of initial conditions gives rise to a front moving to the location of the stationary free boundary in the long run. Thus the stability of the stationary free boundary is shown in an asymptotic sense.

In the sequel we shall deal with the following cases:

$$(1)\quad \begin{cases} \dfrac{\partial u}{\partial t} = \varepsilon^2 \dfrac{\partial^2 u}{\partial x^2} - u + f(x)\, H(u-1) \\[2ex] u|_{t=0} = \psi \end{cases}$$

$$(2)\quad \begin{cases} \dfrac{\partial u}{\partial t} = \varepsilon^2 \dfrac{\partial^2 u}{\partial x^2} - f(x,u) \\[2ex] u|_{t=0} = \psi \quad ; \quad \dfrac{\partial u}{\partial x}\Big|_{x=0} = \dfrac{\partial u}{\partial x}\Big|_{x=1} = 0 \end{cases}$$

$$
(3) \quad \begin{cases} \dfrac{\partial u}{\partial t} = \varepsilon^2 \dfrac{\partial^2 u}{\partial x^2} - \dfrac{u}{1+u} + f(x) \\ u|_{t=0} = \psi \quad ; \quad \dfrac{\partial u}{\partial x}\Big|_{x=0} = \dfrac{\partial u}{\partial x}\Big|_{x=1} = 0 \end{cases}
$$

$$
(4) \quad \begin{cases} \left(\dfrac{\partial u}{\partial t} = \varepsilon^2 \dfrac{\partial^2 u}{\partial x^2} + u - f \right) . \; u = 0 \\ u \le 0 \quad , \quad \dfrac{\partial u}{\partial t} - \varepsilon^2 \dfrac{\partial^2 u}{\partial x^2} + u - f \le 0 \\ u|_{t=0} = \psi \quad ; \quad \dfrac{\partial u}{\partial x}\Big|_{x=-1} = \dfrac{\partial u}{\partial x}\Big|_{x=1} = 0 \end{cases}
$$

The notation H refers to the Heaviside function.

The diffusivity ε^2 is supposed to be small and as a consequence (1) - (4) are singular perturbation problems.

Each of the problems (1) - (4) has an interpretation in terms of applications.
The first problem (1) is a simple model arising in plasma physics, cf. [4], [5]. The second problem (2) is used in population dynamics as a model for competing species, cf. [6], [7], [8]. The problem in (3) models certain enzym reactions, cf. [9], [10] and finally, (4) is a Stephan problem modelling a system with two phases, cf. [11], [12]. The last problem is often formulated in terms of an equivalent variational inequality, cf. [13].

Differences between the cases (1) - (4) are due to the structure of the non-linearity, which have their influence on the structure of an asymptotic approximation as $\varepsilon\downarrow 0$. These differences become already transparent by comparing the flow diagrams in the (x,u) -plane corresponding to the degenerate ($\varepsilon = 0$) equations.

(1) $\varepsilon = 0$

$$\frac{\partial u}{\partial t} = - u + f(x)\, H(u-1)$$

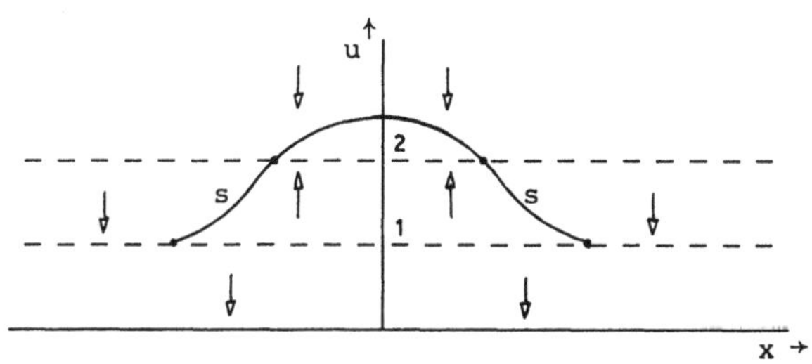

fig. I.(1)

(2) $\varepsilon = 0$

$$\frac{\partial u}{\partial t} = - f(x,u)$$

fig. I.(2)

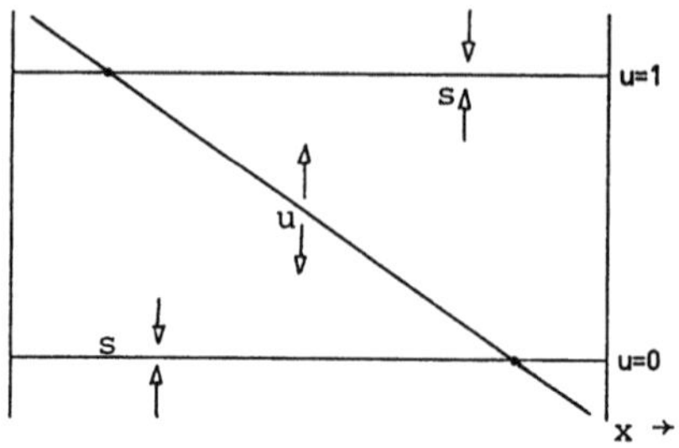

(3) $\varepsilon = 0$

$$\frac{\partial u}{\partial t} = - \frac{u}{1+u} + f(x)$$

fig. I.(3)

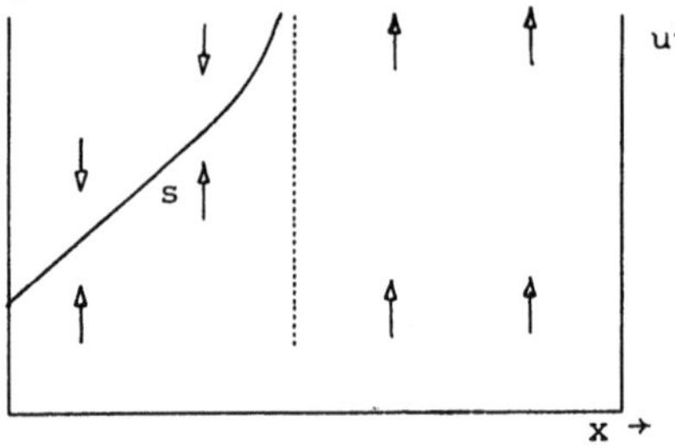

(4) $\varepsilon = 0$

$u \leq 0$,

$$\frac{\partial u}{\partial t} = - u + f$$

where $u < 0$

fig. I.(4)

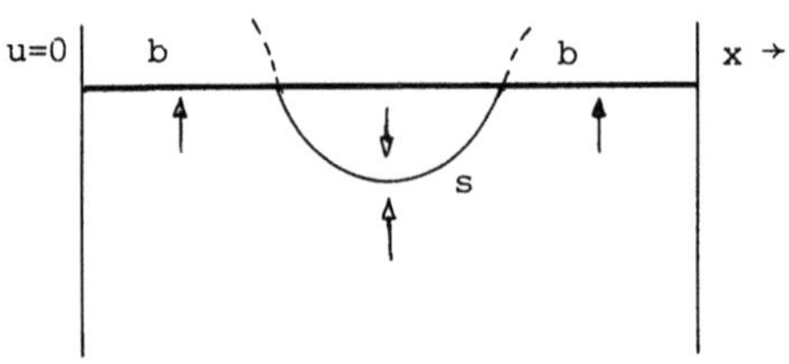

In these figures s, u and b indicate stable, unstable manifolds and barriers, respectively.

In the next sections we shall see how these differences in flow diagrams inherit to the behaviour of the approximation. In each of these sections we shall start by analysing the stationary solution. Next we discuss the formation of fronts from general initial conditions and the dynamics of these fronts on a long time scale.

1. A problem from plasma physics

In case (1) we assume in line with figure I.(1) for the coefficient f:

(1.1) f is sufficiently regular, even and positive,
f is decreasing for $x > 0$ and $f \to 0$ for $x \to \infty$,
$f(0) > 2$

For the initial data ψ we suppose

(1.2) ψ is sufficiently regular, even and positive
ψ is decreasing for $x > 0$ and $\psi \to 0$ for $x \to \infty$
$\psi(0) > 1$, $\psi(f^{-1}(1)) < 1$

It is now clear that the solution of (1) is even. Therefore it is sufficient to consider the region $x \geq 0$. As a matter of fact the conditions (1.1) and (1.2) are stronger than necessary in order to simplify the discussion. Generalisations are left to the reader.

<u>A stationary solution</u> can be approximated as

$$(1.3) \quad U^s = \theta\cdot\{f(x)+\varepsilon^2 f''(x)+\ldots\} + (1-\theta)\cdot\{G_0^s(\zeta)+\varepsilon G_1^s(\zeta)+\ldots\} \quad \text{for } x \geq 0$$

Here the first part of the approximation is the regular part of the expansion, which in 0th order coincides with the non-trivial stable manifold in figure I.(1). The second part is a layer at the <u>unknown</u> free boundary $R^s(\varepsilon) = R_0^s + \varepsilon R_1^s + \ldots$. The free boundary is defined by requiring $U^s = 1$ at $x=R(\varepsilon)$. The layer variable is given by

$$(1.4) \quad \zeta = (x - R^s)/\varepsilon$$

The width of the layer corresponds to a significant stretching of the coordinates, cf. [14]. In the layer the solution jumps with an O(1)-jump to the trivial stable manifold. Next the two local expansions have to be combined to obtain a global approximation. For this gluing process the function θ is used. θ is a C^∞ cut-off function with $\theta(x) = 1$ on $[0, R^s-2\varepsilon^{1/2}]$ and $\theta(x) = 0$ on $[R^s - \varepsilon^{1/2}, \infty]$, $\theta \geq 0$, $\theta' \leq 0$. The region $[R^s-2\varepsilon^{1/2}, R^s- \varepsilon^{1/2}]$ is the overlap region where the two parts of the approximation have to coincide up to a sufficiently high degree.

Analysis of the layer and implementation of the matching conditions at both sides will provide us with both the shape of the layer and the location of the free boundary:

$$(1.5) \quad \frac{d^2G_0^s}{d\zeta^2} - G_0^s = - f(R_0^s)\, H(-\zeta)$$

$$G_0^s(0) = 1 \ , \ \lim_{\zeta\to\infty} G_0^s = 0 \ , \ \lim_{\zeta\to-\infty} G_0^s = f(R_0) \text{ and } G_0^s \text{ is } C^1$$

i.e.

(1.6) $$G_0^s(\zeta) = (1 + e^{\zeta})\, H(-\zeta) + e^{-\zeta}\, H(\zeta)$$

and

(1.7) $$f(R_0^s) = 2$$

Due to our assumptions (1.7) has a unique positive solution $R_0^s = f^{-1}(2)$.

Higher order terms such as G_1^s, R_1^s can be determined in an analogous way.

Note that the location of the free boundary follows from the internal structure of the layer.

The time evolution from initial data ψ on an O(1) time scale is clear from the flow diagram in figure I.(1). Layer formation, i.e. creation of a front takes place at the positive point:

(1.8) $$a_1 = \psi^{-1}(1)$$

A sketch is given below in figure 1-a. In general this front is created at a position O(1) different from the position of the stationary free boundary. It is logical to expect that on a longer time scale the front will move to the stationary position. The dynamics of this process can easily be determined. Let us introduce

(1.9) $$t = \tau/\varepsilon$$

On this $O(1/\varepsilon)$ time scale the solution is approximated as in (1.3), but with the difference that for the location of the moving front we put $R(\tau,\varepsilon) = R_0(\tau) + \varepsilon\, R_1(\tau) + \ldots$.

Now the layer expansion depends on $\zeta = (x - R(\tau,\varepsilon))/\varepsilon$ and on τ and we denote it by $G_0(\zeta,\tau) + \varepsilon\, G_1(\zeta,\tau) + \ldots$.

It is easy to check that this leads to the following equation for G_0

(1.10) $$-\frac{dR_0}{d\tau} \cdot \frac{dG_0}{d\zeta} = \frac{d^2G_0}{d\zeta^2} - G_0 + f(R_0)\, H(-\zeta)$$

Further G_0 has to satisfy the same conditions as in (1.5).

In (1.10) τ merely acts as a parameter.

Again this layer problem provides us with the shape of the layer G_0 and also with the dynamics of the front. A simple calculation yields

$$(1.11) \qquad \frac{dR_0}{d\tau} = 2\left[1 - \frac{2}{f(R_0)}\right] \cdot \left\{1 - \left[1 - \frac{2}{f(R_0)}\right]^2\right\}^{-1}$$

$$R_0(0) = a_1$$

It is clear that R_0^s is a stable stationary point for the equation (1.11). Therefore, the moving front stabilizes in the long run at the position of the stationary free boundary. Altogether this reasoning supports the idea that the stationary solution of (1) is stable.

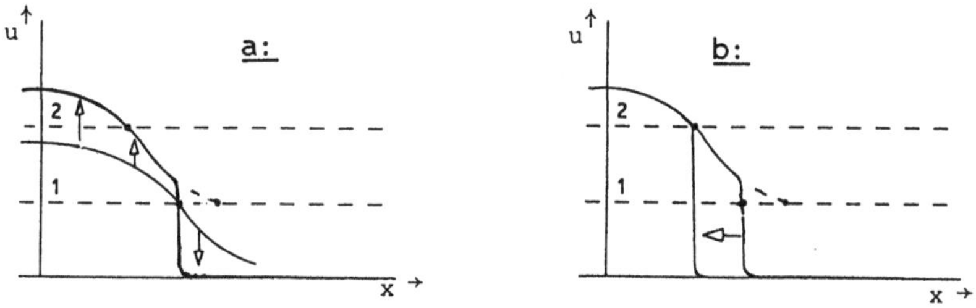

Figure 1

In this case the dynamics of the moving front is governed by an <u>internal driving mechanism.</u>

<u>2. A problem describing competing species</u>

In this section we consider problem (2) where in line with figure I.(2) the non-linearity f satisfies:

(2.1) $f(x,u)$ is sufficiently regular,
the nullset of f consists of $u = 0$, $u = 1$ and $u = a(x)$,
$a(x)$ is sufficiently regular and $a(0)>1$, $a(1)<0$, $a'<0$,
f degenerates on its nullset only in the 2 intersection points $(x_0,0)$ and $(x_1,1)$ where $a(x_0) = 0$, $a(x_1) = 1$,

$$\frac{\partial f}{\partial u}(x,u) > 0 \quad \text{for} \quad u = 0, x \in [0,x_0] \quad \text{and} \quad u = 1,\ x \in [x_1,1]$$

$$\frac{\partial f}{\partial u}(x,u) < 0 \quad \text{for} \quad u = a(x), x \in (x_1,x_0)$$

$F(x) = \int_0^1 f(x,u)du$ is strictly decreasing on (x_1,x_0).

As for the initial data ψ we suppose

(2.2) $\psi(x)$ is sufficiently regular
$0 < \psi < 1$, for all $x \in [0,1]$
the curves $u = a(x)$ and $u = \psi(x)$ intersect in one point where $x = p$ and the intersection is transversal i.e.

$$f(p,\psi(p)) = 0 \quad , \quad \frac{d}{dx} f(x,\psi(x))\Big|_{x=p} < 0$$

A stationary solution can now be found which jumps from 0 to 1 at a free boundary located at $y^s = y_0^s + \varepsilon\, y_1^s + \ldots$. In this case an approximation of the solution has the simple form

$$(2.3)\qquad U^s = G_0^s(\xi) + \varepsilon\, G_1^s(\xi) + \ldots .$$

This approximation consists only of a layer of width ε with a layer variable $\xi = (x-y^s)/\varepsilon$. For the highest order term G_0^s we obtain

$$(2.4)\qquad \frac{d^2G_0^s}{d\xi^2} - f(y_0^s\,,\,G_0^s) = 0$$

$$\lim_{\xi\to-\infty} G_0^s = 0,\ \lim_{\xi\to\infty} G_0^s = 1 \quad \text{and} \quad G_0^s\,(0) = \frac{1}{2}$$

Note that we define the free boundary as the point where the solution equals $\frac{1}{2}$.

Multiplying the equation (2.4) with $\frac{dG_0^s}{d\xi}$ and integrating from $-\infty$ to ∞ we end up with a condition which determines the location of the free boundary

$$(2.5)\qquad F\,(y_0^s) = 0$$

where $F(x)$ was introduced in (2.1). Note that (2.5) has a unique solution $y_0^s \in (x_1, x_0)$.

In this case the location of the free boundary follows again from the internal structure of the layer.

In a standard way one can also derive the shape of the layer but we do not need an explicit formula for $G_0^s(\xi)$ for our purpose.

Let us now consider the time evolution from initial data ψ.

From the flow diagram I.(2) it follows that on an $O(1)$ time scale layer formation, i.e. creation of a front, takes place at the point p where $(p,\psi(p))$ lies on the unstable manifold, see figure 2-a. In general p is $O(1)$ different from y_0^s and, as in the previous case, it is logical to expect a moving front on a longer time scale:

$$(2.6)\qquad t = \tau/\varepsilon$$

To investigate the dynamics of this front we denote its location by $y(\tau,\varepsilon) = y_0(\tau) + \varepsilon\, y_1(\tau) + \ldots$ and we introduce as a notation for the layer $G(\xi,\tau) = G_0(\xi,\tau) + \ldots$ with $\xi = (x-y(\tau,\varepsilon))/\varepsilon$. It follows that the highest order term of the layer has to satisfy

$$(2.7) \qquad \frac{d^2G_0}{d\xi^2} + \frac{dy_0}{d\tau} \cdot \frac{dG_0}{d\xi} - f(y_0, G_0) = 0$$

$$\lim_{\xi \to -\infty} G_0 = 0 \quad , \quad \lim_{\xi \to \infty} G_0 = 1$$

In this problem τ acts merely as a parameter.

It is possible to find a relation between $\frac{dy_0}{d\tau}$ and y_0 from (2.7) using phase plane analysis of the equivalent 1st order system for

$$g = G_0 \quad , \quad v = \frac{dG_0}{d\xi}$$

The equation (2.7) is equivalent to the system

$$\frac{d}{d\xi}\begin{bmatrix} g \\ v \end{bmatrix} = \begin{bmatrix} v \\ -\frac{dy_0}{d\tau} \cdot v + f(y_0, g) \end{bmatrix}$$

This system possesses three stationary points (0,0), $(a(y_0,0),0)$ and (1,0). They correspond to the stable 0-manifold, unstable manifold and stable 1-manifold in figure I.(2). The conditions for $\xi \to \pm\infty$ in (2.7) imply that we look for a heteroclinic orbit connecting (0,0) and (1,0) in the phase plane. It is clear that such a special orbit can only exist for a special value of $\frac{dy_0}{d\tau}$. To see this consider the situation where $y_0 > y_0^s$. Herebelow a sketch of the phase portret is given starting with $\frac{dy_0}{d\tau} = 0$ and next decreasing $\frac{dy_0}{d\tau}$ to negative values.

(i) $\frac{dy_0}{d\tau} = 0$

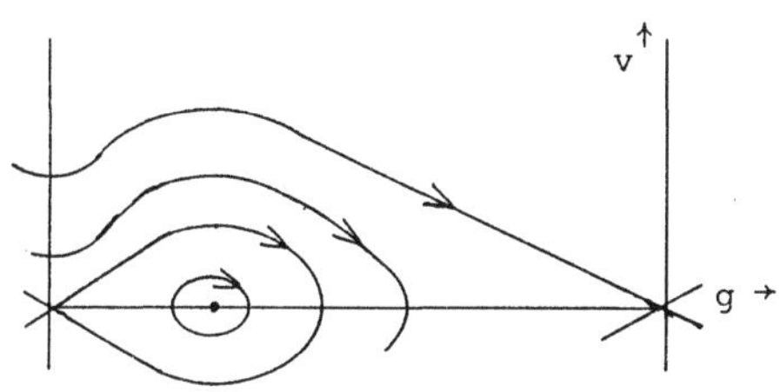

(ii) $\frac{dy_0}{d\tau} < 0$

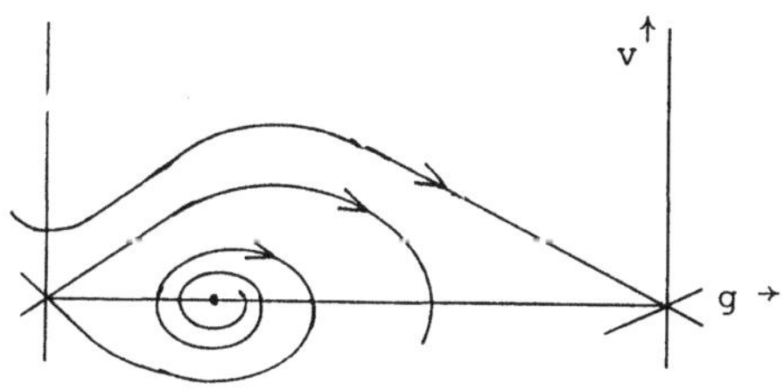

(iii) $\frac{dy_0}{d\tau} < 0$

heteroclinic orbit

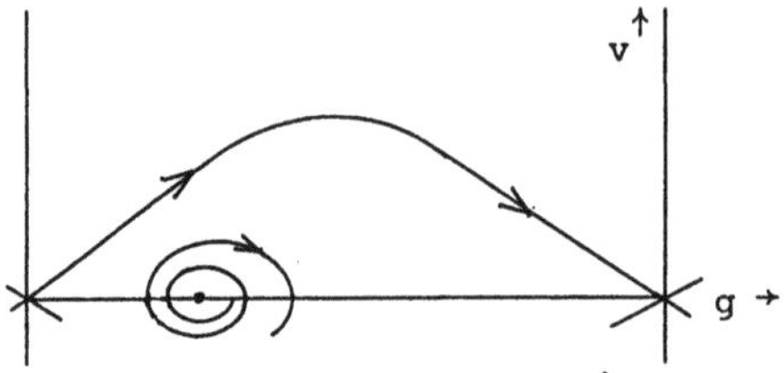

In the phase portret for $\frac{dy_0}{d\tau} = 0$ there are closed orbits around $(a(y_0),0)$ in a region enclosed by a homoclinic orbit connecting (0,0) to itself. For the part of this orbit corresponding to the unstable manifold of (0,0) we introduce the notation $v = V(g,\alpha)$ for $v \geq 0$. Here α abbreviates $\frac{dy_0}{d\tau}$. It is easy to check that

$$\text{(2.8)} \qquad \frac{d}{dg}\left(\frac{1}{2}V^2\right) = -\alpha V + f(y_0,g)$$

Differentiation with respect to α and solving the equation for $\frac{\partial V}{\partial \alpha}$ in terms of V shows that $\frac{\partial V}{\partial \alpha} < 0$ where $V > 0$. Decreasing $\alpha = \frac{dy_0}{d\tau}$ has the effect that closed orbits around $(a(y_0),0)$ become spirals.

At the same time the unstable manifold from (0,0) moves upwards and its intersection point with the line $v = 0$ moves to the right towards 1. At a certain moment this intersection point coincides with (1,0) and the heteroclinic orbit is there. From (2.8) it follows that this happens if

$$\text{(2.9)} \qquad \frac{dy_0}{d\tau} = F(y_0) \,/\, I\left(\frac{dy_0}{d\tau}\right)$$

$$y_0\big|_{\tau=0} = p$$

where $$I(\alpha) = \int_0^1 V(g,\alpha)\,dg$$

The dynamics of the moving front follows from an ODE. Moreover, it is clear that y_0^s is a stable stationary point of this ODE, i.e. the front stabilizes in the long run at its stationary position. The dynamics of the moving front is again governed by an <u>internal</u> driving mechanism.

Figure 2

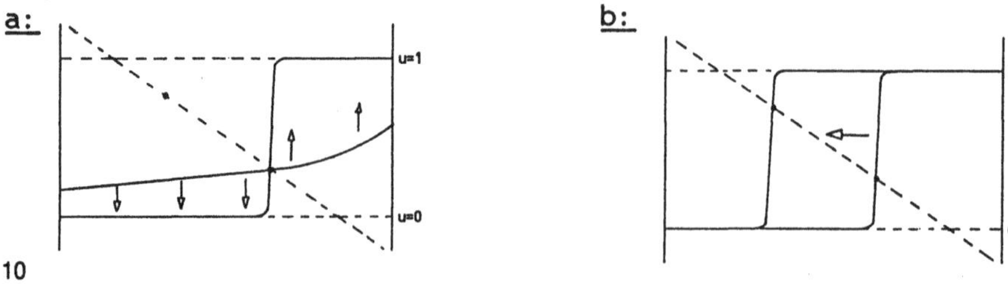

3. A problem from enzym reaction kinetics

Here we consider problem (3), where in line with figure I.(3) the inhomogeneous term f satisfies

(3.1) f is sufficiently regular,
$f > 0$, $f > 1$ for $x > x_1$, $f < 1$ for $x < x_1$
$\int_0^1 f(x)dx < 1$

As for the initial data ψ we suppose, that

(3.2) ψ is sufficiently regular, $\psi \geq 0$

A stationary solution can be approximated as

$$(3.3)\quad U^s = \theta_l \cdot \{U_0^s + \ldots\} + \theta_r \cdot \{\frac{1}{\varepsilon^2} W_0^s + \ldots\} + (1-\theta_r-\theta_l) \cdot \{G_0^s + \ldots\}$$

Here the first and second parts describe the approximation at the left and the right of a free boundary $y^s(\varepsilon) = y_0^s + \ldots$. The third part describes the transition at the free boundary in a layer of width ε. θ_l and θ_r are cut-off functions analogous to θ in (1.3). θ_l equals θ with R^s replaced by y^s. θ_r is found from θ_l reflecting the argument in y^s, i.e. $\theta_r(y^s+z) = \theta_l(y^s-z)$.
It is simple to check, that

$$(3.4)\quad - U_0^s/(1+U_0^s) + f = 0 \quad , \text{ i.e. } \quad U_0^s = f/(1-f)$$

For the leading term in the 2nd part we obtain

$$(3.5)\quad \frac{d^2W_0^s}{dx^2} - 1 + f = 0$$

In addition W_0^s has to satisfy

$$\frac{dW_0^s}{dx}(1) = 0 \quad ; \quad W_0^s(y_0^s) = \frac{dW_0^s}{dx}(y_0^s) = 0$$

The first condition is inherited from the full problem.
The latter two conditions can be understood as matching conditions with the layer at the free boundary.
From the layer equation it follows that $G_0(\xi)$ behaves as $\frac{1}{2}(1-f(y_0^s)) \cdot \xi^2 + \ldots$ as the layer variable ξ tends to $+\infty$.
Hence, in order to match the zeroth and first order Taylor term of W_0^s at y_0^s have to vanish.
An easy calculation shows that

$$W_0^s(x) = \int_{y_0^s}^{x} (x-z)\,(1-f(z))dz$$

and the location of the free boundary follows from

$$(3.6) \qquad \int_{y_0^s}^{1} (1-f(z))dz = 0$$

Note that in this case the location of the free boundary is determined by the outer expansion at the right in contrast with the previous cases.

As for the <u>time evolution from initial data</u> ψ the behaviour at the outer expansion at the right will also be the dominant phenomenon. On an O(1) time scale it follows from the flow diagram that a large jump builds up at $x = x_1$. At the left of x_1 the highest order term of the regular approximation tends to the stable manifold, whereas at the right of x_1 there is no mechanism to stop the growth of the approximation as long as $t = O(1)$ and the approximation is O(1). However, stabilization will take place on a longer time scale $t = \tau/\varepsilon^2$ when the approximation has become large: $O(1/\varepsilon^2)$.

The structure of the approximation on this time scale is analogous to (3.3), but now the terms in the expansion and the free boundary also depend on τ. Therefore, we drop the superscript s in our notation, i.e the moving free boundary body is denoted by $y(\tau,\varepsilon) = y_0(\tau) + \ldots$. The highest order term at the right of the free boundary W_0 has to satisfy

$$(3.7) \qquad \frac{\partial W_0}{\partial \tau} = \frac{\partial^2 W_0}{\partial x^2} - 1 + f \qquad \text{for } \tau \geq 0,\ x \in [y_0(\tau), 1]$$

$$W_0|_{\tau=0} = 0$$

$$\frac{\partial W_0}{\partial x}\Big|_{x=1} = 0 \qquad\qquad W_0\,\Big|_{x=y_0(\tau)} = \frac{\partial W_0}{\partial x}\Big|_{x=y_0(\tau)} = 0$$

As before, the conditions at the moving free boundary can be understood as matching conditions, since analysis of the layer leads to $G_0(\xi,\tau) = \frac{1}{2}(1-f(y,\tau))\cdot\xi^2 + \ldots$ for $\xi \to +\infty$. The problem (3.7) is of Stephan type. W_0 is found from a diffusion equation with an extra condition at the moving free boundary, which determines the dynamics of the moving free boundary. Thus, in this case the dynamics of the moving front is determined in an external way outside the layer. It is a nice exercise to check, that a travelling layer on a time scale $1/\varepsilon$ with a term $\frac{dy_0}{d\tau}$ entering the layer equation as in the previous case, leads to unfeasible layer solutions because of exponential growth for $\xi \to \infty$. A sketch of the dynamic behaviour of the solution is given in the next figure.

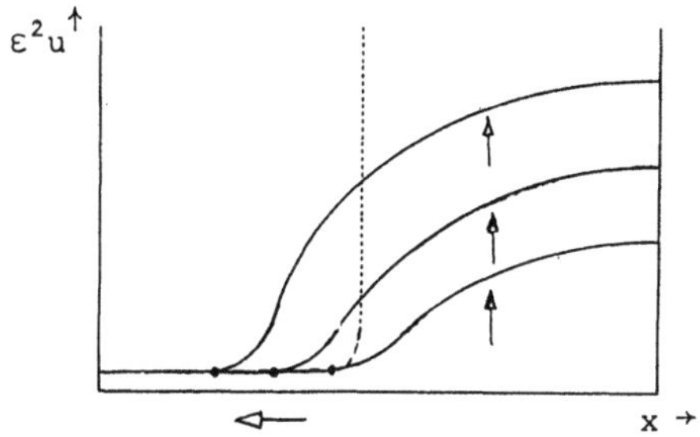

Figure 3

4. A Stephan problem

Let us now consider a sufficiently regular solution of (4). In line with figure I.(4) we assume for the inhomogeneous term f:

(4.1) f is sufficiently regular, f is even
$f < 0$ on $(-\alpha_0, \alpha_0)$, $f > 0$ elsewhere

For the initial data ψ we suppose

(4.2) ψ is sufficiently regular, ψ is even
$\psi < 0$ on $(-x_0, x_0)$ with $x_0 > \alpha_0$, $\psi \equiv 0$ elsewhere
ψ/f increases with $x \in (\alpha_0, x_0)$

In this case the stationary solution is even and it can be approximated for $x \geq 0$ by:

(4.3) $$U^s = \{f + \varepsilon^2 f'' \ldots\} + \{\varepsilon\, G_1^s(\xi) + \ldots\}$$
where $U^s < 0$, i.e. $0 \leq x < y^s(\varepsilon)$

$$U^s = 0 \qquad \text{elsewhere}$$

Here $y^s(\varepsilon) = y_0^s + \varepsilon y_1^s + \ldots$ denotes the free boundary. The first part of the approximation is regular and the second part is a layer correction with a layer variable $\xi = (x - y^s(\varepsilon))/\varepsilon$. Requiring C^1 regularity just as for the exact solution one finds that in highest order the location of the free boundary follows from the outer expansion:

(4.4) $$f(y_0^s) = 0 \quad \text{, i.e. } f_0^s = \alpha_0.$$

In order to satisfy the condition for the derivative at the free boundary the layer correction plays a role. It is easy to verify that

(4.5) $$G_1^s(\xi) = -f'(\alpha_0) \cdot e^{\xi} \quad \text{for } \xi < 0$$

and $y_1^s = 1.$

The evolution from initial data ψ on an O(1) time scale is given approximately by

(4.6) $U = \{U_0 + \ldots\} + \{\varepsilon^2 G_2 + \ldots\}$ where $U < 0$, i.e. $0 \le x < L(t,\varepsilon)$

$U = 0$ elsewhere

In this case the free boundary $L(t,\varepsilon) = L_0(t)+\varepsilon^2 L_2(t) + \ldots$ will travel with O(1) speed, if $t = O(1)$.

The highest order term of the regular expansion satisfies

(4.7) $$\frac{\partial U_0}{\partial t} = -U_0 + f$$

$$U_0|_{t=0} = \psi$$

i.e. U_0 is given by

(4.8) $$U_0(x,t) = (\psi(x)-f(x))\exp(-t) + f(x)$$

The first term in the approximation of the free boundary is found from the nulset of U_0, i.e.

(4.9) $$\frac{\psi}{f}(L_0) = 1 - e^t$$

The dynamics of the front follows in first order from an algebraic equation, which is determined from the regular expansion.
For $t \to \infty$ the moving front $L_0(t)$ given by (4.9) stabilizes at the stationary position where $f = 0$, see figure 4.

The layer at the moving front plays a role only in the next order. From the point of view of asymptotic analysis, however, this layer is very interesting.
The width of the layer is $O(\varepsilon^2)$ and the layer variable is $\xi = (x-L(t,\varepsilon))/\varepsilon^2$.
The first term of the layer becomes:

(4.10) $$G_2(\xi,t) = K_2(t)\exp\left(-\frac{dL_0}{dt}\cdot\xi\right) \qquad \text{for } \xi \le 0$$

Using the C^1 regularity at the free boundary one obtains

(4.11) $$L_2(t) = -\left(\frac{dL_0}{dt}\right)^{-1}$$

An expliciet expression for $K_2(t)$ can also be derived.

Note that $\frac{dL_0}{dt} \to 0$ as t tends to $+\infty$. This means that the decay rate of the exponential function in the layer diminishes for $t \to \infty$. At the same time $L_2(t)$ develops a singularity for $t \to \infty$. The conclusion is that the asymptotic analysis for large time is more complex than anticipated up to now.

Of course, the reason is that a long time scale is involved in the evolution of the layer. This long time scale is necessary to connect the layer of width $O(\varepsilon^2)$ for $t = O(1)$ to the layer of width $O(\varepsilon)$ in the stationary solution, see figure (4). This transition takes place when $\frac{dL_0}{dt}$ becomes of the same order as ε. Since a $L_0(t) = \alpha_0 + \mu \exp(-t) + \ldots$ for $t \to \infty$ it is logical to introduce a new time variable σ by putting

$$(4.12) \qquad t = \ln\left(\frac{1}{\varepsilon}\right) + \sigma$$

Then a layer correction has the form $\varepsilon\, G_1(\xi,\sigma) + \ldots$ with a layer variable $\xi = (x-L(\sigma,\varepsilon))/\varepsilon$ where the location of the free boundary is $L(\sigma,\varepsilon) = \alpha_0 + \varepsilon\, L_1(\sigma) + \ldots \ldots$.
It turns out that G_1 and L_1 have to satisfy a problem of Stephan type themselves.

$$(4.13) \qquad \frac{\partial G_1}{\partial \sigma} - \frac{dL_1}{d\sigma} \cdot \frac{\partial G_1}{\partial \xi} = \frac{\partial^2 G_1}{\partial \xi^2} - G_1$$

$$G_1|_{\xi=0} = -f'(\alpha_0) \cdot L_1 - \psi(\alpha_0)e^{-\sigma}$$

$$\frac{\partial G_1}{\partial \xi}|_{\xi=0} = -f'(\alpha_0)$$

Though this problem is of the same order of complexity as the original problem it is not difficult to verify that asymptotics as required for the matching for $\sigma \to \pm\infty$ are consistent with (4.13):

$$(4.14) \qquad L_1 = \mu\, e^{-\sigma} + \ldots \qquad \text{for } \sigma \to -\infty$$

$$G_1 = -\frac{f'(\alpha_0)}{\mu} e^{-\sigma} \exp(\xi\mu e^{-\sigma}) + \ldots \qquad \text{for } \sigma \to -\infty$$

$$L_1 = 1 + \ldots \qquad \text{for } \sigma \to \infty$$

$$G_1 = -f'(\alpha_0) \cdot \exp(\xi) \qquad \text{for } \sigma \to \infty$$

Anyway it is clear that the transition phenomenon in the width of the layer is due to a complex internal driving mechanism.

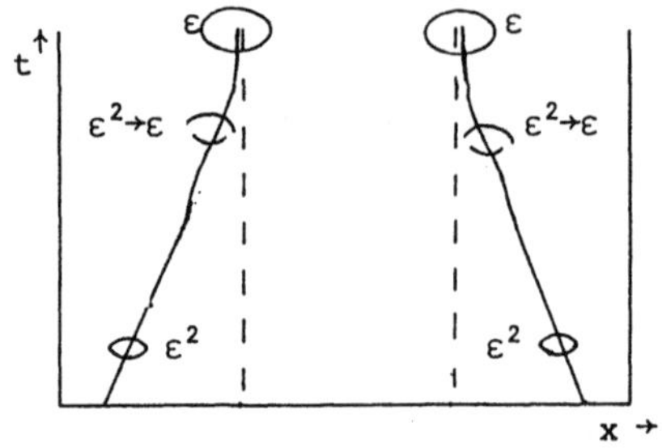

Figure 4

transition in layerwidth

at $t = O(\ln(\frac{1}{\varepsilon}))$

5. References

[1] Buckmaster, J., Ludford, G.S.S.
Theory of Laminar Flames, Cambridge Un. Press , London, 1982.

[2] van Harten, A., Matkowsky, B.J.
A New Model in Flame Theory, SIAM J. Appl. Math., vol 42, 1982.

[3] Garbey, M.
Singular Travelling Phenomena in (degenerate) Parabolic-Hyperbolic Singular Perturbation Problems, preprints, Un. of Valenciennes, 1987.

[4] Eckhaus, W., van Harten, A., Peradzynski, Z.
A Singularly Perturbed Free Boundary Problem Describing a Laser Sustained Plasma, SIAM J. Appl. Math., vol 45, 1985.

[5] Eckhaus, W., van Harten, A., Peradzynski, Z.
Plasma Produced by a Laser in a Medium with Convection and Free Surface Satisfying a Hamilton-Jacobi Equation, Physica D, vol 27, 1987.

[6] Nishiura, Y., Fuji, H.
Stability of Singularly Perturbed Solutions to Systems of Reaction Diffusion Equations, SIAM J. Math. An., vol 18, 1987.

[7] Fife, P.C.
Mathematical Aspects of Reacting and Diffusing Systems, Lect. Notes in Biomath., Springer, Berlin, 1979.

[8] van Harten, A., Burger, E.
Approximate Green Functions as a Tool to Prove Correctness of a Formal Approximation in a Model of Competing and Diffusing Species, Pac. J. of Math., vol 125, 1986.

[9] Brauner, C.M., Eckhaus, W., Garbey, M., van Harten, A.
Asymptotics of a Rather Unusual Type in a Free Boundary Problem, SIAM J. Math. An., vol 18, 1987.

[10] Michaelis, M., Menten, M., Biochem, Z. Vol 49, 1973.

[11] van Harten, A., van Hassel, R.R.
On a Singularly Perturbed Time Dependent Free Boundary Problem, J. of Math. and Appl., to appear.

[12] Moet, H.J.K.
Asymptotic Analysis of Singularly Perturbed Variational Inequalities, thesis, Un. of Utrecht, 1985.

[13] Bensoussan, A., Lions, J.L.
Applications des Inequations Variationelles en Contrôle Stochastique, Dunod, Paris, 1978.

[14] Eckhaus, W.
Asymptotic Analysis of Singular Perturbations, North Holland, Amsterdam, 1979.

Boundary Layer Analysis for Some Problems of Singular Domain Perturbation

H.J. Weinitschke

Institut für Angewandte Mathematik, Universität Erlangen-Nürnberg, D-8520 Erlangen, Fed. Rep. of Germany

This article is concerned with analyzing the effect of small circular holes in a plane domain on the solution of certain nonlinear boundary value problems. Apart from its intrinsic mathematical interest, this problem has various applications. Examples include the stress concentration in thin elastic membranes and shells at the edge of a small hole, and the buckling of plates and shells, which may be of the bifurcation or snapping type. It is usually much simpler to calculate stresses and critical loads for a shell without hole. Therefore, it is of interest to assess asymptotically the effect of the singular domain perturbation on the stresses and on the critical loads. Another application relates to tubular chemical reactors, where internal cooling pipes of small diameter are inserted into the reactor vessel.

The study of this type of problems by asymptotic methods was initiated by LAGERSTROM and CASTEN [1]. A model for incompressible flow at low Reynolds number is given by the boundary value problem

$$y'' + \frac{n-1}{r} y' + yy' = 0 \ , \quad y(\varepsilon) = 0 \ , \quad y(r) \to 1 \ , \quad r \to \infty$$

which is known to have a unique solution $y(r;\varepsilon)$ for any $\varepsilon > 0$ [2]. In the absence of the hole, the solution is $y(r;0) = 1$, the boundary condition at $r = \varepsilon$ being lost in the limit $\varepsilon \to 0$. For small ε, the solution $y(r,\varepsilon)$ will be slightly perturbed except near the hole, where $y = 0$ at $r = \varepsilon$. Hence the convergence of $y(r,\varepsilon)$ for $\varepsilon \to 0$ is nonuniform and we encounter a singular perturbation problem of the layer type.

A different example for this type of domain perturbation is the eigenvalue problem for the Laplacian in a bounded two-dimensional annular domain $D(\varepsilon)$. Assuming Dirichlet boundary conditions at the outer boundary of $D(\varepsilon)$ and at an inner circular boundary of radius ε, the problem is to estimate, asymptotically for $\varepsilon \to 0$, the perturbed eigenvalues $\lambda_n(\varepsilon)$, $n = 1,2,\ldots,$ in terms of the eigenvalues λ_n of the unperturbed domain D. OZAWA has treated this problem by methods of scattering theory [3].

In the past few years, investigation of this type of problems has been continued by WAN, LANGE, and this writer [4,5,6], with a view towards new applications to non-linear problems, whose solutions depend on an additional parameter λ. Of interest are not only *regular* solution branches, but also critical points at $\lambda =$

λ_c , say, where *singularities* such as limit points or bifurcation points occur. As in the eigenvalue problem for the Laplacian, we are interested in asymptotic expansions of $\lambda_c(\varepsilon)$ at singular points.

We do not address the problem here in any generality, rather, we wish to give a summary of our previous work on some specific problems, supplemented by more recent work. We restrict ourselves to axisymmetric problems defined on circular domains with outer radius normalized to one and inner radius ε . The results for small ε are obtained by adapting the method of matched asymptotic expansions to the problems in question. Sections 1 and 2 deal with solutions at regular points; Sections 3 and 4 with solutions at singular points.

1. Nonlinear Membrane Problems

The starting point for the work on elastic membranes [4] was a representation theorem, expressing the solution of the finite deflection annular membrane problem explicitly in terms of a solution of the circular membrane problem [7]. Here, it is assumed that either the radial stress σ_r or the radial displacement u is precribed at the outer edge, while the inner edge is assumed free of traction, that is, $\sigma_r(\varepsilon) = 0$. Within the FOPPL nonlinear membrane model, the dimensionless stress $\sigma_r = cy$ in a shallow membrane of revolution with a hole at the apex, subjected to uniform normal pressure, is determined by the following boundary value problem (BVP) [8].

$$y'' + \frac{3}{x}y' + \frac{2}{y^2}\left(1 - \frac{\varepsilon^2}{x^2}\right)^2 = \mu(r) \qquad 0 < \varepsilon < x < 1$$
$$y(\varepsilon) = 0 \; , \quad y(1) = S > 0 \;\; \underline{\text{or}} \;\; y'(1) + (1-\nu)y(1) = H \tag{1.1}$$

where μ is a measure for the curvature. The solution of (1.1) for $\mu = 0$ is given by [7]

$$y(x) = (1-\varepsilon^2)^{1/3}\left(1 - \frac{\varepsilon^2}{x^2}\right)z(\xi) \qquad \xi = \left(\frac{x^2-\varepsilon^2}{1-\varepsilon^2}\right)^{1/2} \tag{1.2}$$

where $z(\xi)$ is the solution of the circular membrane problem satisfying $z'(0) = 0$ and

$$z(1) = (1-\varepsilon^2)^{-4/3}S \;\; \underline{\text{or}} \;\; z'(1) + [\nu - \varepsilon^2(1+\nu)]z(1) = (1-\varepsilon^2)^{-1/3}H \; . \tag{1.3}$$

Clearly $z(\xi)$ is the solution for the unperturbed problem with $z(0) \neq 0$. Let $\sigma_{\theta,0}$ be the circumferential stress corresponding to this limit solution, then it is a consequence of (1.2),(1.3) that

$$\lim_{\varepsilon\to 0} S_c(\varepsilon) = 2 \; , \qquad S_c(\varepsilon) := \sigma_\theta(\varepsilon)/\sigma_{\theta,0}(0) \; . \tag{1.4}$$

S_c is called the stress concentration factor. The result (1.4) was conjectured by SCHWERIN in 1929. What cannot be obtained from (1.2),(1.3) is how S_c depends on ε , for small ε .

In a singular perturbation solution of (1.1), which--unlike the solution (1.2), (1.3)--is not restricted to uniform pressure and $\mu = 0$, there will be an inner solution $Y(s;\varepsilon)$ near $x = \varepsilon$, obtained by setting $s = (x/\varepsilon) - 1$, which satisfies

$$\ddot{Y} + \frac{3}{1+s}\dot{Y} + \varepsilon^2 \frac{2}{Y^2}\left(1 - \frac{1}{(1+s)^2}\right)^2 = \varepsilon^2\mu , \qquad Y(0;\varepsilon) = 0 \qquad \dot{Y} = \frac{dY}{ds} . \tag{1.5}$$

For $\varepsilon \to 0$ the solution is $Y_0(s) = C_0(1 - (1+s)^{-2})$ which is nothing but the factor $1 - \varepsilon^2/x^2$ in (1.2). Expanding the solution $Y(s;\varepsilon)$ in an asymptotic power series in ε^2 , and matching Y for $s \to \infty$ with an appropriate outer solution expansion $z_0(x) + \varepsilon^2 z_1(x) + \ldots$, the following asymptotic formula for $S_c(\varepsilon)$ is obtained

$$S_c(\varepsilon) = 2[1 + \frac{\varepsilon^2}{C_0^3}(C_0^2 C_1 - \frac{1}{2} + \frac{1}{4} C_0^2 \mu) + O(\varepsilon^4)] \qquad \varepsilon \to 0 \tag{1.6}$$

where $C_0 = z_0(0)$ and $C_1 = z_1''(0)/2$. For details of these calculations, we refer to [4]. We have applied similar techniques to more general nonlinear elastic membrane models.

2. Shells of Revolution

For the sake of brevity, we shall consider the special case of a shallow spherical shell under uniform normal load, clamped at the outer edge, with zero radial traction and bending moment at the hole. The basic equations then reduce to the BVP

$$Lf = -\mu^2 g + fg + 2\gamma , \quad Lg = \mu^2 f - \frac{1}{2} f^2 , \quad Ly := y'' + \frac{3}{x} y'$$
$$\varepsilon f'(\varepsilon) + (1+\nu) f(\varepsilon) = g(\varepsilon) = 0 , \quad f(1) = g'(1) + (1-\nu) g(1) = 0 \tag{2.1}$$

where μ^2 = curvature, γ = load intensity, $f(x)$ = angular deflection and $g(x)$ = radial stress resultant. In the limit case of a shell closed at the apex, the boundary conditions at $x = \varepsilon$ change to $f'(0) = g'(0) = 0$. Hence both f and g exhibit a boundary layer near $x = \varepsilon$, which can be analyzed again via the stretching $s = (x/\varepsilon) - 1$, leading to equations for $F(s;\varepsilon) = f(x(s);\varepsilon)$ and $G(s;\varepsilon) = g(x(s);\varepsilon)$ analogous to (1.5). There is one basic difference here: While the leading terms of an asymptotic expansion of F and G are of the form $A + B/(1+s)^2$, as in $Y_0(s)$, the solutions $F_n(s), G_n(s)$, $n \geq 1$ contain logarithmic terms not present in the solutions $Y_n(s)$ of the membrane problem. It is clear that the outer solutions will also contain logx-terms. It was shown in [5] that in order to match inner and outer solutions to all orders, the inner expansions $F(s;\varepsilon)$ and $G(s;\varepsilon)$ must be of the form

$$Z(s;\varepsilon) = Z_0(s) + \varepsilon^2 \log\varepsilon \bar{Z}(s) + \varepsilon^2 Z_1(s) + \varepsilon^4 \log^2\varepsilon Z_2(s)$$
$$+ \varepsilon^4 \log\varepsilon Z_3(s) + \varepsilon^4 Z_4(s) + \ldots \tag{2.2}$$

where Z stands for F or G. As a consequence, we obtain the following stress concentration factors for the radial stress and moment resultants

$$
\begin{aligned}
S_N(\varepsilon) &= \frac{N_\theta(\varepsilon)}{N_{\theta,0}(0)} = 2\{1 + \frac{1}{2}(\varepsilon^2 \log \varepsilon) \frac{B_0^2}{C_0} \frac{\nu-1}{\nu+1} + \frac{\varepsilon^2}{C_0}[C_2 + \frac{1}{2} B_0^2 h_1(\nu)] + o(\varepsilon^2) \\
S_M(\varepsilon) &= \frac{M_\theta(\varepsilon)}{M_{\theta,0}(0)} = 2\{1 + (\varepsilon^2 \log\varepsilon) \frac{1+\nu}{1-\nu}(\frac{\gamma}{B_0} - C_0 \frac{\nu}{\nu+1}) \\
&\qquad + \frac{\varepsilon^2}{B_0} [B_2 - \frac{1}{4} (\gamma + B_0 C_0 h_2(\nu)) + o(\varepsilon^2)
\end{aligned}
\tag{2.3}
$$

where $B_0 = f_0(0)(1+\nu)/(\nu-1)$, $C_0 = g_0(0)$, $h_1 = (8\nu+3\nu^2)/8(2+\nu)^2$, $h_2 = (1+3\nu)/2(1+\nu)$, and $f_0(x)$, $g_0(x)$ are the solutions of the BVP for the unperturbed shell. It is seen that calculation of first order correction terms in the expressions for S_N and S_M is much simpler here than in the membrane problem. All we need to solve is the BVP for the shell without hole. This solution is the leading term outer solution for the annular shell and determines B_0 and C_0. The constants B_2 and C_2 are found from higher order terms, but they do not contribute to the $\varepsilon^2\log\varepsilon$-terms in (2.3). The preceding analysis assumes that γ is below the buckling load. In the following sections, we consider some cases of singular points.

3. Singular Perturbation of Eigenvalues

Given a nonlinear BVP on a circular domain, describing a radially symmetric situation,

$$
\begin{aligned}
&\Delta u + a_0(r)u_r + \lambda^2\rho(r)u + f(r,u) = 0 , \qquad f(r,0) = 0 \\
&u(\varepsilon,\theta) = 0 , \qquad \beta_0 u(1,\theta) + \beta_1 u_r(1,\theta) = 0 ,
\end{aligned}
\tag{3.1}
$$

for $u = u(r,\theta)$, with polar coordinates r,θ $a_0(r)$ and $\rho(r)$ are assumed analytic for $|r| \leq 1$. In addition to the solution $u \equiv 0$, (3.1) generally has nontrivial solutions which bifurcate from $u \equiv 0$ at the eigenvalues of the linearized problem. Looking first for radially symmetric solutions $u = u(r)$, (3.1) can be reduced, upon linearization, to the eigenvalue problem

$$
u'' + \frac{1}{r} a(r)u' + \lambda^2\rho(r)u = 0 , \quad u(\varepsilon) = \beta_0 u(1) + \beta_1 u'(1) = 0 \tag{3.2}
$$

where $a(0) = 1$, $a(r)$ analytic. In the absence of the ε-perturbation, we have $u'(0) = 0$ instead of $u(\varepsilon) = 0$. The resulting eigenvalues and - functions are denoted by λ_n and $U_n(r)$, respectively. We wish to find the eigenvalues $\lambda_n(\varepsilon)$ and eigenfunctions $u_n(r;\varepsilon)$ of (3.2) in the form

$$
\begin{aligned}
\lambda_n(\varepsilon) &= \lambda_n + \delta_1(\varepsilon)\alpha_1 + \delta_2(\varepsilon)\alpha_2 + \dots \\
u_n(r;\varepsilon) &= U_n(r) + \mu_1(\varepsilon)h_1(r) + \mu_2(\varepsilon)h_2(r) + \dots .
\end{aligned}
\tag{3.3}
$$

where the α_k and $h_k(r)$ also depend on n. In order to determine the asymptotic scales $\delta_k(\varepsilon)$ and $\mu_k(\varepsilon)$, we first consider the special case $a(r) = \rho(r) = 1$ with $u(1) = 0$ (vibration of an annular membrane), where the $u_n(r;\varepsilon)$ are given by $J_0(\lambda_n(\varepsilon)r) + c_n Y_0(\lambda_n(\varepsilon)r)$ and the eigenvalues $\lambda_n(\varepsilon)$ are the roots of $J_0(\lambda\varepsilon)Y_0(\lambda) - J_0(\lambda)Y_0(\lambda\varepsilon) = 0$. Expanding this transcendental equation for small ε, one finds [5]

$$\lambda_n(\varepsilon) = \lambda_n + \frac{\pi}{2}\frac{Y_0(\lambda_n)}{J_1(\lambda_n)}(\log\frac{1}{\varepsilon})^{-1} + O((\log\frac{1}{\varepsilon})^{-2}) \qquad \varepsilon \to 0 \tag{3.4}$$

where the unperturbed eigenvalues λ_n are the zeros of $J_0(\lambda)$. This suggests that in the expansion (3.3) for the more general problem (3.2), we should have $\delta_k(\varepsilon) = (\log 1/\varepsilon)^{-k}$. This can indeed be confirmed by matching inner and outer solutions of (3.2), which also yields $\mu_k(\varepsilon) = \delta_k(\varepsilon)$ as well as appropriate equations for the successive determination of α_k and $h_k(r)$, $k = 1,2,\ldots$ [5].

For unsymmetric eigenfunctions of the type $v_n(1)\cos m\theta$, we find the following expansions for the eigenvalues

$$\lambda_{n,m}(\varepsilon) = \lambda_{n,m} + \alpha_{1,m}\varepsilon^{2m} + \alpha_{2,m}\varepsilon^{2m+2}\log\varepsilon + O(\varepsilon^{2m+2}) . \qquad \varepsilon \to 0$$

It is seen that the symmetric eigenvalues $\lambda_n(\varepsilon)$ are considerably more sensitive to the domain perturbation than the unsymmetric eigenvalues $\lambda_{n,m}(\varepsilon)$.

The second example concerns the bifurcation buckling of a circular plate subjected to radial thrust and weakened by a central hole of radius ε. We assume ε to be so small that buckling is determined by radially symmetric deformations. The basic equations are then obtained from (2.1) by setting $\mu = 0$ and incorporating the radial load at the outer edge, which yields

$$Lf + \lambda^2 f = fg , \qquad Lg = -\frac{1}{2}f^2 , \tag{3.5}$$

the boundary conditions being the same as in (2.1). The linearized problem (3.5) can be solved in terms of Bessel functions, the eigenvalues $\lambda_n(\varepsilon)$ are the roots of the equation

$$\begin{aligned}&[J'_{-\beta}(\tfrac{\lambda\varepsilon}{r(\varepsilon)}) + \nu\frac{r(\varepsilon)}{\lambda\varepsilon}J_{-\beta}(\tfrac{\lambda\varepsilon}{r(\varepsilon)})]J_\beta(\tfrac{\lambda}{r(\varepsilon)})\\ &\quad - [J'_\beta(\tfrac{\lambda\varepsilon}{r(\varepsilon)}) + \nu\frac{r(\varepsilon)}{\lambda\varepsilon}J_\beta(\tfrac{\lambda\varepsilon}{r(\varepsilon)})]J_{-\beta}(\tfrac{\lambda}{r(\varepsilon)}) = 0\end{aligned} \tag{3.6}$$

where $r(\varepsilon) = (1-\varepsilon^2)^{1/2}$ and $\beta = [1 + \lambda^2\varepsilon^2/(1-\varepsilon^2)]^{1/2}$. This equation is somewhat awkward to solve as the unknown λ also appears in β. A singular perturbation solution has been carried out from which an asymptotic expansion of $\lambda_n(\varepsilon)$ can be obtained. For the lowest eigenvalue, which determines the buckling load, we find

$$\lambda_1(\varepsilon) = \lambda_1 + \frac{\varepsilon^2}{J_2(\lambda_1)}[\frac{\pi}{2}\lambda_1^2 Y_1(\lambda_1) - \lambda_1 J_1'(\lambda_1)] + O(\varepsilon^4\log\varepsilon) , \qquad \varepsilon \to 0 \tag{3.7}$$

where λ_1 is the lowest eigenvalue of the unperturbed problem (= circular plate). The details will be reported elsewhere.

4. Singular Perturbation of Limit Points

The problems to be considered in this section originate from a BVP of type (3.1) except that $f(r,0) \neq 0$. More precisely, it is assumed that a radially symmetric solution branch $U(r,\lambda)$ of the unperturbed problem has a simple quadratic limit point at $\lambda = \lambda_0$. The objective is to examine the effect of the domain perturbation on λ_0, that is, to estimate $\lambda_0(\varepsilon) - \lambda_0$ for small ε, where $\lambda_0(\varepsilon)$ is the limit point of the perturbed problem. As the terms $a_0 u_r$ and ρu in (3.1) are not essential in this problem, we drop them and also set $\beta_1 = 0$.

The limit point λ_0 is determined from the extended system

$$U'' + \frac{1}{r} U' + \lambda_0 f(r,U) = 0 , \qquad U'(0) = U(1) = 0 \qquad (4.1)$$

$$V'' + \frac{1}{r} V' + \lambda_0 f_u(r,U)V = 0 , \qquad V'(0) = V(1) = 0 , \qquad V(0) = 1 . \qquad (4.2)$$

The perturbed problem is defined by the same differential equations, but with boundary conditions

$$u(\varepsilon;\varepsilon) = u(1;\varepsilon) = 0 , \qquad v(\varepsilon;\varepsilon) = v(1;\varepsilon) = 0 , \qquad \lim_{\varepsilon \to 0} v(r;\varepsilon) = V(r) \qquad (4.3)$$

$$r \text{ fixed, } 0 < r \leq 1 .$$

Here we have denoted the solution of the perturbed problem by $u(r;\varepsilon)$, $v(r;\varepsilon)$. The construction of asymptotic expansions is slightly more complicated because parts of the BVP for $u(r;\varepsilon)$ and for $v(r;\varepsilon)$ must be solved simultaneously. A detailed analysis of the interplay between these two BVP's results in the correct form of the inner and outer expansions for u and v.

Rewriting the first equation of (4.1) for $u(r;\varepsilon)$ in terms of the inner variable $s = r/\varepsilon$, we have

$$\frac{d^2u}{ds^2} + \frac{1}{s}\frac{du}{ds} + \varepsilon^2 \lambda_0(\varepsilon) f(\varepsilon s,u) = 0 , \qquad s > 1 \qquad (4.4)$$

and the condition $u = 0$ at $s = 1$. For smooth f we deduce that the inner expansion for u at the limit point has the form

$$u(r;\varepsilon) = \delta(\varepsilon) \log s (U(0) + o(1)) . \qquad \delta(\varepsilon) = (\log \frac{1}{\varepsilon})^{-1} . \qquad (4.5)$$

A similar argument shows that $v(r;\varepsilon)$ must be of the form (4.5), with $V(0) = 1$. Assuming outer expansions for $u(r;\varepsilon)$ and $v(r;\varepsilon)$ as in (3.3), the main problem then is finding the asymptotic scales $\delta_k(\varepsilon)$, $\mu_k(\varepsilon)$ and $\nu_k(\varepsilon)$ in the expansions of u, v and $\lambda_0(\varepsilon)$, respectively, from matching outer and inner expansions. The result for $\lambda_0(\varepsilon)$ is

$$\lambda_0(\varepsilon) = \lambda_0 + \alpha_1\delta(\varepsilon) + \alpha_2\delta^2(\varepsilon) + \dots , \qquad \varepsilon \to 0 \tag{4.6}$$

where α_1 is determined from the solution $U(r)$, $V(r)$ of the unperturbed problem. The complete expansions for $u(r;\varepsilon)$ and $v(r;\varepsilon)$ are given in [5], together with a scheme for the successive calculation of α_1 , α_2 , etc., and the coefficient functions in the inner and outer expansions for u and v .

This analysis has been extended to simple cubic limit points, which arise in problems where $f = f(r,U;\mu)$. At certain critical values μ_0 of the second parameter μ , two simple quadratic limit points λ_1 and λ_2 may coalesce into a simple cubic limit point λ_0 . The values $\lambda_0(\varepsilon)$ and $\mu_0(\varepsilon)$ of the perturbed problem are obtained by solving the extended system (see, for example [9])

$$\begin{aligned}
&u'' + \frac{1}{r}u' + \lambda_0 f(r,u;\mu_0) = 0 && u(\varepsilon) = u(1) = 0 \\
&v'' + \frac{1}{r}v' + \lambda_0 f_u(r,u;\mu_0)\, v = 0 && v(\varepsilon) = v(1) = 0 \\
&w'' + \frac{1}{r}w' + \lambda_0 f_u(r,u;\mu_0)\, w = -\lambda_0 f_{uu}(r,u;\mu_0)v^2 , && w(\varepsilon) = w(1) = 0
\end{aligned} \tag{4.7}$$

and two normalizing conditions which rule out $v = w = 0$, for example $[v,v] = 1$ and $(v,w) = 0$, where $[.,.]$ and $(.,.)$ are appropriate inner products. In the present problem, one may assume $V(0) = W(0) = 1$ for the unperturbed problem and replace the inner product conditions by lim-conditions as in (4.3). The essential result of the asymptotic analysis is that both $\lambda_0(\varepsilon)$ and $\mu_0(\varepsilon)$ have expansions of the form (4.6), with the scale $\delta(\varepsilon) = 1/\log(1/\varepsilon)$.

5. Numerical Results

The various asymptotic solutions have been compared extensively with high accuracy numerical solutions. Some representative examples are discussed in this section. Table 1 presents stress concentration factors for annular mambranes, $S = 0.5$ and for plates, $\mu = 0$. The accuracy of the asymptotic results for membranes is better than that for plates, as to be expected from comparing (1.6) and (2.3).

Table 1 Stress Concentration Factors

Membrane: $S_C(\varepsilon)$			Plate:		$S_N(\varepsilon)$		$S_M(\varepsilon)$	
ε	Asympt.	Numer.	ε	γ	Asympt.	Numer.	Asympt.	Numer.
0.20	1.9962	1.9973	0.10	10^2	2.2529	2.1951	1.7261	1.7054
0.10	1.9990	1.9991	0.10	10^3	2.0449	2.0423	2.0619	2.5105
0.05	1.9998	1.9998	0.01	10^2	2.0078	2.0078	1.9905	1.9905
0.01	2.0000	2.0000	0.01	10^3	2.0029	2.0029	1.9644	1.9641

In Table 2 the estimate (3.4) is compared with the exact Bessel function solution for the vibrating membrane, while Table 3 shows a comparison between (3.6) and (3.7) for the annular plate buckling problem. Finally, Table 4 contains asymp-

Table 2 Vibrating Membrane

ε	2 Term Asympt.	3 Term Asympt.	Numer. $\lambda_1(\varepsilon)$
0.100	3.075	3.299	3.314
0.050	2.920	3.053	3.064
0.010	2.740	2.796	2.801
0.001	2.628	2.653	2.655

Table 3 Annular Plate

ε	2 Term Asympt.	Numer. $\lambda_1(\varepsilon)$
0.20	3.613	3.743
0.10	3.752	3.792
0.05	3.812	3.821
0.01	3.831	3.831

Table 4 Limit Points $\lambda_0(\varepsilon)$

ε	2 Term Asympt.	3 Term Asympt.	Numer. $\lambda_0(\varepsilon)$
0.100	3.204	3.645	3.891
0.040	2.861	3.220	3.200
0.010	2.602	2.777	2.761
0.001	2.401	2.460	1.469

totic approximations to the limit point problem treated in Section 4 for the special case $f(r,u) = e^u$. In this case, the unperturbed problem (4.1),(4.2) has a closed form solution

$$\lambda_0 = 2 , \qquad U(r) = \log\left(\frac{2}{1+r^2}\right)^2 , \qquad V(r) = \frac{1 - r^2}{1 + r^2} . \tag{5.1}$$

The approximation (4.6) for $\lambda_0(\varepsilon)$ is given by [5]

$$\lambda_0(\varepsilon) = 2 + 2\log 4(\log 1/\varepsilon)^{-1} + 2.961(\log 1/\varepsilon)^{-2} + O((\log 1/\varepsilon)^{-3}) \tag{5.2}$$

as $\varepsilon \to 0$. We conclude from Table 4 that even for very small values of ε the effect of the domain perturbation is quite remarkable, for example, of $\varepsilon = 0.001$, $\lambda_0(\varepsilon)$ is about 23% in excess of λ_0. It may be recalled that the BVP (4.1) with $f = e^u$ is a simplified model for an exothermic chemical reaction in an infinite circular cylindrical reactor. The ε-perturbation corresponds to a cooling rod of radius ε.

Adknowledgement

The author would like to thank Elizabeth Ong and Rodney Sinclair for assistance in the numerical calculations. The manuscript was completed while the author was a Visiting Professor of the University of Washington, Seattle, USA.

References

1. P.A. Lagerstrom and R.G. Casten, Basic concepts underlying singular perturbation techniques, SIAM Rev. 14, 63-120 (1972).
2. J. Kevorkian and J.D. Cole, Perturbation Methods in Applied Mathematics, Springer, New York, Heidelberg, Berlin (1981).

3. S. Ozawa, Singular variation of domains and eigenvalues of the Laplacian, Duke Math. Journ. 48, 767-778 (1981).
4. F.Y.M. Wan and H.J. Weinitschke, Boundary layer solutions for some nonlinear elastic membrane problems, ZAMP 38, 79-91 (1987).
5. H.J. Weinitschke and C.G. Lange, Asymptotic solutions for finite deformation of thin shells of revolution with a small circular hole, Quart. Appl. Math. 45, 401-417 (1987).
6. C.G. Lange and H.J. Weinitschke, Singular perturbations of singular points with application to tubular chemical reactors, to appear.
7. H. Grabmüller and H.J. Weinitschke, Finite displacements of annular elastic membranes, J. Elasticity 16, 135-147 (1986).
8. E. Reissner, Rotationally symmetric problems in the theory of thin elastic shells, Proc. 3rd U.S. National Congress of Applied Mechanics, 51-69 (1958).
9. H.J. Weinitschke, On the calculation of limit and bifurcation points in stability problems of elastic shells, Int. J. Solids Structures 21, 79-95 (1985).

Global Behaviour of Very Thin Cellular Structures. Applications to Networks

D. Cioranescu[1] *and J. Saint Jean Paulin*[2]

[1]CNRS – Lab. Analyse Numérique, 4 place Jussieu, F-75252 Paris Cedex 05, France
[2]CNRS – Lemta, 2 rue de la Citadelle, B.P. 850, F-54011 Nancy, France

We consider a very thin network made of periodically distributed wires. Such a network is a particular case of thin cellular large spaces structures, which are characterized by the fact that the material is distributed along layers (honeycomb structures) or along beams (reinforced structures). The thickness of the material is small compared with the dimension of the period. In the case of networks we have also to take into account their small thickness which makes them plate-like structures.

We study here a thermal problem in a network and our aim is to give the dependence of the temperature on the three small parameters describing this structure : the plate-thickness, the period and the thickness of the material in the elementary cell.

1. Statement of the Problem. Variational Formulation.

Let us consider the network defined by figure 1 ; it is contained in the domain $\omega \times (-\frac{e}{2}, \frac{e}{2})$, $\omega = (0,\ell_1) \times (0,\ell_2)$. As the thickness e of the network is small compared with the period we suppose that

$$e = \varepsilon\, \eta ,$$

η being a small parameter. We have periodicity in x_1 and x_2 and the cells are homothetic to a fixed cell $\mathcal{Y} = Y \times (-\frac{e}{2}, \frac{e}{2})$, $Y = (0,1) \times (0,1)$ in the x_1 and x_2-directions. The ratio is $\varepsilon : 1$.

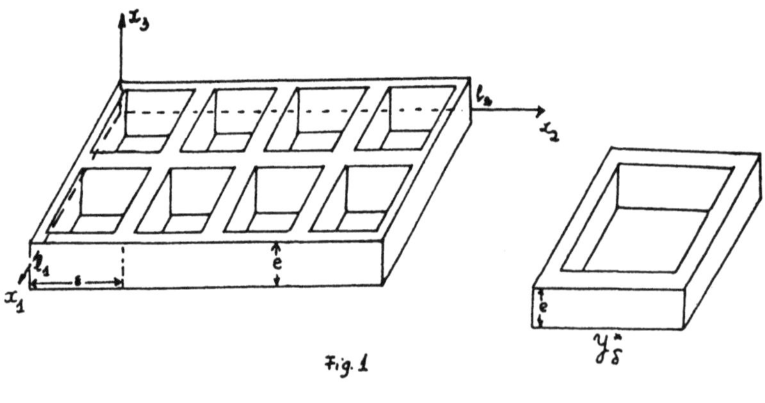

Fig. 1

It is supposed that the holes do not meet the boundary $\partial\omega$. We denote by Y_δ^* the part occupied by the material in Y:

$$Y_\delta^* = \{y \in Y \,,\ \mathrm{dist}(y,\partial Y) \leq \frac{\delta}{2}\}.$$

The cross section $\omega^{\varepsilon\delta}$ consists in beams of thickness $\varepsilon\delta$. In the sequel we also use the notation:

$$\mathcal{Y}_\delta^* = Y_\delta^* \times]-\frac{1}{2}, \frac{1}{2}[.$$

Consider the thermal problem:

$$\begin{aligned}
&-\frac{\partial}{\partial x_i}(a_{ij}\frac{\partial u_e^{\varepsilon\delta}}{\partial x_j}) = f_e && \text{in } \omega^{\varepsilon\delta} \times (-\frac{e}{2}, \frac{e}{2}) \\
&a_{3j}\frac{\partial u_e^{\varepsilon\delta}}{\partial x_j} n_3 = g_e^{\pm} && \text{on } \omega^{\varepsilon\delta} \times \{\pm\frac{e}{2}\} \\
&a_{\alpha j}\frac{\partial u_e^{\varepsilon\delta}}{\partial x_j} n_\alpha = 0 && \text{on the lateral boundary of holes} \\
&u_e^{\varepsilon\delta} = 0 && \text{on } \partial\omega \times (-\frac{e}{2}, \frac{e}{2}).
\end{aligned} \tag{1}$$

In the following the indices i, j, k take values in $\{1,2,3\}$ and α, β, γ in $\{1,2\}$. The constant conductivity coefficients a_{ij} are independent of e, ε, δ and satisfy:

$$\exists\, m > 0 \text{ such that } a_{ij}\,\xi_i\,\xi_j \geq m\,\xi_i\,\xi_i \qquad \forall\, \xi_i \in \mathbb{R}. \tag{2}$$

Let us suppose that f_e and $g_e^{\pm}$ are smooth functions (for instance C^1).

Under these assumptions problem (1) has a unique solution $u_e^{\varepsilon\delta} \in H^1(\omega^{\varepsilon\delta} \times (-\frac{e}{2}, \frac{e}{2}))$. To give the dependence of $u_e^{\varepsilon\delta}$ on ε, η and δ we shall take successively $\varepsilon \to 0$ with η and δ fixed, $\eta \to 0$ with δ fixed and finally $\delta \to 0$. The first step is a homogenization process. The other steps use perturbation methods.

In order to work on a fixed domain independent of $e = \varepsilon\,\eta$ we make a dilatation in the x_3-direction:

$$z_2 = x_\alpha \,,\quad z_3 = \frac{x_3}{e} = \frac{x_3}{\varepsilon\,\eta} \qquad z_3 \in [-\frac{1}{2}, \frac{1}{2}].$$

In this change of variable any function $\varphi_e(x_1,x_2,x_3)$ will be transformed into the function $\varphi_\eta(z_1,z_2,z_3) = \varphi_e(z_1,z_2,ez_3)$. We use the notation

$$\Omega^{\varepsilon\delta} = \omega^{\varepsilon\delta} \times (-\frac{1}{2}, \frac{1}{2}) \;,\; \Gamma_{\pm}^{\varepsilon\delta} = \omega^{\varepsilon\delta} \times \{ \pm \frac{1}{2} \} \;,\; \Omega = \omega \times (-\frac{1}{2}, \frac{1}{2}) .$$

Introduce the functional space :

$$V^{\varepsilon\delta} = \{ v \in H^1(\Omega^{\varepsilon\delta}) \,,\, v = 0 \text{ on } \partial\omega \times (-\frac{1}{2}, \frac{1}{2}) \} .$$

Then the system (1) after dilatation, can be written in the variational form :

$$u^{\varepsilon\delta} \in V^{\varepsilon\delta}$$

$$\int_{\Omega^{\varepsilon\delta}} [a_{\alpha\beta} \frac{\partial u_\eta^{\varepsilon\delta}}{\partial z_\beta} \frac{\partial v}{\partial z_\alpha} + \varepsilon^{-1} \eta^{-1} (a_{\alpha 3} \frac{\partial u_\eta^{\varepsilon\delta}}{\partial z_3} \frac{\partial v}{\partial z_\alpha} + a_{3\beta} \frac{\partial u_\eta^{\varepsilon\delta}}{\partial z_\beta} \frac{\partial v}{\partial z_3}) + \varepsilon^{-2} \eta^{-2} a_{33} \frac{\partial u_\eta^{\varepsilon\delta}}{\partial z_3} \frac{\partial v}{\partial z_3}] dz = \int_{\Omega^{\varepsilon\delta}} f_\eta \, v \, dz + e^{-1} \int_{\Gamma_\pm^{\varepsilon\delta}} g_\eta^{\pm} \, v \, dz_1 \, dz_2 \quad \forall v \in V^{\varepsilon\delta} . \tag{3}$$

2. Limit for $\varepsilon \to 0$. Homogenization.

We have the following result

Theorem 1.

There exists an extension operator $P^{\varepsilon\delta} \in \mathcal{L}(H^1(\Omega^{\varepsilon\delta}) \,,\, H^1(\Omega))$ such that :

$$\varepsilon \, \eta \, P^{\varepsilon\delta} u_\eta^{\varepsilon\delta} \xrightarrow[\varepsilon \to 0]{} u^{\delta\eta} \quad \text{in } H^1(\Omega) \text{ weak} \tag{4}$$

The function $u^{\delta\eta}$, independent of z_3, can be identified to a function, still denoted $u^{\delta\eta}$, $u^{\delta\eta} \in H_0^1(\omega)$ which satisfies the homogenized system :

$$- q_{\alpha\beta}^{\delta\eta} \frac{\partial^2 u^{\delta\eta}}{\partial z_\alpha \partial z_\beta} = |\text{meas } Y_\delta^*| \, (g_\eta^+ + g_\eta^-) \qquad \text{in } \omega . \tag{5}$$

The homogenized coefficients are given by

$$q_{\alpha\beta}^{\delta\eta} = - \int_{-1/2}^{1/2} \int_{Y_\delta} \int [a_{\gamma\beta} \frac{\partial w_\alpha^{\delta\eta}}{\partial y_\gamma} + \eta^{-1} a_{3\beta} \frac{\partial w_\alpha^{\delta\eta}}{\partial y_3}] \, dy \,, \tag{6}$$

and the functions $w_\alpha^{\delta\eta}$ are solutions of the system

$$- \frac{\partial}{\partial y_\beta} (a_{\gamma\beta} \frac{\partial w_\alpha^{\delta\eta}}{\partial y_\gamma}) - \eta^{-1} \frac{\partial}{\partial y_\beta} (a_{3\beta} \frac{\partial w_\alpha^{\delta\eta}}{\partial y_3}) - \eta^{-1} \frac{\partial}{\partial y_\beta} (a_{\gamma 3} \frac{\partial w_\alpha^{\delta\eta}}{\partial y_\gamma}) - \eta^{-2} \frac{\partial}{\partial y_3} (a_{33} \frac{\partial w_\alpha^{\delta\eta}}{\partial y_3}) = 0 \quad \text{in } Y_\delta^* \times (-\frac{1}{2}, \frac{1}{2})$$

$$[a_{\gamma j} \frac{\partial w_\alpha^{\delta\eta}}{\partial y_\gamma} + \eta^{-1} (a_{3j} \frac{\partial w_\alpha^{\delta\eta}}{\partial y_3})] \, n_j = 0 \quad \text{on } \partial[Y_\delta^* \times (-\frac{1}{2}, \frac{1}{2})] \setminus \partial[Y \times (-\frac{1}{2}, \frac{1}{2})]$$

$$w_\alpha^{\delta\eta} + y_\alpha \text{ is periodic in } y_1 \text{ and } y_2 \,,\, \int_{-1/2}^{1/2} \int_{Y_\delta} w_\alpha^{\delta\eta} \, dy = 0 . \tag{7}$$

Remark.

As a rule, there is no summation in repeated upper indices $\delta\eta$.

Sketch of the proof.

We use here standard homogenization methods, for details on these methods we refer the reader to : BENSOUSSAN-LIONS-PAPANICOLAOU [1].

We begin by giving a priori estimates on $u_\eta^{\varepsilon\delta}$. They are a consequence of (3) and of the fact that the Poincaré constant of the domain $\Omega^{\varepsilon\delta}$ can be estimated independently of ε. We have, with C independent of ε and η:

$$\| u_\eta^{\varepsilon\delta} \|_{H^1(\Omega^{\varepsilon\delta})} \leq C\,\varepsilon^{-1}\,\eta^{-1}$$

$$\left| \frac{\partial u_\eta^{\varepsilon\delta}}{\partial z_3} \right|_{L^2(\Omega^{\varepsilon\delta})} \leq C \tag{8}$$

Then, by using extension operators in perforated domains (see BRIZZI-CHALOT [3], CIORANESCU-SAINT JEAN PAULIN [5]), we have the convergence (4) and moreover, from the second estimate (8) it follows :

$$\frac{\partial u^{\delta\eta}}{\partial z_3} = 0 \tag{9}$$

To find the limit equation we first introduce the flux

$$\sigma_i^\varepsilon = a_{i\beta} \frac{\partial u_\eta^{\varepsilon\delta}}{\partial z_\beta} + \varepsilon^{-1}\,\eta^{-1}\,a_{i3} \frac{\partial u_\eta^{\varepsilon\delta}}{\partial z_3} .$$

Its extension by zero in the holes is denotes $\tilde{\sigma}_i^\varepsilon$. A consequence of (8) is

$$\varepsilon\,\eta\,\tilde{\sigma}_i^\varepsilon \xrightarrow[\varepsilon \to 0]{} \sigma_i \quad \text{in} \quad L^2(\Omega) \text{ weak}.$$

We give now the limit equation satisfied by σ_i by using the same technique as that of CAILLERIE [4], which is based on a special choice of test functions for cylindrical domains.

Let $\varphi \in \mathcal{D}(\Omega)$ and take $v = \int_0^{z_3} \varphi(z_1,z_2,s)ds$ as test function in (3), it follows :

$$\varepsilon^2\,\eta^2 \int_\Omega \tilde{\sigma}_\alpha^\varepsilon \frac{\partial v}{\partial z_\alpha} dz + \varepsilon\,\eta \int_\Omega \tilde{\sigma}_3^\varepsilon\,\varphi\,dz = \varepsilon^2\,\eta^2 \int_{\Omega^{\varepsilon\delta}} f_\eta\,v\,dz + \varepsilon\,\eta \int_{\Gamma_\pm^{\varepsilon\delta}} g_\eta^\pm\,v\,dz_1\,dz_2$$

and by passing to the limit for $\varepsilon \to 0$ we get

$$\sigma_3 = 0 . \tag{10}$$

Now, take in (4) v independent of z_3 (so v can be identified to a $H_0^1(\omega)$ function), by passing to the limit we have :

$$-\frac{\partial}{\partial z_\alpha}\left[\int_{-1/2}^{1/2} \sigma_\alpha \, dz_3\right] = \text{meas } Y_\delta^* \ l\,(g_\eta^+ + g_\eta^-) \qquad \text{in } \omega . \tag{11}$$

It remains to express σ_α in terms of $u^{\delta\eta}$. To do it we apply the variational homogenization method of TARTAR [7]. Set

$$w_\alpha^\varepsilon(z_1, z_2, z_3) = \varepsilon\, w_\alpha^{\varepsilon\eta}\left(\frac{z_1}{\varepsilon}, \frac{z_2}{\varepsilon}, z_3\right)$$

where $w_\alpha^{\delta\eta}$ is defined by (7). It is easy to check that w_α^ε is solution of the adjoint problem to (3) with null right handside terms. Define now the extension of w_α^ε to $\mathbb{R}^n$ by :

$$\mathcal{Q}^\varepsilon w_\alpha^\varepsilon (z_1, z_2, z_3) = - z_\alpha + \varepsilon P^\varepsilon \left[\left(w_\alpha^{\delta\eta} + y_\alpha\right)\left(\frac{z_1}{\varepsilon}, \frac{z_2}{\varepsilon}, z_3\right)\right],$$

where P^ε is the extension operator introduced above. Set now :

$$\xi_{\alpha i}^\varepsilon = a_{\gamma i}\frac{\partial w_\alpha^\varepsilon}{\partial z_\gamma} + \varepsilon^{-1}\eta^{-1} a_{3i}\frac{\partial w_\alpha^\varepsilon}{\partial z_3} .$$

One verifies that its extension by zero, denoted $\tilde{\xi}_{\alpha i}^\varepsilon$, satisfies :

$$-\left(\tilde{\xi}_{\alpha\beta,\beta}^\varepsilon + \varepsilon^{-1}\eta^{-1}\tilde{\xi}_{\alpha 3,3}^\varepsilon\right) = 0 \ \text{ in } \Omega . \tag{12}$$

Moreover we have the convergences :

$$\mathcal{Q}^\varepsilon w_\alpha^\varepsilon \longrightarrow - z_\alpha \ \text{ in } H^1(\Omega) \ \text{ weak},$$

$$\int_{-1/2}^{1/2} \tilde{\xi}_{\alpha\beta}^\varepsilon \, dz_3 \longrightarrow - q_{\alpha\beta}^{\delta\eta} \ \text{ in } L^2(\Omega) \text{ weak}, \tag{13}$$

where $q_{\alpha\beta}^{\delta\eta}$ are the constants defined by (6).

Let $\varphi \in H_0^1(\omega)$, take $v = \varphi\, \mathcal{Q}^\varepsilon w_\alpha^\varepsilon$ in (3) multiply (12) by $\varphi P^\varepsilon u_\eta^{\varepsilon\delta}$ and integrate by parts. Recalling the convergences (4), (13) and the relations (9), (10)

$$\int_{-1/2}^{1/2} \sigma_\alpha \, dz_3 = q_{\alpha\beta}^{\delta\eta}\frac{\partial u^{\delta\eta}}{\partial z_\beta}$$

and thus, from (11) we have the homogenized equations (5).

3. Limit for $\eta \to 0$.

We fix δ in (5). Take $\eta \to 0$. We have :

Theorem 2.

Let $u^{\delta\eta}$ be solution of homogenized equation (5). Assume that

$$g_\eta^+ + g_\eta^- \longrightarrow g^+ + g^- \text{ in } L^2(\omega) \text{ weak}.$$

Then :

$$u^{\delta\eta} \underset{\eta\to 0}{\longrightarrow} u^\delta \text{ in } H_0^1(\omega) \text{ weak}, \tag{14}$$

where u^δ satisfies the limit equation :

$$- q_{\alpha\beta}^\delta \frac{\partial^2 u^\delta}{\partial z_\alpha \partial z_\beta} = |\text{meas } Y_\delta^*| (g^+ + g^-) \qquad \text{in } \omega. \tag{15}$$

The coefficients $q_{\alpha\beta}^\delta$ are given by :

$$q_{\alpha\beta}^\delta = - \int_{Y_\delta} (a_{\gamma\beta} - \frac{a_{\gamma 3} a_{3\beta}}{a_{33}}) \frac{\partial w_\alpha^\delta}{\partial y_\gamma} dy_1 \, dy_2, \tag{16}$$

with w_α solution of the system :

$$w_\alpha^\delta + y_\alpha \in W(Y^*) = \{\psi \in H^1(Y_\delta^*), \ \psi \text{ periodic}, \ \int_{Y_\delta} \psi \, dy = 0\}$$

$$\int_{Y_\delta} (a_{\gamma\delta} - \frac{a_{\gamma 3} a_{3\beta}}{a_{33}}) \frac{\partial w_\alpha^\delta}{\partial y_\gamma} \frac{\partial \varphi}{\partial y_\beta} dy = 0, \quad \forall \varphi \in W(Y^*). \tag{17}$$

Sketch of the proof.

In (8), the constant C is independent of η, hence, there is a subsequence $u^{\delta\eta}$ satisfying (14). To pass to the limit in equation (5), one has to investigate the behaviour of $w_\alpha^{\delta\eta}$ as $\eta\to 0$. We again make use of Caillerie's choice of test functions. Multiply the system (7) defining $w_\alpha^{\delta\eta}$ by $w_\alpha^{\delta\eta} + y_\alpha$. By applying Poincaré-Wirtinger inequality (see BREZIS [2]) it follows :

$$\| w_\alpha^{\delta\eta} \|_{H^1(Y_\delta)} \leqslant C$$

$$\left| \frac{\partial w_\alpha^{\delta\eta}}{\partial y_3} \right|_{L^2(Y_\delta^*)} \leqslant C\eta \tag{18}$$

and therefore (up to a subsequence) :

$$w_\alpha^{\delta\eta} \underset{\eta\to 0}{\longrightarrow} w_\alpha^\delta \text{ in } H^1(Y_\delta^*) \text{ weak} \tag{19}$$

and w_α^δ is independent of y_3. Moreover $w_\alpha^\delta + y_\alpha \in W(Y^*)$.

Multiply now (7) by the test-function $\eta\psi$ where :

$$\psi(y_1, y_2, y_3) = \int_0^{y_3} \varphi(y_1, y_2, s) ds, \quad \varphi \in \mathcal{D}(\mathcal{Y}).$$

Due to a priori estimates (15) we deduce that :

$$a_{\gamma 3} \frac{\partial w_\alpha^\eta}{\partial y_\gamma} + \eta^{-1} a_{33} \frac{\partial w_\alpha^\eta}{\partial y_3} \qquad 0 \quad \text{in } L^2(\mathcal{Y}_\delta^*) \text{ weak}. \tag{20}$$

If we take now in (7) as test function $\varphi \in W(\mathcal{Y}_\delta^*)$, φ independent of y_3, we obtain

$$\int_{Y_\delta^*} \frac{a_{3\beta}}{a_{33}} (a_{\gamma 3} \frac{\partial w_\alpha^\eta}{\partial y_\gamma} + \eta^{-1} a_{33} \frac{\partial w_\alpha^\eta}{\partial y_3}) \frac{\partial \varphi}{\partial y_\beta} dy + \int_{Y_\delta^*} (a_{\gamma\beta} - \frac{a_{\gamma 3} a_{3\beta}}{a_{33}}) \frac{\partial w_\alpha^\eta}{\partial y_\gamma} \frac{\partial \varphi}{\partial y_\beta} dy = 0$$

where we pass to the limit by using (20), which gives (17). We check that the matrix $(a_{\gamma\beta} - (a_{\gamma 3} a_{3\beta} / a_{33}))_{\gamma,\beta}$ is positive definite. To obtain the limit coefficients (16) we pass to the limit directly in formulas (16) by using the convergences (19) and (20). Moreover, the problem (15) is well-posed as it can be verified that $(q_{\alpha\beta}^\delta)_{\alpha,\beta}$ is positive-definite too.

4. Limit for $\delta \to 0$.

To study the limit of u^δ for $\delta \to 0$, we apply our former results on reinforced and honeycomb special strucutres (see CIORANESCU-SAINT JEAN PAULIN [6]). We have :

Theorem 3.

If $\delta \to 0$:

$$u^\delta \longrightarrow u^* \quad \text{in } H_0^1(\omega) \text{ weak} \tag{21}$$

where u^* is the solution of the problem :

$$- q_{\alpha\beta}^* \frac{\partial^2 u^*}{\partial z_\alpha \partial z_\beta} = 2(g^+ + g^-) \quad \text{in } \omega . \tag{22}$$

The coefficients $q_{\alpha\beta}^*$ are given by :

$$q_{\alpha\beta}^* = 0 \quad \text{if} \quad \alpha \neq \beta$$
$$q_{\underline{\alpha\alpha}}^* = \frac{1}{A_{\underline{\alpha\alpha}}} \quad \text{(no summation on } \alpha \text{)} \tag{23}$$

where $(A_{ij})_{ij} = (a_{ij})_{ij}^{-1}$.

Proof.

The main steps of the proof are the following (for details see [6]). First, from system (17) we have the a priori estimate

$$|\nabla w_\alpha^\delta|_{L^2(Y_\delta^*)} \leq C \delta^{1/2} \tag{24}$$

which used in (16) gives :

$$\delta^{-1} q^{\delta}_{\alpha\beta} \rightarrow q^{*}_{\alpha\beta} .$$

Due to the periodicity we can translate the domain of integration Y^*_δ in formulas (16) and work on Y^0_δ :

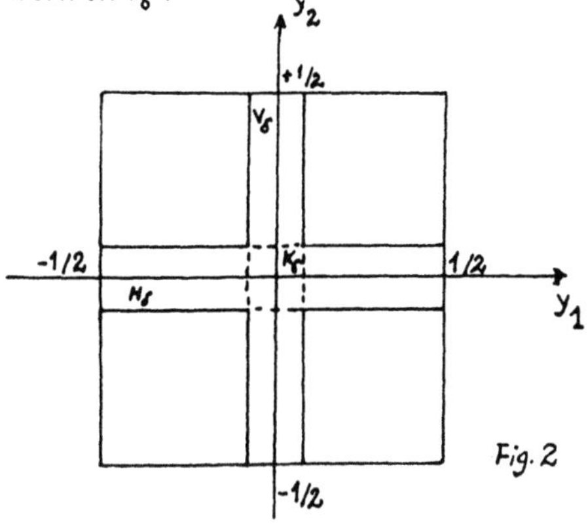

Fig. 2

Now decompose Y^0_δ in the horizontal bar H_δ , the vertical bar V_δ and the central square K_δ . We change by affine transformations H_δ , V_δ and K_δ into $Y^0 =]-1/2, 1/2[\times]-1/2, 1/2[$. For instance by $t_1 = y_1$, $t_2 = y_2/\delta$ we transform H_δ into Y^0 and any function $\varphi(y_1,y_2)$ into $\varphi_H(t_1,t_2) = \varphi(t_1,\delta t_2)$. From (24) we have :

$$\frac{\partial w^{\delta}_{\alpha,H}}{\partial t_1} \longrightarrow h^{\alpha}_1 \quad , \quad \delta^{-1} \frac{\partial w^{\delta}_{\alpha,H}}{\partial t_2} \longrightarrow h^{\alpha}_2 \quad \text{in } L^2(Y^0) \text{ weak ,} \tag{25}$$

with (from the periodicity) $\int_{Y^0} h^{\alpha}_1 \, dt = - \delta_{\alpha 1}$ (where δ_{ij} is the Kronecker symbol).

A similar transformation on V_δ leads to

$$\delta^{-1} \frac{\partial w^{\delta}_{\alpha,V}}{\partial t_1} \longrightarrow v^{\alpha}_1 \quad , \quad \frac{\partial w^{\delta}_{\alpha,V}}{\partial t_2} \longrightarrow v^{\alpha}_2 \quad \text{in } L^2(Y^0) \text{ weak ,} \tag{26}$$

where $\int_{Y^0} v^{\alpha}_2 \, dt = - \delta_{\alpha 2}$.

Therefore, from formulas (16) we have :

$$q^{*}_{\alpha\beta} = - b_{\gamma\beta} \int_{Y^0} (h^{\alpha}_{\gamma} + v^{\alpha}_{\gamma}) dt \tag{27}$$

where

$$b_{\gamma\beta} = [a_{\gamma\beta} - \frac{a_{\gamma 3} \; a_{3\beta}}{a_{33}}] .$$

It remains to give the explicit values of $\int_{Y^0} h_2^\alpha \, dt$ and $\int_{Y^0} v_1^\alpha \, dt$. Let $\varphi = \varphi(y_2)$, $\varphi \in \mathcal{D}(-1/2, 1/2[)$. Multiplying (17) by φ , integrating by parts, passing to the limit for $\delta \to 0$ and using (25) and (26) we get :

$$\int_{Y^0} (b_{12} h_1^\alpha + b_{22} h_2^\alpha) dt = 0 .$$

Analogous calculation with φ independent on y_2 gives :

$$\int_{Y^0} (b_{11} v_1^\alpha + b_{21} v_2^\alpha) dt = 0 .$$

Using these relations in (27) we have formulas (23).

References.

1. A. Bensoussan, J.L. Lions, G. Papanicolaou. Asymptotic analysis for periodic structures. North-Holland, Amsterdam, 1978.
2. H. Brezis. Analyse fonctionnelle. Masson, Paris, 1983.
3. R. Brizzi, J.P. Chalot. Homogénéisation dans des ouverts à frontière fortement oscillante, Thèse de 3ème cycle, Université de Nice, 1978.
4. D. Caillerie. Etude de quelques problèmes de perturbation en théorie de l'élasticité et de la conduction thermique. Thèse d'Etat, Université Pierre et Marie Curie, Paris, 1982.
5. D. Cioranescu, J. Saint Jean Paulin. J. of Math. Anal. Appl. 71, 2 (1979), 590-607.
6. D. Cioranescu, J. Saint Jean Paulin. J. Math. Pures et Appl. 65 (1986), 403-422.
7. L. Tartar. in Functional Analysis and Numerical Analysis. Proc. of the Japan-France Seminar 1976. Ed. by Fujita,Japan Society for the Promotion of Science, 1978, 469-482.

Nonlinear Diffusion in Stellar Magnetic Loops

F. Verhulst

Mathematisch Instituut, Rijksuniversiteit Utrecht,
P.O. Box 80.010, NL-3508 TA Utrecht, The Netherlands

The description of a hot plasma, confined to a magnetic loop in a stellar corona, leads to singularly perturbed nonlinear reaction-diffusion equations with rather unusual side conditions. We formulate the problems and we survey a number of preliminary qualitative and quantitative results.

1. Formulation of the Problem and Astrophysical Background

Observations from outside the atmosphere of the earth have revealed that the outer atmosphere of the sun, the corona, and of other stars is highly structured. One of the features is the presence of a hot plasma confined to relatively persistent magnetic loops.

A description of observations, physical mechanisms and modelling aspects has been given by Pakkert et al. [1]; the reader is referred to [1] for the astrophysical references. In these models the quantities to be determined are the pressure P and the temperature T in the loop,

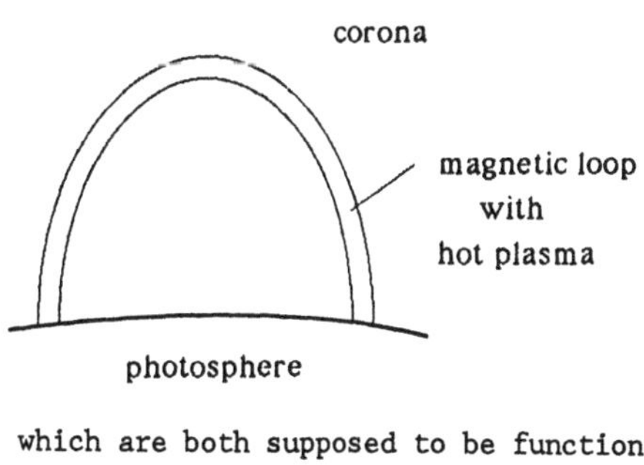

Figure 1
Coronal loop.

which are both supposed to be functions of time t and the spatial variable z; because of the strong magnetic field, the loop can be described adequately by one space variable. The dominant mechanisms are thermal diffusion, heating, radiation losses and, in some stars, radiation pressure.

The energy balance equation reads

$$\frac{5}{2}\frac{P}{T}\frac{\partial T}{\partial t} - \frac{\partial P}{\partial t} = \epsilon \frac{P}{T}\frac{\partial}{\partial z}\left(T^{3/2} P \frac{\partial T}{\partial z}\right) + \alpha - P^2 \chi(T) \tag{1}$$

where $t \geq 0$, $0 \leq z \leq 1$.

The parameter ϵ is small and positive, it is partly determined by the coefficient of thermal conductivity. The first term on the righthand side represents the thermal diffusion, its explicit form depends on the characteristics of the hot, rarefied plasma. The heating rate α is a positive constant (taken a constant because of lack of a well-founded physical theory). The function $\chi(T)$ represents the radiative losses, it increases quickly and monotonically until $T = \tau_1$, a temperature typical for regions near the bottom of the corona, it decreases monotonically for $T > \tau_1$. In [1] and [2] $\chi(T)$ takes the form

$$\chi(T) = \beta_0\, \epsilon^{-6}\, T^5 \quad , \quad T_0 \leq T \leq \tau_1$$

$$\chi(T) = \beta_1\, T^{-5} \quad , \quad T_0 \geq \tau_1$$

where β_0 and β_1 are constants independent of ϵ and which are such that $\chi(T)$ is continuous.

The base of the loop corresponds with $z = 0$, the summit with $z = 1$. Assuming hydrodynamic equilibrium, the continuity equation decouples whereas the momentum equation reduces to

$$\frac{\partial P}{\partial z} = -\lambda \tag{2}$$

with λ a positive constant, $0 < \lambda \leq 1$. All quantities have been normalised with characteristic pressure, temperature, length scale etc. Equation (2) holds for the sun and most stellar coronae. In some stellar coronae however, radiation pressure is effective; an example is the star τ Sco. In such a case the momentum equation (2) is replaced by

$$\frac{\partial P}{\partial z} = -\lambda + S(T) \tag{3}$$

In the case of τ Sco the radiation pressure is governed by

$$S(T) = \delta\, T^{15} \quad , \quad T \leq \tau_2$$

$$S(T) = \delta\, T^{-4} \quad , \quad T \geq \tau_2$$

with δ a constant independent of ϵ, $\tau_1 < \tau_2$ and $\delta\, \tau_2^{15} = \delta\, \tau_2^{-4}$ (continuity).

In the time-dependent case, system (1,2) or (1,3) is supplied with initial conditions in the form of a smooth initial state T(z) and P(z). The boundary conditions are

$$T(0,t) = T_0 \quad , \quad \frac{\partial T}{\partial z}(1,t) = 0 \tag{4}$$

with T_0 a positive constant and $0 < T_0 < \tau_1$. Another condition arises from the assumption that the total mass of the loop is constant, which in the original coordinate system means that the loop length is a constant:

$$\int_0^1 \frac{T}{P}\, dz = L \tag{5}$$

2. Singular Perturbation Analysis of the Solar Loop Problem.

In this section we shall develop a quantitative description of the static solutions of system (1,2) based on [1] and [2].

First we are trying to use a regular expansion near $z = 0$, $T_0 \le T \le \tau_1$. Putting $\epsilon = 0$ we obtain

$$\alpha - P^2\, \chi(T) = 0 \tag{6}$$

as a first formal relation between P and T. Using (6) we find that the temperature does not rise enough with z to produce solutions of practical interest; also we can not meet the loop length condition (5).

a. The alternative is that a boundary layer exists near $z = 0$. We scale $T = \epsilon^{\nu}\, \bar{T}$, $z = \epsilon^{\sigma}\, \zeta$, $0 \le \nu \le 1$, $\sigma > 0$ in (1) and find a significant degeneration

$$1 + \tfrac{3}{2}\nu - 2\sigma = -6 + 5\nu.$$

The boundary layer equation becomes to lowest order

$$\frac{1}{\bar{T}} \frac{d}{d\zeta} (\bar{T}^{3/2} \frac{d\bar{T}}{d\zeta}) - \beta_0\, \bar{T}^5 = 0, \tag{7}$$

which means that in this boundary layer only conduction and radiative losses play a part. Integration of (7) produces in the original variables the integral

$$\epsilon^7\, T^3\, (\frac{dT}{dz})^2 - \frac{4}{17}\, \beta_0\, T^{17/2} = c \tag{8}$$

with c a constant. The phase-plane has been presented in Figure 2. The size of the boundary layer depends on T_0 and τ_1; if these constants do not depend on ϵ we have

$\nu = 0$, $\sigma = 7/2$.

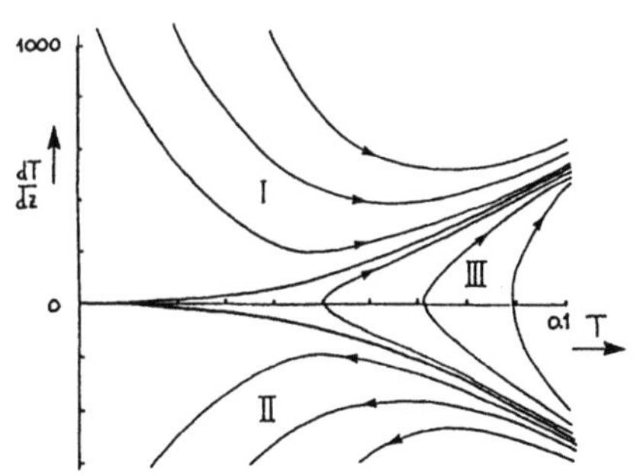

Figure 2, Phase-plane for boundary layer near $z = 0$; $\tau_1 = 0.1$, $\epsilon = 2.5 \times 10^{-2}$

Solutions of type II (Fig. 2) are not interesting as T decreases, solutions of type I are monotonically increasing, type III solutions can be decreasing first with z, after which they are increasing until $T = \tau_1$.

b. The second boundary layer.

The solutions are arriving at $T = \tau_1$ with a steep temperature gradient so we need a second boundary layer in z to describe the solutions. Repeating the scaling we find now the significant degeneration

$$1 + \tfrac{3}{2}\nu - 2\sigma = -5\nu.$$

Conduction and radiative losses are still dominant. The boundary layer equation becomes to lowest order

$$\frac{1}{\bar{T}} \frac{d}{d\zeta} (\bar{T}^{3/2} \frac{d\bar{T}}{d\zeta}) - \beta_1 \bar{T}^{-5} = 0 \tag{9}$$

with integral (in the original variables)

$$\epsilon \tfrac{1}{4} T^3 (\frac{dT}{dz})^2 + \tfrac{1}{3} \beta_1 T^{-3/2} = d, \tag{10}$$

d a constant. The phase-plane is presented in Figure 3; starting with a positive temperature gradient at $T = \tau_1$, the gradient remains positive

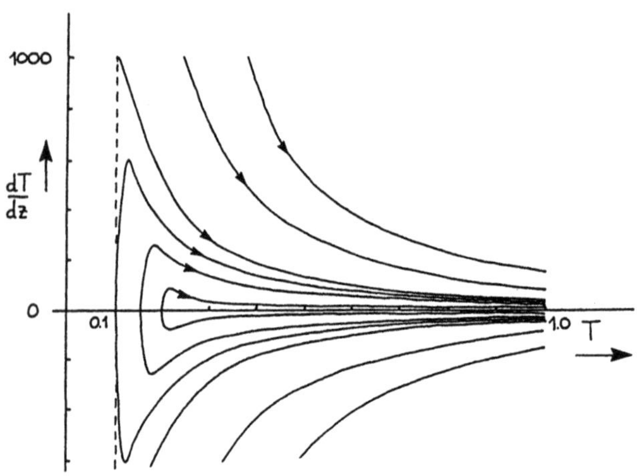

Figure 3, Phase-plane for second boundary layer near $\tau_1 = 0.1$, $\epsilon = 2.5 \times 10^{-2}$.

but it decreases towards zero. At the same time the temperature increases monotonically until the radiation losses are becoming dominated by the heating. The transition takes place in a neighbourhood of a temperature T_h given by

$$\alpha = \beta_1 P_0^2 T_h^{-5}$$

c. From this stage on conduction and heating dominate, there is no spatial boundary layer; T becomes large with ϵ. The integral describing the behaviour of temperature and pressure is (see [1])

$$\epsilon P^2 \left(\frac{dT}{dz}\right)^2 T^3 + \frac{4}{7} \alpha T^{7/2} = e \quad (11)$$

with e a constant and from (2)

$$P(z) = P_0 - \lambda z \quad (12)$$

where $P_0 \geq \lambda$ is a constant.

We can integrate (11) again (see [1]); it follows that at $z = 1$ we have the maximum temperature T_m given by

$$T_m = \left(\frac{7e}{4\alpha}\right)^{2/7}.$$

The expressions obtained in this section have been used to construct static solutions by a shooting procedure. One chooses P_0 and T_0 to obtain a solution satisfying (4) and (5) by varying P_0; see [2].

3. Existence of Static Solutions.

We reformulate the static solar loop problem as follows. In (12) P_0 is considered as an unknown parameter, to be determined. Next we set

$$w = T^{5/2} \quad (13)$$

and introduce (Euler transformation)

$$y = \log P = \log (P_0 - \lambda z) \quad (14)$$

as a new spatial coordinate. In the static case (1) becomes with (13,14)

$$\frac{2}{5} \epsilon \lambda^2 w_{yy} + \alpha w^{2/5} - e^{2y} w^{2/5} \chi(w^{2/5}) = 0 \quad (15)$$

for $\log(P_0 - \lambda) < y < \log P_0$. The boundary conditions (4) become

$$w_y(\log(P_0-\lambda)) = 0, \quad w(\log P_0) = T_0^{5/2} \quad (16)$$

Condition (5) takes the form

$$\int_{\log(P_0-\lambda)}^{\log P_0} w(y)^{2/5} \, dy = \lambda L \quad (17)$$

Hulshof et al. [3] show that for a rather general class of radiative loss functions $\chi(T)$ of the type we are interested in and for ϵ sufficiently small, (15) with the conditions (16,17) has a decreasing solution w(y) (T increasing with z).

In [3] one has furthermore the estimates

$$P_0 = O(\epsilon^{-2/7}), \quad T_0 = O(\epsilon^{-2/7}).$$

The proof has been based on the analysis of a certain operator which has been obtained by formally integrating (15) and which has contraction properties as $\epsilon \to 0$.

4. Time-dependent Solutions.

In the solar coronal loop problem (1,2) we may discern at least two possibilities:

a. $P(t,0) = $ constant (or constant to first order in ϵ);

b. $P(t,0) = f(t)$ which yields with (1) $P(t,z) = f(t) - \lambda z$; the function $f(t)$ has to be determined such that $P(t,z) \geq 0$, $t \geq 0$, $z \in [0,1]$.

Nothing is known about the existence of solutions of (1,2) in the cases a or b with conditions (4,5) and appropriate initial values. The case a should be easier as P can be used again as an independent variable (section 3).

A number of quantitative results have been presented in [1]; in that paper however, the existence of two boundary layers near the base of the loop in the static case is ignored. The analysis in [1] points towards an increasing T near $z = 1$ and subsequently the building of a transition layer in z from high to lower temperature; the transition layer then moves with time towards lower regions of the loop (smaller values of z).

The results are very preliminary and the time-dependent case is basically still an open problem.

5. The τ Sco Loop Problem.

In the case of OB-stars like τ Sco radiation pressure plays a part in the corona. The momentum equation (3) contains S(T) which increases steeply with T until $T = \tau_2$, after which S(T) decreases monotonically. In the case of τ Sco the functions $\chi(T)$ and S(T) have been given in section 1.

In this section we shall sketch how to construct static solutions. The equations are

$$\epsilon \frac{P}{T} \frac{d}{dz}(T^{3/2} P \frac{dT}{dz}) + \alpha - P^2 \chi(T) = 0 \tag{18}$$

$$\frac{dP}{dz} = -\lambda + S(T) \tag{19}$$

The analysis is based on [4] and [5].

We expect very different behaviour, qualitatively and quantitatively, as with increasing T the terms on the righthandside of (19) may balance, causing the solutions to be oscillating. A singular perturbation analysis as in section 2 runs as follows.

a. For $T_0 \leq T \leq \tau_1$ a boundary layer in z exists with the same structure as in section 2a. To first order we recover (8).

b. We also recover the second boundary layer in z which exists for values of T near and above τ_1. With our choice of parameters we recover to first order (10).

c. In a region of values of T near τ_2 we have different behaviour: there is no boundary layer in z but a temperature regime which causes conduction and heating to balance. After scaling and looking for significant degenerations we find to a first approximation

$$\frac{d\bar{P}}{dz} = \delta\, \bar{T}^{15} \tag{20}$$

$$\epsilon\, \frac{\bar{P}}{\bar{T}} \frac{d}{dz}(\bar{P}\, \bar{T}^{3/2} \frac{d\bar{T}}{dz}) + \alpha = 0 \tag{21}$$

We drop the bars, put $T^{65/2} = u$ and transform as in (14)

$$P = e^y \ , \ \frac{d}{dy} = P \frac{d}{dP} \text{ (Euler transformation).}$$

System (20,21) then reduces to

$$\frac{2}{65}\, \epsilon\, \delta^2 \frac{d^2u}{dy^2} + \alpha\, u^{2/65} = 0 \tag{22}$$

We can integrate (22) to find

$$\frac{1}{65}\, \epsilon\, \delta^2 \left(\frac{du}{dy}\right)^2 + \frac{65}{67}\, \alpha\, u^{67/65} = \text{constant,} \tag{23}$$

which can be integrated again to produce an implicit relation for u(y). Rewritten in the original variables, (23) becomes

$$\epsilon\, \frac{1}{4}\, P^2\, T^{33} \left(\frac{dT}{dz}\right)^2 + \frac{1}{67}\, \alpha\, T^{67/2} = \text{constant} \tag{24}$$

We may use (24) or (23), its phase-plane and the relation for u(y) to characterise the behaviour of the solutions in this domain.

d. A similar situation exists in a region of values of T above τ_2. The reader may verify that radiation pressure becomes less effective at higher temperature, pressure decreases, conduction and heating balance. The behaviour of the solutions becomes as described in section 3c.

The singular perturbation analysis of this section can be combined with a numerical analysis of the equations (18,19). We find that the solutions show repeated transitions of the domains described in c and d, resulting in oscillatory behaviour of the solutions. In Figs. 4 and 5 we present a typical case of solutions T(z) and P(z).

Finally we remark that it is possible to obtain two constant stationary solutions of

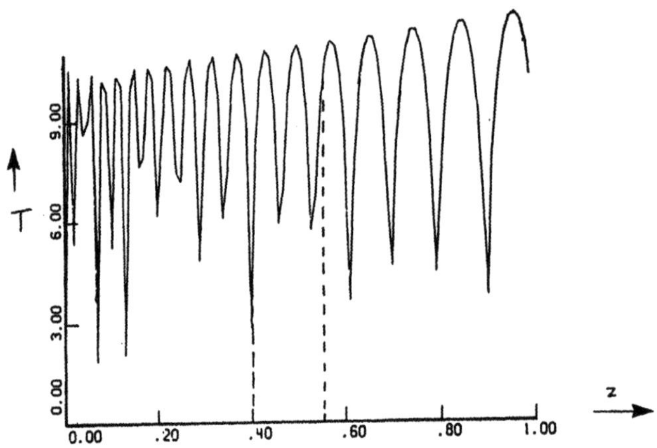

Figure 4. T(z) is the case with radiation pressure; the parameters apply to the τ Sco case. (fig. from [4]).

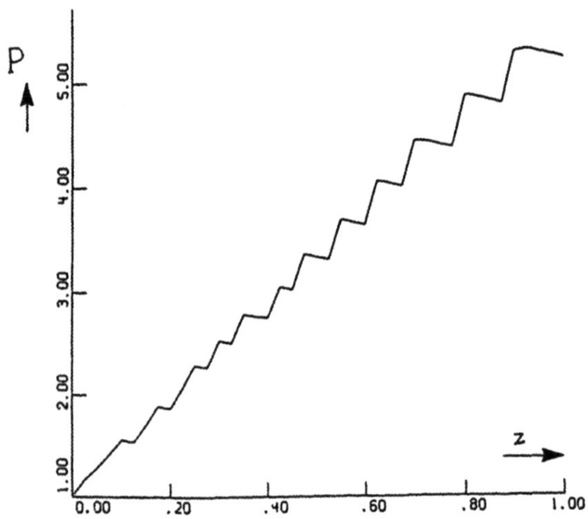

Figure 5. P(z) in the case with radiation pressure; the parameters apply to the τ Sco case (fig. from [4]).

system (1,3). In these cases the loop length or total mass condition (5) has a fixed L but the solutions turn out to be physically relevant. An analysis of these solutions is in progress.

References

1. J.W. Pakkert, P.C.H. Martens and F. Verhulst:
The thermal stability of coronal loops by nonlinear diffusion asymptotics, Astronomy & Astrophysics, 179 (1987), pp.285-299.

2. F. Verhulst and P. Zegeling: Boundary layers in coronal loops, submitted to Astronomy & Astrophysics.

3. J. Hulshof, D. Terman and F. Verhulst: Existence of stationary solutions in the coronal loop problem, Report 88-06 Delft (1988), submitted to Nonlinear Analysis.

4. P. Zegeling, graduate thesis, Rijksuniversiteit Utrecht (1987).

5. P. Zegeling, J. Kuijpers and F. Verhulst, to be published.

Perturbation Solutions of Magnetomechanical Signal Processing Devices

*G.A. Maugin and A.N. Abd-Alla**

Université Pierre et Marie Curie, Laboratoire de Modélisation en Mécanique, Tour 66, 4 Place Jussieu, F-75252 Paris Cedex 05, France
*On leave from the University of Sohag, Egypt

Magnetomechanical materials, such as ferromagnetic elastic polycrystals exhibiting magnetostrictive couplings can be used in a variety of signal processing devices although not to the same extent as piezoelectric materials. The nonlinear characteristics of these devices may be either beneficial or undesirable. Perturbation and multiple-scale techniques are used to study, in order of increasing complexity, the propagation of bulk nonlinear magnetoelastic waves and the generation of harmonics, resonators exhibiting the phenomenon of anisochronism, and magnetoacoustic nonlinear surface waves exhibiting mode couplings.

1. PREREQUISITES

Magnetostriction is the production of a stress even in the applied magnetic field. It is, per se, a nonlinear effect in contradistinction, say, with piezoelectricity [1]-[2]. It is only one type of electromagnetomechanical couplings in deformable solids [3]-[4]. This contribution presents a theoretical study of nonlinear effects which can be useful or undesirable in transducers, resonators and S.A.W. systems using magnetostrictive couplings [5]. The mathematical tools used have become conventional. They pertain essentially to perturbation methods at a more or less sophisticated level (direct expansion in a small parameter, multiplescale technique, Galerkin representation, coupled-amplitude equations, numerical simulations [6]). Before focussing on these nonlinear effects we need recall the basic nonlinear, rotationally invariant continuum equations of magneto strictive insulators that we have developed elsewhere [7] and give the essential linear-response characteristics about which perturbations are eventually studied.

A. - Equations of nonlinear magnetoelasticity [7]

We consider elastic insulators in the framework of quasi-magnetostatics. Nonlinearities may have elastic, magnetic and mixed origins. Of necessity all equations are written down in the material description so that all boundaries are fixed even for nonlinear strains. The equations consist of the balance equations of motion, Maxwell equations and the associated boundary conditions together with the nonlinear, coupled, stress and magnetic-induction constitutive equations. We have thus (for a domain D_o of regular boundary ∂D_o equipped with outward unit normal **N**) :

. Field equations in D_o :

$$\rho_o \frac{\partial^2 u_i}{\partial t^2} = \frac{\partial T^t{}_{Ki}}{\partial X_K} \ , \ \frac{\partial B_K}{\partial X_K} = 0 \ , \ H_K = -\partial \phi / \partial X_K \tag{1.1}$$

. Boundary conditions at ∂D_o :

$$N_K [T^t_{Ki}] = 0 \ , \ N_K [B_K] = 0 \ , \ [\phi] = 0 \tag{1.2}$$

. Constitutive equations

$$T^t_{Ki} = \delta_{Ri} \ (C_{KRMN} \ U_{M,N} + \gamma'_{KRMNPQ} \ U_{M,N} \ U_{P,Q}$$

$$+ \delta_{KRMNPQAB}\ U_{M,N}\ U_{P,Q}\ U_{A,B} + B'''_{KRMN}\ H_M\ H_N$$

$$+ B'''_{KRMNPQ}\ H_M H_N\ u_{P,Q} + \text{h.o.t}) \tag{1.3}$$

$$B_K = \mu_{KL}\ H_L - B^{\circ}_{KLMN}\ H_L\ U_{M,N} + \frac{1}{6}\ \chi_{KLMN}\ H_L\ H_M\ H_N$$

$$- \lambda^{0}_{KRMNPQ}\ H_R\ U_{M,N}\ U_{P,Q} + \text{h.o.t} \tag{1.4}$$

where upper and lower case Latin indices refer to material and present Cartesian coordinates, respectively. The symbols ρ_0, $u_i = \mathcal{X}_i - X_K\ \delta_{Ki}$, $\mathcal{X}_i$, B_K, H_K, $\emptyset$ and T^t_{Ki} denote, respectively, the matter density at K_R (the reference configuration), the elastic displacement, the nonlinear motion, the "Lagrangian" magnetic induction and field, the magnetostatic scalar potential and the total Piola-Kirchhoff nonsymmetric stress tensor. δ_{Ki} is a shifter. C_{KLMN} and μ_{KL} are tensors of elasticity coefficients of second order and of magnetic permeabilities, χ_{KLMN} is a tensor of magnetic susceptibilities of the fourth order, while Υ_{KRMNPQ} and $\delta_{KRMNPQAB}$ are effective tensors of elasticity coefficients of the third and fourth orders, B'''_{KRMN} and B°_{KLMN} are effective tensors of magnetostriction coefficients of the first order and B'''_{KRMNPQ} and λ^0_{KRMNPQ} are effective tensors of magnetostriction of the second order. The last two of these are neglected in the following analysis. The material has been assumed centrosymmetric and, obviously, does not exhibit any natural piezomagnetism (a seldom encountered coupling).

B. - Comments on linear solutions

Simple prerequisites for a better understanding of nonlinear wave and vibration phenomena are the consideration of bulk and surface wave propagation in the linear and linearized theory and the usual response of an elastic resonator. That is, we are interested in three of the usually considered "configurations" in electro or magneto-mechanical devices [8], which are sketched in Figure 1.

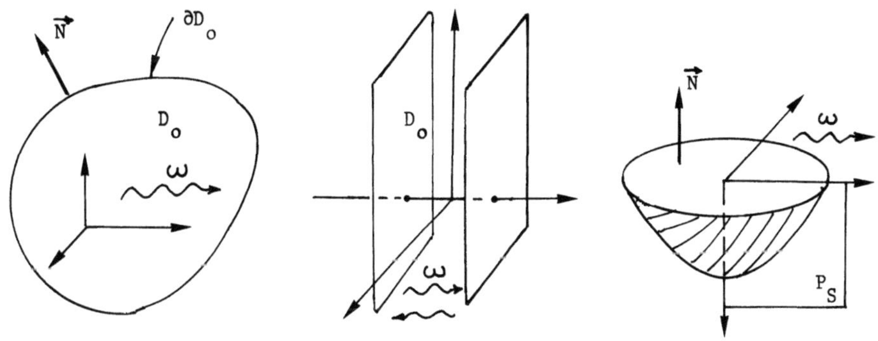

(a) bulk wave (b) elastic resonator (c) surface wave

Figure 1 - Studied "configurations"

Fully linear equations. - In the absence of initial magnetic field, eqns. $(1.1)_1$ and $(1.1)_2$ fully uncouple and read, on account of eqns. (1.3) and (1.4),

$$\rho_0 \ddot{u}_R - C_{KRMN}\ u_{M,NK} = 0 \tag{1.5}$$

$$\mu_{KL}\ \emptyset_{L,K} = 0 \tag{1.6}$$

yielding, separately, a problem of linear elasticity for an anisotropic solid on the one hand, and an anisotropic potential problem on the other hand. The first of these, when studying free bulk oscillations, gives elastic eigenmodes which provide an orthonormal basis $\{1_R^\alpha\ ;\ \alpha=1,2,3\}$ which can be used to project any other wave problem (this is assumed herein after). For instance, on this basis, eqn. (1.5) gives the mode equations

$$\rho_0\, \ddot{u}^\alpha = m_\alpha\, u^\alpha_{,\mathfrak{X}\mathfrak{X}} \qquad \alpha = 1,2,3 \tag{1.7}$$

where $\mathfrak{X}$ is a curvilinear coordinate in the direction of propagation.

Linearized equations in presence of a spatially uniform magnetic field. The example (1.5)-(1.6) exhibits, in fact, a typical feature of magnetostriction : this coupling is operative only in the presence of a bias magnetic field H°. Indeed, if we now linearize the system (1.1) and (1.3)-(1.4) about an initial configuration of zero displacement but non zero, spatially uniform magnetic field $H^\circ_K = -\phi_K$, instead of eqns. (1.5)-(1.6) we are led to the quations

$$\begin{aligned} &\rho_0 \ddot{u}_R - C_{KRMN}\, u_{M,NK} + F'_{KRN}\,(H^\circ)\phi_{,NK} = 0 \\ &\mu_{KL}\,\phi_{,LK} + F^*_{KMN}\,(H^\circ)\, u_{M,NK} = 0 \quad , \end{aligned} \tag{1.8}$$

where F'_{KRN} and F^*_{KLN} are third-order tensorial coefficients - not necessarily equal-which depend on H° and go to zero with it. Equations (1.8) and the associated boundary conditions at $D_\circ$,

$$\begin{aligned} &N_K[C_{KLMN}\, u_{M,N} - F'_{KLN}\,\phi_N] = 0 \\ &N_K[\mu_{KL}\,\phi_{,L} + F^\circ_{KMN}\, u_{M,N}] = 0 \qquad [\phi] = 0 \quad , \end{aligned} \tag{1.9}$$

provide a coupled linear problem of the same type as linear piezoelectricity [2]. In particular, the system (1.8) in nondimensional notations, projected onto the above-mentioned basis $\{1_R^\alpha\}$, provides "one-dimensional" problems for each α-component, of the type (commas for differentiation have been omitted)

$$u_{tt} - u_{xx} + \bar{\beta}_1 \phi_{xx} = 0 \quad , \quad \phi_{xx} + \bar{\beta}_2 u_{xx} = 0 \tag{1.10}$$

or

$$u_{tt} - \frac{1}{K^2}\, u_{xx} = 0 \quad , \quad K^2 \simeq 1 - \varepsilon_m \ , \ 0 < \varepsilon_m \equiv \bar{\beta}_1 \bar{\beta}_2 \ll 1, \tag{1.11}$$

where ε_m may be called a magnetoacoustic coupling coefficient. It has for effect to reduce the elastic speed by a very small amount [9].

For surface waves (Figure 1c), the system (1.8)-(1.9) is of the same type as the corresponding one in piezoelectricity when the free surface $X_2 = 0$ is not earthed, so that one has to simultaneously solve for the external field and match internal $(X_2>0)$ and èxternal solutions $(X_2<0)$ at the flat boundary $X_2 = 0$. Because of anisotropy, surface modes, in general, cannot be split and one is left with a four-mode solution of the generalized Rayleigh-wave type in the form

$$\begin{aligned} u_K &= \sum_{\alpha=1}^{4} C_K^\alpha \exp(-k_\circ \chi_\alpha X_2)\, \exp[i(\omega t - k_\circ X_1)] \\ \phi &= \sum_{\alpha=1}^{4} G^\alpha \exp(-k_\circ \chi_\alpha X_2)\, \exp[i(\omega t - k_\circ X_1)] \end{aligned} \tag{1.12}$$

for $X_2 > 0$. This is true for a general orientation of $\underline{H}^\circ$. However, for centrosymmetric cubic crystals and $\underline{H}^\circ$ along the propagation axis X_1, on obtains a magneto-elastic Rayleigh wave with a displacement polarized parallel to the sagittal plane P_S, while for $\underline{H}^\circ$ set orthogonal to P_S, a split of the general problem (1.12) is indeed achieved, yielding a pure elastic Rayleigh wave (polarized in the sagittal plane) and a shear-horizontal (SH) magnetoelastic wave (u_S, ϕ). This possibility is exploited in the nonlinear problem.

Finally, in the resonator configuration (Figure 1b)), the linear version (1.5) provides stationnary-wave solutions (of a purely elastic nature, frequency normalized to one)

$$u = U_0 \sin t \sin x \quad , \; h = (2q+1)\frac{\pi}{2} = \frac{\pi}{2}, \frac{3\pi}{2}, \frac{5\pi}{2}, \ldots \tag{1.13}$$

The following sections reconsider the above three problems with the purpose of placing in evidence nonlinear effects due to magnetostriction and elastic nonlinearities in the presence of a bias magnetic field. We note that the magnetostriction of uniformly magnetized bodies, along with thermoelasticity, is one example of an elasticity theory with internal strains [1].

2. - NONLINEARITIES AND BULK WAVES

A. - Progressive waves

We use the monomode hypothesis [2] and this allows us to extract from the system (1.1) - (1.2) a nonlinear model problem (for one displacement component u - not necessarily longitudinal - and the magnetic scalar potential ϕ) in adimensional form as

$$\begin{aligned} &u_{tt} - u_{xx}(1 + 2\gamma u_x + 3\delta\, u_x^2) = \beta_1(\phi_x^2)_x \\ &\phi_{xx} - \beta_2(\phi_x u_x)_x + \frac{1}{6}\chi(\phi_x^3)_x = 0 \end{aligned} \tag{2.1}$$

in spacetime (x,t), where γ and δ account for elastic nonlinearities, χ for magnetic nonlinearities and β_1 and β_2 for magnetostriction [compare eqns. (1.10)]. Small parameters ε_1 and ε_2 can be introduced in such a way that

$$\begin{aligned} &\varepsilon_1 \simeq \gamma u_x \simeq \varepsilon_1\beta_1 \quad , \; u_x = O(10^{-4}) \\ &\varepsilon_2 \simeq \beta_2\, u_x \simeq \chi = O(\varepsilon_1) \end{aligned} \tag{2.2}$$

A straightforward expansion of the solution (u,ϕ) à la Poincaré in terms of ε_1 and $\varepsilon_2 = O(\varepsilon_1)$ such that (here H° is set along the direction of propagation)

$$\begin{aligned} &u(x,t) = \varepsilon_1 u^{(o)}(x,t) + \varepsilon_2\, u^{(1)}(x,t) + \ldots \\ &\phi_x(x,t) = -H^{(o)} + \varepsilon_2\, \phi_x^{(1)}(x,t) + \varepsilon_2^2 \phi_x^{(2)}(x,t) + \ldots \end{aligned} \tag{2.3}$$

for the boundary and initial conditions

$$u(x=0,t) = U_\circ(1-\cos\omega t) \; , \; \phi_x(x,0) = -H^\circ = \text{const.}, \quad \forall x \quad , \tag{2.4}$$

yields a hierarchy of problems that we do not reproduce here. The source condition $(2.4)_1$ is essentialy accounted for in the $u^{(o)}$ solution and additional conditions at the source (see a recent work for a similar problem in elasticity and piezoelectricity [10]) allows one to find the missing "integration constants" at higher orders. In any case, the solution obtained exhibits the production of harmonics of unphysically spatially growing amplitude with distance, restricting this type of solution to a near-field approximation as is common in nonlinear acoustics. A uniformly valid far-field solution for (2.1) with boundary initial conditions (2.4) is to be looked for using a multiple-scale technique with slow (space) variable $s = \varepsilon x$, $\varepsilon = \varepsilon_1 = O(\varepsilon_2)$:

$$\begin{aligned} &u(x,t) = \tilde{u}(x,s,t) \quad , \; \phi_x = \tilde{\phi}_x(x,s,t) \\ &\tilde{u} = \varepsilon_1\tilde{u}^{(o)}(x,s,t) + \ldots \quad , \; \tilde{\phi}_x = -H^{(o)} + \varepsilon_2\tilde{\phi}_x^{(1)} + \ldots \end{aligned} \tag{2.5}$$

With a source term $(2.4)_1$, the secularity condition at order one imposes, after expansion in Fourier series of the implicit solution, that $u^{(o)}$ be of the form ($\varepsilon_m = \bar{\beta}_1\,\bar{\beta}_2 \equiv \beta_1\,\beta_2 H^{(o)2}$)

$$u^{(o)} = U_\circ\left[1 - \sum_{n=1}^{\infty} \frac{2J_n(ns/L_B^\circ)}{n^2 s/L_B} \cos(n\Psi) - L_B^{-1}\varepsilon_m\, x\right] \tag{2.6}$$

where $\Psi = t - x$, $L_B = (\gamma U_o)^{-1}$ is the breaking distance for shock-wave formation, J_n is the n-th Bessel function of the first kind and ε_m has previously been defined. The solution (2.6) consists of an infinity of multiple harmonics of the fundamental vibration Ψ with amplitudes which have a pseudo-periodic behavior along the propagation direction (this is the celebrated Fubini-Ghiron solution of nonlinear acoustics [2],[11]) and of a term due to magnetostriction which creates a spatially uniform strain in agreement with a previously made remark. Such an additional term may also exist in pure elasticity depending on the initial conditions [12].

B. - Stationary waves

In the case of the resonator (Figure 1b) of Thickness 2h, with H° directed throughout the thickness, one has to solve eqns. (2.1) for $|x| < h$ and Maxwell equation $\phi_{xx} = 0$ for $|x| > h$ on account of the matching conditions at $|x| = h$:

$$[-\phi_x + \beta_2 \phi_x u_x + \frac{1}{6} \chi \phi_x^3] = 0 \quad , \qquad [\phi] = 0 \quad , u_x = 0 \tag{2.7}$$

Since the thickness 2h is usually much smaller than L_B one is tempted to consider a straightforward expansion in terms of $\varepsilon_1 = O(\varepsilon_2)$ for the above set problem. At order zero one obtains a pure elastic solution such as (1.13). Tuning the resonator on the first mode (q = 0) and computing the first order solution, one exhibits an additional contribution at the fundamental frequency (here equal to one) and this new term, combined with the zeroth-order solution, yields a shift in the original wave number (originally, one)by the amount (see [13])

$$\delta_A = \varepsilon_m \tag{2.8}$$

This is the so-called anisochronism of the resonator. Here it is due to magnetostriction. An additional anisochronism due to pure nonlinear elasticity appears at the next order of approximation [2],[14].

Another way to look at the nonlinear magnetoelastic resonator problem is to use a Galerkin representation followed by a multiple-scale technique in the manner of Googerdy and Peddleson [15] (see also [2], pp. 84-89). To that purpose the thickness of the resonator is normalized to the spatial internal $(0,\pi)$. The functions sin nx then provide a natural basis for elastic stationary waves and the nonlinear magnetoelastic solution can be projected on it. For the sake of simplicity the nonlinear contributions with coefficients δ and χ are discarded in eqns. (2.1) which, therefore, reduce to

$$\left.\begin{aligned} &u_{tt} - u_{xx}(1 + 2\gamma u_x) = \beta_1 (\phi_x^2)_x \\ &\phi_{xx} - \beta_2 (\phi_x u_x)_x = 0 \end{aligned}\right\} \text{ for } 0 < x < \pi \tag{2.9}$$

$$\phi_{xx} = 0 \qquad \text{for } x < 0,\ x > \pi$$

$$\left.\begin{aligned} &[\phi_x - \beta_2 \phi_x u_x] = 0 \\ &[H^\circ] = 0 \quad , \quad u(x,t) = 0 \end{aligned}\right\} \text{ at } x = 0 \quad , \quad x = \pi \tag{2.10}$$

with the initial conditions

$$u(x,0) = \sin x, \quad u_t(x,0) = 0 \tag{2.11}$$

The natural Galerkin representations are

$$u(x,t) = \sum_{n=1}^{\infty} f_n(t) \sin nx \qquad \phi_x(x,t = - H^{(o)} + \sum_{n=1}^{\infty} \Phi_n(t) \cos nx \tag{2.12}$$

Substituting from this into (2.9) - all other equations will be satisfied - and accounting for the orthonormality of the basis functions we obtain infinite sets of nonlinear ordinary differential equations for the f_n's and the Φ_n's :

$$\ddot{f}_n + n^2 f_n + 2\beta_1 H^{(o)} n\Phi_n = \gamma \sum_{m=1}^{\infty} m^2 f_m[(m+n)f_{m+n} + (m-n)(f_{n-m}-f_{m-n})]$$

$$+ \beta_1 \sum_{m=1}^{\infty} \Phi_n [(m+n)\Phi_{m+n} + (m-n)(\Phi_{n-m} - \Phi_{m-n})] \quad , \tag{2.13$_1$}$$

$$n\Phi_n + \beta_2 H^{(o)} n f_n = \frac{1}{2} \sum_{m=1}^{\infty} m f_m [(m+n)\Phi_{m+n} + (n-m)\Phi_{n-m}]$$

$$- \frac{1}{2} \sum_{m=1}^{\infty} m^2 f_m [\Phi_{m+n} - \Phi_{m-n} - \Phi_{n-m}] \; . \tag{2.13$_2$}$$

This system is now solved by a <u>two-time scale</u> technique <u>with</u>

$$\xi = t \quad , \quad \eta = \varepsilon t \quad , \quad \varepsilon = \varepsilon_1 = O(\varepsilon_2)$$

$$\begin{aligned} f_n(\xi,\eta) &= \varepsilon_1 f_{no}(\xi,\eta) + \varepsilon_1^2 f_{n1}(\xi,\eta) + \ldots \\ \Phi_n(\xi,\eta) &= \varepsilon_2 \Phi_{no}(\xi,\eta) + \varepsilon_2^2 \Phi_{n1}(\xi,\eta) + \ldots \end{aligned} \tag{2.14}$$

Only first two orders are considered in the resulting hierarchy of equations. Symbolically, we have :

$$\begin{aligned} &\frac{\partial^2 f_{no}}{\partial \xi^2} + n^2 f_{no} = 0 \\ &\Phi_{no} = - \beta_2 n H^{(o)} f_{no} \end{aligned} \tag{2.15}$$

then

$$\begin{aligned} &\frac{\partial^2 f_{n1}}{\partial \xi^2} + n^2 f_{n1} = g(f_{mo}, \Phi_{no}) \\ &n\Phi_{n1} = - \beta_2 H^{(o)} n^2 f_{n1} + h(f_{mo}, \Phi_{no}) \end{aligned} \tag{2.16}$$

and so on. We see that the Φ_{ni}'s are determined by the f_{ni}'s and thus one seeks

$$f_{no} = C_n(\eta) \cos [\alpha_n(\eta) - n\xi]. \tag{2.17}$$

The <u>secularity</u> condition for (2.16) imposes restrictions on the C_n's and α_n's. For instance, for a <u>two-mode process</u>, n = 1,2, and we obtain the following conditions :

$$\frac{dC_1}{d\eta} + \frac{1}{4} C_1 C_2 = 0 \quad , \quad \frac{dC_2}{d\eta} - \frac{1}{16} C_1^2 = 0 \tag{2.18}$$

and

$$\frac{d\alpha_1}{d\eta} = \frac{1}{4} C_2 - \varepsilon_m \quad , \quad \frac{d\alpha_2}{d\eta} = - \frac{1}{16} \frac{C_1^2}{C_2} - 4\varepsilon_m \tag{2.19}$$

Equations (2.18) integrate immediately to

$$C_1 = \text{sech}(\eta/8) \quad , \quad C_2 = \frac{1}{2} \tanh(\eta/8) \; . \tag{2.20}$$

Then it is readily checked that

$$u_{10} = \text{sech}\,(\eta/8)\cos[\xi + \delta_A\eta - \ln\cosh(\eta/8)]\sin x \qquad (2.21)$$

so that the anisochronism of the resonator, $\Delta\omega/\omega_0$ (with $\omega_0 = 1$), is obtained as

$$\frac{\Delta\omega}{\omega_0} \simeq \varepsilon\delta_A - \varepsilon\frac{t}{128} + O(\varepsilon^4 t^3) \quad , \qquad (2.22)$$

and it involves a time effect which was overlooked in the foregoing simpler analysis.

3. - NONLINEARITIES AND SURFACE WAVES

Most of the time the surface-wave problem remains multimode and this obviously generates quite an amount of difficulty in the nonlinear regime (for sufficiently powerful input signals). This problem has been examined in some detail in Ref.[16] and in great detail in Ref. [17] to which we refer the reader. Here we shall content ourselves with a very few indications. One can first try a straight forward expansion of the full multimode problem for eqns. $(1.1)_{1-2}$ for $X_2 > 0$, $(1.1)_2$ for $X_2 < 0$ and the boundary conditions (1.2) at $X_2 = 0$ with $[\varepsilon_1 = O(\varepsilon_2)]$

$$u_K = \varepsilon_1 u_K^{(o)} + \varepsilon_1^2 u_K^{(o)} + \dots$$
$$\phi_{,K} = -H^{(o)}_K + \varepsilon_2 \phi_{,K}^{(1)} + \varepsilon_2^2 \phi_{,K}^{(2)} + \dots \qquad (3.1)$$

In this manner, with a zeroth-order solution shown to be purely elastic [compare (1.12) which is already magnetoelastic] written as

$$u_K^{(o)} = U_1 \sum_{\alpha=1}^{3} C_K^{\alpha} \exp(-k_o \chi_\alpha X_2)\exp[i(\omega t - k_o X_1)] \quad , \qquad (3.2)$$

it is found that, to first order, magnetostriction and nonlinear elasticity cause a production of second harmonic which varies linearly with distance while magnetostriction alone causes an anisochronism (here change in the propagation velocity of the ω-component) which is independent of travelled distance. In fact, we have for the ω-component

$$u_K^{(\omega)} = U(X_1)\sum_{\alpha=1}^{3} C_K^{\alpha} \exp[i\omega(t - \frac{X_1}{V(X_1)} - \frac{n_2^{\alpha} X_2}{V_o})] \qquad (3.3)$$

with

$$U = U_1(1 + k_o U_1 e_M X_1) \quad , \quad V = V_o(1 + U_1 f_M) \qquad (3.4)$$

where e_M and f_M are computable quantities. In contrast, elastic nonlinearities require to go one order further in the expansion and yield instead of (3.4) (see [14] or [2]) :

$$U = U_1(1 + k_o^4 U_1^2 e_E X_1^2) \quad , \quad V = V_o(1 + k_o^3 U_1^2 f_E X_1) \qquad (3.5)$$

Since surface waves can travel over long distances it may also be salient to envisage a uniformly valid far-field solution. To that purpose a two-space scale technique is used with slow space variables $\xi = \varepsilon X_1$, $\eta = \varepsilon X_2$, $\varepsilon = \varepsilon_1 = O(\varepsilon_2)$ with the expansions

$$u_K(X_1,X_2,t) = \tilde{u}_K(X_1,X_2,\xi,\eta,t) = \varepsilon_1 \; \tilde{u}_K^{(o)} + \ldots$$
$$\phi_{,K}(X_1, X_2, t) = - \tilde{H}_K (X_1, X_2, \xi, \eta, t) = - H^{(o)}_K + \varepsilon_2 \tilde{\phi}_{,K}^{(1)} + \ldots \tag{3.6}$$

The zeroth order system is satisfied identically by taking $u_K^{(o)}$ as a linear combination of harmonics with slowly varying amplitudes in the form

$$\tilde{u}_K^{(o)} = \frac{1}{2} U_1 \sum_{n=1}^{\infty} \sum_{\alpha=1}^{3} [C_K^\alpha A_n^\alpha (\zeta) \exp (in\psi^\alpha) + c.c]$$
$$\psi^\alpha = \omega t - k_o X_1 - k_o \lambda^\alpha X_2 \quad , \quad \zeta = \xi + \mu_n^\alpha \eta \tag{3.7}$$

Thus there is some freedom left to this solution through the functional dependence of the A_n^α where μ_n^α is some kind of depth parameter to be determined for the α-th component of the n-th harmonic. At the next-order of approximation $u_K^{(1)}$ is shown to consist of several contributions one of which being the general solution of the homogeneous form of the equations satisfied by $u_K^{(1)}$, hence a contribution of the same form as (3.7) with functions B_n^α replacing A_n^α. The total solution ($u_K^{(1)}$, $\phi^{(1)}$), including also contributions "forced" by $u_K^{(o)}$, must check the boundary conditions. These conditions yield a linear inhomogeneous system for the B_n^α, which is singular having zero determinant. The compatibility for solving this then produces the looked for condition imposed on the A_n^α 's in the form of nonlinear complex ordinary differential equations akin to the coupled-amplitude equations of the nonlinear optics of absorbing transparent media (compare |18|). For instance, with H° set orthogonal to the sagittal plane and a centrosymmetric cubic crystal, we know that the linear problem splits in two problems, one of which involving an SH magnetoelastic mode (see comments in Section 1). We consider this is still practically valid at order ε in the nonlinear case, so that we can envision a monomode (SH) process coupled to the magnetic field. Limiting ourselves to the interactions between the fundamental and the second harmonic in the solution (3.7) there remain two complex amplitudes A_1 and A_2 for $\tilde{u}_3^{(o)}$ and the nonlinear ODE's governing these amplitudes are shown to read

$$\frac{dA_1}{d\zeta} = - \Gamma_{121} A_1^* A_2 - \Gamma_1 A_1 \; , \quad \frac{dA_2}{d\zeta} = \bar{\Gamma}_{211} A_1^2 - \Gamma_2 A_2 \quad , \tag{3.8}$$

where the Γ's are complex and Γ_1 and Γ_2 vanish with $H^{(o)}$. If the latter two are zero, eqns. (3.8) in fact admit exact solutions with the behaviour $|A_2| \propto \tanh (A\zeta)$, $|A_1| \propto \operatorname{sech} (A\zeta)$ - as in an index - matching medium in optics. For Γ_1 and Γ_2 nonzero, i.e. in the presence of magnetostriction ($H^{(o)}$ is necessary for that) there are no analytic solutions but a numerical solution shows that magnetostriction modifies the known exact solution only for distances much longer than the breaking distance L_B computed from the nonlinear elastic behaviour (the pure elastic case was approached in [19] and treated by means of asymptotic expansions [20]). This concludes the present contribution which shows how modern methods of asymptotic analysis can help characterize the essential nonlinearities of magnetomechanical devices.

REFERENCES

1. G.A.MAUGIN, in: Electromagnetic Interactions in Elastic Solids , ed.H. Parkus , pp.243-324 (Springer-Verlag , Vienna, 1979).

2. G.A.MAUGIN: Nonlinear Electromechanical Effects and Applications - A Series of Lectures (World Scientific , Singapore, 1985).

3. G.A.MAUGIN(Editor): The Mechanical Behavior of Electromagnetic Solid Continua (North-Holland, Amsterdam, 1984).

4. G.A.MAUGIN: Continuum Mechanics of Electromagnetic Solids (North-Holland, Amsterdam, 1988).

5. V.M.RISTIC: Principles of Acoustic Devices (J.Wiley-Interscience , New York , 1983).

6. A.H.NAYFEH and D.T.MOOK: Nonlinear Oscillations (J.Wiley-Interscience , New York , 1979).

7. A.N.ABD-ALLA and G.A.MAUGIN: J.Acoust.Soc.Amer., 82, 1746-1752 (1987).

8. G.A.MAUGIN, B.COLLET and J.POUGET: in: Nonlinear Wave Propagation in Solids, ed.T.W.Wright , Vol.AMD-77 ,pp.57-84 (A.S.M.E., New York, 1986).

9. G.A.MAUGIN and A.HAKMI: J.Acoust.Soc.Amer., 76, 826-840 (1984).

10. N.DAHER and G.A.MAUGIN:"Intermodulation and Generation of Elastic and Piezoelectric Waves in Anisotropic Solids" (submitted for publication, 1988), also , same authors ,in:Elastic Wave Propagation ,eds.M.F.McCarthy and M.Hayes,(North-Holland, Amsterdam, 1988).

11. R.T.BEYER: Nonlinear Acoustics (Dept of the Navy , US Government Printing Office, Washington, D.C., 1974).

12. J.H.CANTRELL, W.T.YOST and P.LI: Phys.Rev., 35B, 9780-9782 (1987).

13. A.N.ABD-ALLA and G.A.MAUGIN:"Harmonic Generation and Anisochronism in Magnetostrictive Materials" (submitted for publication, 1988).

14. M.PLANAT: Thesis , L.P.M.O.-C.N.R.S., Besançon, France (1984).

15. A.GOOGERDY and J.PEDDLESON Jr.,"An Evaluation of the Galerkin Method for the Solution of Nonlinear Wave-Propagation Problems", Commun.to the 21st Society of Engineering Science Meeting (Blacksburg, Va., U.S.A., October 1984).

16. A.N.ABD-ALLA and G.A.MAUGIN: in :Recent Developments in Surface Acoustic Waves ,eds. D.F.Parker and G.A.Maugin (Euromech 226, Nottingham, U.K.;1987), Springer Series on Wave Phenomena (Springer-Verlag, Berlin,Heidelberg, 1988).

17. A.N.ABA-ALLA and G.A.MAUGIN:"Linear and Nonlinear Surface Waves on Magnetostrictive Substrates" (submitted for publication, 1988).

18. G.C.BALDWIN: An Introduction to Nonlinear Optics (Plenum Press, New York , 1969).

19. E.G.LEAN and C.C.TSENG : J.Appl.Phys., 41, 3912-3917 (1970).

20. M.PLANAT: J.Appl.Phys., 57, 4911-4915 (1985).

An Asymptotic Theory for a System of Weakly Nonlinear Wave Equations

W.T. van Horssen

Department of Mathematics and Computer Science,
Delft University of Technology,
Julianalaan 132, NL-2628 BL Delft, The Netherlands

1. Introduction

In this paper an asymptotic theory for a class of initial-boundary value problems for systems of nonlinearly coupled wave equations is presented. The theory implies the well-posedness of the problem in the classical sense and the validity of formal approximations on long time-scales. As an application of the theory an initial-boundary value problem for a system of nonlinearly coupled wave equations is studied using a two-timescales perturbation method. From an aero-elastic analysis it is shown that this initial-boundary value problem may be regarded as a model describing the galloping oscillations of overhead transmission lines.

2. A Model of the Galloping Oscillations of Overhead Transmission Lines

Galloping can be defined as a low frequency, large amplitude phenomenon involving an almost purely vertical oscillation of single conductor lines on which for instance ice has accreted. The frequencies involved are so low that the assumption can be made that the aerodynamic forces are as in steady flow. Another consequence of these low frequencies is that structural damping may be neglected. In severe cases galloping may give rise to conductor damage due to impact of conductor lines. The usual conditions causing galloping (see also [6]) are those of incipient icing in a stable atmospheric environment implying uniform (but not necessarily high velocity) airflows.

Recently in [1] an oscillator with two degrees of freedom has been considered to describe the oscillations of a rigid circular cylinder with a small ice ridge in a uniform airflow. In that approach a system of two coupled, ordinary differential equations is obtained, describing the displacements of he cylinder in two directions. And in [4] a cylinder-shaped transmission line has been considered to describe the vertical displacement of the conductor due to galloping. In this paper the vertical as well as the horizontal displacements of a circular cylinder-shaped conductor with a small ice ridge will be considered in a horizontal airflow. In order to describe the galloping oscillations a right-handed coordinate system is set up where one of the endpoints of the conductor is the origin. Through this point three mutually perpendicular axes

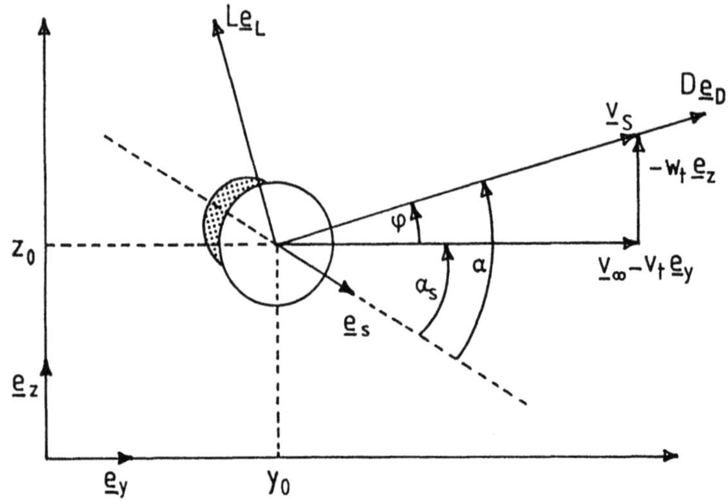

Figure 2.1. Cross-section of the circular conductor with ice ridge

(the x-, y-, and z-axis) are drawn, where the z-axis coincides with the direction of gravity. The endpoints of the conductor are supposed to be in (0,0,0) and (1,0,0), where 1 is the distance between the pylons. To model galloping a cross-section (perpendicular to the x-axis) of the conductor is considered. It is assumed that all cross-sectional shapes are identical and symmetric. Along the axis of symmetry of a cross-section a vector $\underline{e}_s$ is defined to be directing away from the ice ridge and starting in the centre of the cross-section. In fig. 2.1 the centre of the cross-section is considered to be at $x = x_0$, $y = y_0$ and $z = z_0$ with $0 < x_0 < 1$.

Let $v(x_0,t)$ respectively $w(x_0,t)$ denote the y-coordinate respectively the z-coordinate of the centre of the cross-section at $x = x_0$ and time t. It is assumed that every cross-section perpendicular to the x-axis oscillates in a plane parallel to the (y,z)-plane. Furthermore, the torsion of the conductor is not taken into account. Let the static angle of attack α_s (assumed to be constant and identical for all cross-sections) be the angle between $\underline{e}_s$ and the uniform airflow $\underline{v}_\infty$. In this uniform airflow with flow velocity $\underline{v}_\infty = v_\infty \underline{e}_y$ $(v_\infty > 0)$ the conductor may oscillate due to the lift force $L\underline{e}_L$ and the drag force $D\underline{e}_D$. It should be noted that the drag force $D\underline{e}_D$ has the direction of the virtual wind-velocity $\underline{v}_s$ and that the lift force $L\underline{e}_L$ has a direction perpendicular to $\underline{v}_s$. Now the conductor is considered to be a one-dimensional continuum in which the only interaction between different parts is a tension T, which for all times is assumed to be constant along the conductor. The equations describing the horizontal and the vertical motion of the conductor are given by

$$\rho_c A v_{tt} - TA\frac{\partial}{\partial x}\left[v_x(1+v_x^2+w_x^2)^{-1/2}\right] = D\cos\varphi - L\sin\varphi, \tag{2.1}$$

$$\rho_c A w_{tt} - TA\frac{\partial}{\partial x}\left[w_x(1+v_x^2+w_x^2)^{-1/2}\right] = \rho_c A g + D\sin\varphi + L\cos\varphi, \tag{2.2}$$

where D respectively L are the magnitudes of the drag respectively lift force acting on the conductor per unit length of the conductor, ρ_c the mass-density of the conductor (including the small ice ridge), φ the angle between $\underline{v}_\infty - v_t\underline{e}_y$ and $\underline{v}_s$, and g the gravitational acceleration. The magnitudes D and L of the aerodynamic forces are given by $D = \frac{1}{2}\rho_a d\, c_D(\alpha)v_s^2$ and $L = \frac{1}{2}\rho_a d\, c_L(\alpha)v_s^2$, where ρ_a is the density of the air, d the diameter of the cross-section of the circular part of the conductor, α the angle between $\underline{e}_s$ and $\underline{v}_s$, $v_s^2 = (v_\infty - v_t)^2 + w_t^2$, and $c_D(\alpha)$ and $c_L(\alpha)$ the quasi-steady drag- and lift-coefficients, which may be obtained from windtunnel measurements. For a certain range of values v_∞ some characteristic results from wind-tunnel experiments are given in fig. 2.2.

According to the Den Hartog criterion [3] a two-dimensional section is aerodynamically unstable if $c_D(\alpha) + \frac{d}{d\alpha}c_L(\alpha) < 0$. From fig. 2.2 it follows that this condition is likely to be satisfied for some interval in α with $\alpha_0 < \alpha < \alpha_2$. For a certain range of values of α (including those values which satisfy the Den Hartog criterion) the drag- and lift-coefficients $c_D(\alpha)$ and $c_L(\alpha)$ can be approximated by $c_{D0} + c_{D1}(\alpha-\alpha_1) + c_{D2}(\alpha-\alpha_1)^2 + c_{D3}(\alpha-\alpha_1)^3$ and $c_{L1}(\alpha-\alpha_1) + c_{L2}(\alpha-\alpha_1)^2 +$

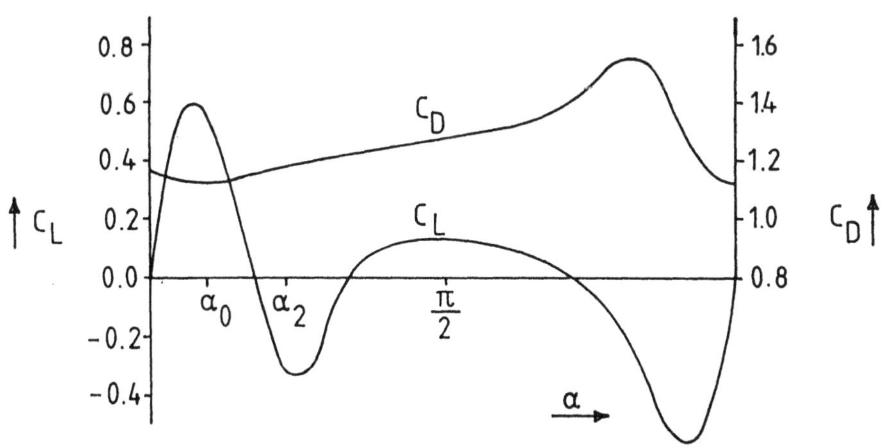

Figure 2.2. Typical variation of the drag- and lift-coefficients c_D and c_L with angle of attack for a symmetric profile with small icy nose

$+ c_{L3}(\alpha-\alpha_1)^3$ with $\alpha_0 < \alpha_1 < \alpha_2$ and $c_L(\alpha_1) = 0$. Since galloping is a low frequency oscillation it may be assumed that $|v_t| < v_\infty$ and $|w_t| < v_\infty$. The right-hand sides of the equations (2.1) and (2.2) can be considered as functions of $v_t v_\infty^{-1}$ and $w_t w_\infty^{-1}$, and so these right-hand sides can be expanded in Taylor series near $v_t v_\infty^{-1} = 0$ and $w_t w_\infty^{-1} = 0$. Since the amplitudes of the oscillations are small compared to the length of the conductor and since galloping is an oscillation described by the lower modes of vibration, it may be assumed that $|v_x| < 1$ and $|w_x| < 1$. And so, the left-hand sides of the equations (2.1) and (2.2) can be expanded in Taylor series near $v_x = 0$ and $w_x = 0$. Neglecting terms of degree four and higher, applying some transformations, introducing dimensionless variables, assuming $\alpha_s = \alpha_1 + O(\epsilon)$, the equations (2.1) and (2.2) become up to order ϵ (for detailed calculation see [5]):

$$\bar{v}_{\bar{t}\bar{t}} - \bar{v}_{\bar{x}\bar{x}} = \epsilon \left\{ a_{10}\bar{v}_{\bar{t}} + a_{01}\bar{w}_{\bar{t}} + a_{20}\bar{v}_{\bar{t}}^2 + a_{11}\bar{v}_{\bar{t}}\bar{w}_{\bar{t}} + a_{02}\bar{w}_{\bar{t}}^2 + a_{03}\bar{w}_{\bar{t}}^3 \right\} \quad , \tag{2.3}$$

$$\bar{w}_{\bar{t}\bar{t}} - \bar{w}_{\bar{x}\bar{x}} = \epsilon \left\{ b_{01}\bar{w}_{\bar{t}} + b_{11}\bar{v}_{\bar{t}}\bar{w}_{\bar{t}} + b_{02}\bar{w}_{\bar{t}}^2 + b_{03}\bar{w}_{\bar{t}}^3 \right\} \quad , \tag{2.4}$$

where $\epsilon = \dfrac{\rho_a\, d\, l\, v_\infty}{2\pi A(\rho_c T)^{1/2}}$ and $0 < \epsilon << 1$, and where $a_{10}, a_{01}, \ldots, b_{03}$ are constants of $O(1)$ determined by the aerodynamic lift- and drag-coefficients, and where $\bar{v}$, $\bar{w}$, $\bar{t}$ and $\bar{x}$ are the dimensionless variables related to respectively v, w, t and x.

If a conductor with fixed endpoints is considered the boundary conditions $\bar{v}(0,\bar{t}) = \bar{v}(\pi,\bar{t}) = \bar{w}(0,\bar{t}) = \bar{w}(\pi,\bar{t}) = 0$ are obtained. In section 5 the partial differential equations (2.3) and (2.4) subject to these Dirichlet boundary conditions and the initial conditions $\bar{v}(\bar{x},0) = V_0 \sin(m\bar{x})$, $\bar{v}_{\bar{t}}(\bar{x},0) = V_1 \sin(m\bar{x})$, $\bar{w}(\bar{x},0) = W_0 \sin(n\bar{x})$ and $\bar{w}_{\bar{t}}(\bar{x},0) = W_1 \sin(n\bar{x})$ will be studied. It should be noted that monochromatic initial values are considered, because the galloping oscillations often affect only a single mode of vibration.

3. The Well-posedness of the Problem

The following weakly nonlinear initial-boundary value problem for a (with respect to x and t) twice continuously differentiable and vector-valued function $\underline{u}(x,t;\epsilon) = (u_1(x,t;\epsilon), u_2(x,t;\epsilon), \ldots, u_n(x,t;\epsilon))^T$ will now considered ($1 \leq n < \infty$):

$$\underline{u}_{tt} - \underline{u}_{xx} + \epsilon\, \underline{F}(u;\epsilon) = \underline{0}, \qquad 0 < x < \pi,\ t > 0, \tag{3.1}$$

$$\underline{u}(x,0;\epsilon) = \underline{u}_0(x;\epsilon), \qquad 0 < x < \pi, \tag{3.2}$$

$$\underline{u}_t(x,0;\epsilon) = \underline{u}_1(x;\epsilon), \qquad 0 < x < \pi, \tag{3.3}$$

$$\underline{u}(0,t;\epsilon) = \underline{u}(\pi,t;\epsilon) = \underline{0}, \qquad t \geq 0, \tag{3.4}$$

where $0 < |\epsilon| \leq \epsilon_0 << 1$ and $\underline{F}(\underline{u};\epsilon)(x,t) \equiv f(x,t,\underline{u}(x,t;\epsilon),\ \underline{u}_t(x,t;\epsilon),\ \underline{u}_x(x,t;\epsilon);\epsilon)$. In order to obtain a classical solution (that is, a solution which is two times continuously differentiable with respect to x and t) the nonlinearity $\underline{F}$ and the initial values $\underline{u}_0$ and $\underline{u}_1$ have to satisfy certain smoothness conditions (see [5]). To prove existence and uniqueness in the classical sense of the solution of the initial-boundary value problem (3.1)-(3.4) an equivalent system of coupled integral equations will be used. In order to derive this system of integral equations the initial-boundary value problem is transformed into an initial value problem by extending the vector-valued functions $\underline{u}$, $\underline{f}$, $\underline{u}_0$ and $\underline{u}_1$ in x to odd and 2π-periodic functions. Then, assuming that the solution of the initial value problem is twice continuously differentiable, an integral representation for the solution of the initial value problem can be obtained. Using reflection principles the so-obtained system of integral equations can be rewritten as a system of integral equations on the semi- infinite strip $0 \leq x \leq \pi$, $0 \leq t < \infty$, yielding

$$\underline{u}(x,t;\epsilon) = \frac{\epsilon}{2} \int_0^t \int_0^\pi G(\xi,\tau;x,t)\ \underline{F}(\underline{u};\epsilon)(\xi,\tau)\ d\xi\ d\tau + \underline{u}_1(x,t;\epsilon), \tag{3.5}$$

where $G(\xi,\tau;x,t)$ is a (nxn) diagonal-matrix. In fact, G can be identified as the Green's function for the differential operator $\frac{\partial^2}{\partial t^2} - \frac{\partial^2}{\partial x^2}$ and the Dirichlet boundary conditions. It is also worth noticing that $\underline{u}_1(x,t;\epsilon)$ is the solution of linear initial-boundary value problem (3.1)-(3.4) with $\underline{F} \equiv \underline{0}$. Now it can easily be shown that the system of integral eqautions (3.5) and the initial-boundary value problem (3.1)-(3.4) are equivalent if twice continuously differentiable solutions exist. By applying Banach's fixed point theorem (in the space of twice continuously differentiable functions) to the system of integral equations (3.5) it can be shown that a unique, twice continuously differentiable solution of the system of integral equations exists on Ω_L: $0 \leq x \leq \pi$ and $0 \leq t \leq L|\epsilon|^{-1}$, where L is a constant independent of ϵ. It can also be shown that the solution of the system of integral equations depends continuously on the initial values. And so, the well-posedness of the initial-boundary value problem has been established on a time-scale of order $|\epsilon|^{-1}$.

4. The Validity of Formal Approximations

Since the initial-boundary value problem contains a small parameter ϵ perturbation methods may be applied to construct approximations of the solution. In most perturbation methods for weakly nonlinear problems a function is constructed that satisfies the differential equation(s) and the initial conditions up to some order depending on the small parameter ϵ. Such a function is usually called a formal approximation. It requires an additional analysis to show that this formal approximation is an asymptotic approximation as ϵ tends to zero. Therefore suppose that on Ω_L a twice continuously differentiable function $\underline{v}(x,t;\epsilon)$ is constructed satisfying

$$\underline{v}_{tt} - \underline{v}_{xx} + \epsilon\, \underline{F}(\underline{v};\epsilon) = |\epsilon|^m \underline{c}_1(x,t;\epsilon), \qquad m > 1, \tag{4.1}$$

$$\underline{v}(x,0;\epsilon) = \underline{u}_0(x;\epsilon) + |\epsilon|^{m-1}\underline{c}_2(x;\epsilon), \qquad 0 < x < \pi, \tag{4.2}$$

$$\underline{v}_t(x,0;\epsilon) = \underline{u}_1(x;\epsilon) + |\epsilon|^{m-1}\underline{c}_3(x;\epsilon), \qquad 0 < x < \pi, \tag{4.3}$$

$$\underline{v}(0,t;\epsilon) = \underline{v}(\pi,t;\epsilon) = \underline{0}, \qquad 0 \leq t \leq L|\epsilon|^{-1}, \tag{4.4}$$

where $\underline{u}_0$, $\underline{u}_1$, $\underline{F}$, $\underline{c}_1$, $\underline{c}_2$ and $\underline{c}_3$ satisfy certain smoothness conditions (see [5]). From the previous section it follows that the initial-boundary value problem for $\underline{v}$ has a unique, twice continuously differentiable solution on a time-scale of $O(|\epsilon|^{-1})$. This initial-boundary value problem can then be transformed into the equivalent system of integral equations

$$\underline{v}(x,t;\epsilon) = \frac{\epsilon}{2}\int_0^t\int_0^\pi G(\xi,\tau;x,t)\ \underline{F}(\underline{v};\epsilon)(\xi,\tau)\ d\xi\ d\tau\ +$$

$$- \frac{|\epsilon|^m}{2}\int_0^t\int_0^\pi G(\xi,\tau;x,t)\ \underline{c}_1(\xi,\tau;\epsilon)\ d\xi\ d\tau + \underline{v}_\ell(x,t;\epsilon), \tag{4.5}$$

where $\underline{v}_\ell(x,t;\epsilon)$ is the solution of the linear initial-boundary value problem (4.1)-(4.4) with $\underline{F} \equiv \underline{0}$ and $\underline{c}_1 \equiv 0$. By subtracting the system of integral equations (4.5) for $\underline{v}$ from the system of integral equations (3.5) for $\underline{u}$ the following estimate (in c^2-norm) can be obtained: $\|\underline{u} - \underline{v}\| = O(|\epsilon|^{m-1})$ for $0 \leq x \leq \pi$ and $0 \leq t \leq L|\epsilon|^{-1}$. So, for $m > 1$ this estimate implies the asymptotic validity (as $\epsilon \to 0$) of the formal approximation $\underline{v}$.

5. An Asymptotic Approximation of the Solution of a System of Nonlinear Wave Equations

In this section the partial differential equations (2.3) and (2.4) subject to the initial conditions and the Dirichlet boundary condition mentioned at the end of section 2 will be considered. For arbitrary m, n, a_{10}, a_{01}, ..., b_{03}, V_0, ..., W_1 an asymptotic approximation (as ϵ tends to zero) of the solution of the initial-boundary value problem will be constructed. In view of computational difficulties whenever one assumes an infinite series representation for the solution of the nonlinear initial-boundary value problem, one may alternatively investigate the problem in the characteristic coordinates $\sigma = x-t$ and $\xi = x+t$. In this approach the initial-boundary value problem is replaced by an initial value problem. This replacement requires to extend the dependent variable $\underline{u}(x,t;\epsilon) = (\bar{v}(\bar{x},\bar{t};\epsilon), \bar{w}(\bar{x},\bar{t};\epsilon))^T$, the right-hand sides of the equations (2.3) and (2.4) as well as the initial values in x to odd and 2π-periodic functions. In constructing an approximation of the solution of this initial value problem a two-timescales perturbation method will be used, since the straightforward perturbation expansion causes secular terms. Applying the two-timescales perturbation method $\underline{u}(x,t)$ is supposed to be a function of $\sigma = x-t$, $\xi = x+t$ and $\tau = \epsilon t$. By putting $\underline{u}(x,t) = \underline{\tilde{u}}(\sigma,\xi,\tau) = (\tilde{v}(\sigma,\xi,\tau), \tilde{w}(\sigma,\xi,\tau))^T$ an initial value problem for $\underline{\tilde{u}}$ is obtained. Furthermore, it is assumed that $\tilde{v}$ and $\tilde{w}$ may be approximated by the formal perturbation expansions $v_0(\sigma,\xi,\tau) + \epsilon v_1(\sigma,\xi,\tau) + \ldots$ respectively $w_0(\sigma,\xi,\tau) + \epsilon w_1(\sigma,\xi,\tau) + \ldots$. By substituting these approximations into the equations of the initial value problem for $\underline{\tilde{u}}$, and after equating the coefficients of like powers in ϵ, it follows from the power 0 of ϵ that v_0 and w_0 should satisfy:

for $-\infty < \sigma < \xi < \infty$ and $\tau > 0$: $v_{0_{\sigma\xi}} = 0$, $w_{0_{\sigma\xi}} = 0$, and

for $-\infty < \sigma = \xi < \infty$ and $\tau = 0$:

$$v_0(\sigma,\xi,\tau) = V_0 \sin(m\sigma),$$
$$w_0(\sigma,\xi,\tau) = W_0 \sin(n\sigma),$$
$$-v_{0_\sigma}(\sigma,\xi,\tau) + v_{0_\xi}(\sigma,\xi,\tau) = V_1 \sin(m\sigma) \text{ and}$$
$$-w_{0_\sigma}(\sigma,\xi,\tau) + w_{0_\xi}(\sigma,\xi,\tau) = W_1 \sin(n\sigma).$$

From the power 1 of ϵ it follows that v_1 and w_1 also have to satisfy a linear initial value problem, where the right-hand sides of the partial differential equations depend on v_0 and w_0. It turns out that $v_0(\sigma,\xi,\tau) = h_0(\sigma,\tau)+ k_0(\xi,\tau)$ and $w_0(\sigma,\xi,\tau) = f_0(\sigma,\tau)+ g_0(\xi,\tau)$, where h_0, k_0, f_0 and g_0 have to satisfy: $h_0(\sigma,0)+ k_0(\sigma,0) = V_0 \sin(m\sigma)$, $-h_0'(\sigma,0)+ k_0'(\sigma,0) = V_1 \sin(m\sigma)$, $f_0(\sigma,0)+ g_0(\sigma,0)=$ $= W_0 \sin(n\sigma)$, $-f_0'(\sigma,0) + g_0'(\sigma,0) = W_1 \sin(n\sigma)$, $k_0(\sigma,\tau) = -h_0(-\sigma,\tau)$, $h_0(\sigma,\tau) =$ $= h_0(\sigma+2\pi,\tau)$, $g_0(\sigma,\tau) = -f_0(-\sigma,\tau)$ and $f_0(\sigma,\tau) = f_0(\sigma+2\pi,\tau)$ for $-\infty< \sigma< \infty$ and $\tau\geq 0$ (the prime ' denotes differentiation with respect to the first argument). It turns out that h_0 and f_0 are still undetermined with respect to τ. Furthermore, it should be noticed that the equations for v_0, w_0, v_1 and w_1 have been derived under the assumption that v_0, w_0, v_1, w_1 and their derivatives up to order two are O(1). These boundedness conditions on v_0, w_0, v_1 and w_1 determine the behaviour of h_0 and f_0 with respect to τ. It turns out that f_0 and h_0 have to satisfy:

$$2f_{0_{\sigma\tau}} - b_{01} f_{0_\sigma} - b_{03} f_{0_\sigma}^3 - \frac{3b_{03}}{2\pi} f_{0_\sigma} \int_0^{2\pi} f_{0_\theta}^2(\theta,\tau)d\theta = 0, \text{ and} \tag{5.1}$$

$$2h_{0_{\sigma\tau}} - a_{10} h_{0_\sigma} - a_{01} f_{0_\sigma} - a_{03} f_{0_\sigma}^3 - \frac{3a_{03}}{2\pi} f_{0_\sigma} \int_0^{2\pi} f_{0_\theta}^2(\theta,\tau)d\theta = 0. \tag{5.2}$$

In [2, 4] an equation similar to equation (5.1) has been solved. If the method introduced in [2] is applied to equation (5.1) one obtains after some calculations $f_0(\sigma,\tau)$, and so $w_0(\sigma,\xi,\tau)= f_0(\sigma,\tau)- f_0(-\xi,\tau)$. It turns out that w_0 and v_0 are given by:

$$w_0(\sigma,\xi,\tau) = \frac{\lambda(\tau)}{n\phi^{1/2}(\tau)}\left\{\arcsin\left[\left[\frac{\tilde{W}_n\phi(\tau)}{1+\tilde{W}_n\phi(\tau)}\right]^{1/2}\sin(\alpha+n\sigma)\right]\right.$$

$$\left. - \arcsin\left[\left[\frac{\tilde{W}_n\phi(\tau)}{1+\tilde{W}_n\phi(\tau)}\right]^{1/2}\sin(\alpha-n\xi)\right]\right\}, \text{ and}$$

$$v_0(\sigma,\xi,\tau)= \frac{1}{2}\exp\left[\frac{a_{10}}{2}\tau\right]\left\{V_0 \sin(m\sigma)+ \frac{1}{m} V_1 \cos(m\sigma)+ V_0 \sin(m\xi)- \frac{1}{m} V_1 \cos(m\xi)\right\} +$$

$$+ \frac{a_{03}}{b_{03}}\left\{w_0(\sigma,\xi,\tau)- \frac{1}{2}\exp\left[\frac{a_{10}}{2}\tau\right]\left[W_0 \sin(n\sigma)+ \frac{1}{n} W_1 \cos(n\sigma)+ W_0 \sin(n\xi) +\right.\right.$$

$$\left.\left. - \frac{1}{n} W_1 \cos(n\xi)\right]\right\}+ \frac{1}{2}\left[a_{01}- \frac{a_{03}}{b_{03}} b_{01}+ \frac{a_{03}}{b_{03}} a_{10}\right]\int_0^{\tau} \exp\left[\frac{a_{10}}{2}(\tau-\tau')\right] w_0(\sigma,\xi,\tau')\, d\tau'$$

with $\sigma= x- t$, $\xi= x+ t$, $\tau= \epsilon t$, $\tilde{W}_n= n^2W_0^2+ W_1^2$, $\cos\alpha= nW_0\tilde{W}_n^{-1/2}$, $\sin\alpha= W_1\tilde{W}_n^{-1/2}$, $\lambda(\tau)= 4\, m^{-3}(\tau) \exp\left[\frac{b_{01}\tau}{2}\right]$, $\phi(\tau)= \tilde{W}_n^{-1}\, m(\tau)(m(\tau)-2)$, $m^8(\tau)- \frac{8}{7} m^7(\tau)= 2^6\tilde{W}_n b_{03}b_{01}^{-1}(1-\exp(b_{01}\tau))+ \frac{3.2^8}{7}$. Now the linear initial value problem for v_1 and w_1 can be solved and it can be shown that v_0, v_1, w_0, w_1 and their derivatives up to order two are of $O(1)$ for $-\infty < x < \infty$ and $0 \le t \le L|\epsilon|^{-1}$. So, the assumptions under which the equations for v_0, v_1, w_0 and w_1 have been derived, are justified. So far a vector-valued function $\underline{u}_A(x,t;\epsilon) \equiv (v_0(x-t,x+t,\epsilon t)+ \epsilon v_1(x-t,x+t,\epsilon t),\ w_0(x-t,x+t,\epsilon t) + \epsilon w_1(x-t,x+t,\epsilon t))^T$ has been constructed. By applying the asymptotic theory presented in section 3 and 4 it can be shown that $\|\underline{u}-\underline{u}_A\|= O(\epsilon)$ for $0\le x\le \pi$ and $0\le t\le L|\epsilon|^{-1}$. It can also be shown that $\underline{u}_0(x,t;\epsilon)\equiv (v_0(x-t,x+t,\epsilon t),\ w_0(x-t,x+t,\epsilon t))^T$ is an $O(\epsilon)$ asymptotic approximation (as $\epsilon\to 0$)

for $0 \leq x \leq \pi$ and $0 \leq t \leq L|\epsilon|^{-1}$: $\|\underline{u}-\underline{u}_0\| = \|\underline{u}-\underline{u}_A+\underline{u}_A-\underline{u}_0\| \leq \|\underline{u}-\underline{u}_A\| + \|\underline{u}_A-\underline{u}_0\| = O(\epsilon)$. These estimates enable us to approximate the ultimate amplitudes of the galloping oscillations of overhead transmission lines.

6. Acknowledgement

This research-project was supported by the Netherlands Foundation for Mathematics SMC with financial aid from the Netherlands Organization for the Advancement of Pure Research (ZWO).

7. References

1. C.G.A. van der Beek, A.H.P. van der Burgh: Nieuw Archief voor Wiskunde, No. 2, p. 207-225 (July, 1987)
2. S.C. Chikwendu, J. Kevorkian: SIAM Journal on Applied Mathematics 22, p. 235-258 (1972)
3. J.P. den Hartog: Mechanical Vibrations, 4th edition, (McGraw-Hill, New York, 1956)
4. W.T. van Horssen: SIAM Journal on Applied Mathematics, to appear
5. W.T. van Horssen: Report 87-78 of the Faculty of Technical Mathematics and Computer Science, Delft University of Technology, The Netherlands (1987)
6. A. Simpson: Sci. Prog. Oxford 68, p. 285-308 (1983)

Part II

Instability, Bifurcations and Transition to Chaos

Quasi-Codimension 3 Bifurcation of Invariant T^2 Tori for Maps

G. Iooss and J.E. Los

Laboratoire de Mathématiques, U.A. 168, Université de Nice, Parc Valrose, F-06034 Nice, France

1. Introduction.

Bifurcations of invariant tori of dimension higher than one for families of maps, or higher than two for families of vector fields in dissipative systems is a puzzling problem. There are experimental evidences of the existence of such tori, for instance in Bénard convection problem [Go-Be] and in Taylor Couette problem [GRS].

A known way to obtain analytically successive bifurcations leading to high dimensional invariant tori is to study high codimension singularities. For instance, if we consider the very popular codimension 2 problem of a two parameters family of vector fields with a fixed point at the origin, such that at criticality, two pairs of complex conjugate eigenvalues are on the imaginary axis (Hopf-Hopf interaction), then, under non strong resonance conditions ([Gu-Ho] §7.5, [Io-La],[Io81]) one can easily show, depending on some coefficients, that invariant T^3-tori can bifurcate from the secondary branch of invariant T^2-tori. In fact, this bifurcation is only clear on the truncated normal form of the vector fields. If we consider a generic vector field family , one could only prove that the T^2-tori and T^3-tori families exist outside of a little neighborhood (called a "black hole" in [Io.81]) in the parameter space where precisely bifurcation occurs for the normal form. This is due to the lack of normal hyperbolicity in this neighborhood as in [Chen],[Los1].

In the present work, we show that such codimension 2 problems have to be unfolded, using an additional parameter, in the neighborhood of the bifurcation set (a manifold for the truncated normal form). In this three dimensional parameter space, there is now a Cantor set of points where a nice bifurcation, such as assumed in [Ch-Io] and [Sell], occurs. This leads to a Cantor set of little paraboloïdal regions where we know what happens for the dynamics (fig 1), and a complementary region ("bubbles") where we don't know yet.

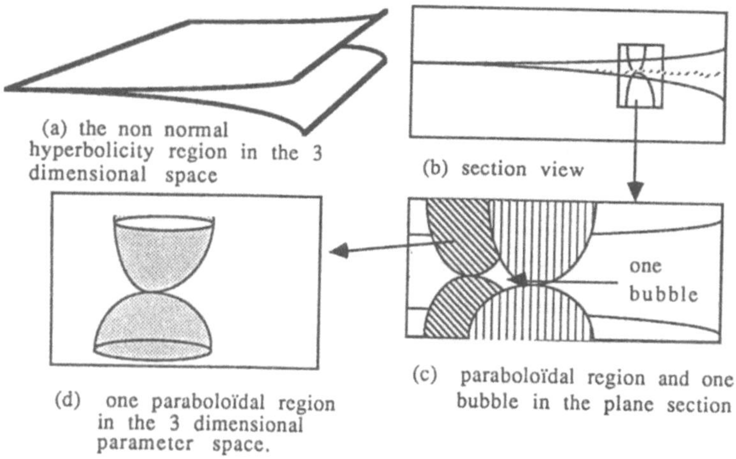

Figure 1 : *The parameter space with the Cantor set of paraboloïdal regions and the bubbles.*

A strong analogue of such a situation with one parameter less, was obtained by A.Chenciner [Chen] in the neighborhood of a degenerate Hopf bifurcation for maps, where such bubbles and parabolic regions in the parameter plane occur, near the curve of circle saddle-node bifurcation , obtained with the normal form. In the same spirit, J.Los [Los1,2] obtained also bubbles and parabolic regions in a codimension 2 problem involving a doubling of invariant $\mathbb{T}^1$-tori. A similar phenomenon was also obtained by Braaksma and Broer [Bra-B] (see also [B-H-T]) in the study of a quasi-periodically forced Hopf bifurcation, leading to a bifurcation $\mathbb{T}^n \to \mathbb{T}^{n+1}$. Their study is however simplified by the fact that the flows on the $\mathbb{T}^n$-tori are linear in such a problem while this is not the case in our study for generic families. Another result of the same type was obtained by Scheurle [Sch] in the study of *reversible* vector fields where he shows how high dimensional invariant tori bifurcate.

The techniques developped below are directly adapted from the one used by Herman [Bost] in his proof of the invariant curve theorem. This one is the main tool in the works of Chenciner [Chen] and Los [Los1] for bifurcation problems. Namely, to prove the persistence of $\mathbb{T}^1$-tori, we extensively use the strong implicit function theorem in Fréchet spaces in the version of R.S.Hamilton [Ham]. The translated curve theorem of Rüssman [Russ] in the Herman formulation [Her83],[Bost], or its adaptations [Chen],[Los1] are now completed with a quasi-conjugacy theorem for matrices near a constant one. An analogous result, for vector fields, and in the case when matrices are close to a diagonal one with simple eigenvalues on the imaginary axis, was obtained by J.Moser [Mos]. He was the first to show the necessity of adding parameters to obtain the quasi-conjugacy to a constant. A general result [Io-Lo] gives the exact number of additional parameters, necessary in any case for the quasi-conjugacy and this allows our system to satisfy the conditions described in [Ch-Io]. Finally it can be shown more generally (see [Io-Lo])that bifurcation to n-dimensional invariant tori occurs genericaly on a Cantor set which is defined by a set in the parameter space having a n-dimensional positive measure. So we say that in some sense it is "quasi-generic".

2. Normal Hyperbolicity regions.

A classical codimension 1 bifurcation for maps is such that one considers a family of maps in $E = \mathbb{R}^2$:

(1) $Z \to Z' = F_\mu(Z), \quad Z \in E,$ μ being a real parameter,

where F is C^∞ in a neighborhood of 0, $F_0(0) = 0$ and $D_Z F_0(0) = T_0$ has 2 simple eigenvalues $\lambda_0 = e^{2i\pi\omega_0}$ and $\bar{\lambda}_0$. If the following non strong resonant conditions are realized : $\omega_0 \neq 0, 1/2, 1/3, 2/3, 1/4, 3/4 \mod 1$ then Sacker showed in 1964 [Sac] that generically there happens a bifurcation of the fixed point into an invariant closed curve, attracting if it bifurcates supercritically, and otherwise repelling. The argument used for proving such a result is called *normal hyperbolicity*. It is mainly based on a fixed point argument in a metric space of C^k-bounded functions.

Let us now consider in $E = \mathbb{R}^4$(for the general case when E is a Banach space, see [Io-Lo]), the case of two interacting bifurcations of the previous type, i.e. we have again a family of maps F_μ (1) , μ being a set of real parameters in $\mathbb{R}^k$ and such that T_0 has 4 simple eigenvalues :

$$\lambda_0, \bar{\lambda}_0, \lambda_1, \bar{\lambda}_1, \text{ where } \lambda_j = e^{2i\pi\omega_j}, \quad j=0,1.$$

The normal form of the unfolding of such a mapping was given by Jost and Zehnder [Jo-Ze] using 2 complex parameters (μ_0,μ_1). Provided the following non strong resonance condition :

(2) $m_0\,\omega_0 + m_1\,\omega_1 \in \mathbb{Z}$, has no solution for $|m_0| + |m_1| \leqslant N$,

is fulfilled, the truncated normal form P_μ of the map F_μ may be written at order N-2 as follows :

$$(3)\qquad \begin{cases} z_0' = \lambda_0 z_0[1+\mu_0+\psi_0(\mu,|z_0|^2,|z_1|^2)]; \\ z_1' = \lambda_1 z_1\,[1+\mu_1+\psi_1(\mu,|z_0|^2,|z_1|^2)]\ ;\ \mu=(\mu_0,\mu_1)\in\mathbb{C}^2; \end{cases}$$

where ψ_0 and ψ_1 are polynomials in their two last arguments, depending smoothly on μ, each of them have valuation larger than one and degree $\left[\frac{N-3}{2}\right]$ in $(|z_0|^2, |z_1|^2)$. The integer N will be fixed latter.

2.1. Normal hyperbolicity region for the primary bifurcation.

Since we are only interested in the $\mathbb{T}^1\to\mathbb{T}^2$-bifurcation let us write the normal form (3) in an easier form for computation .

We can define a primary bifurcating invariant curve for P_μ which is a circle Γ_0 provided that $N\geq 5$. This circle is defined by :

(4) $\qquad z_1=0, \qquad z_0=\rho_0\, e^{2i\pi\theta_0},\ \theta_0\in\mathbb{T}^1$, where ρ_0 is defined by

$$1 = |\,1+\mu_0+\psi_0(\mu,\rho_0^{\,2},0)|\,,$$

and where we assume bifurcation to be supercritical ($\mu_{or}=\mathrm{Re}(\mu_0)>0$).

The restriction of P_μ on Γ_0 is simply a rotation :

$$(5)\qquad \theta_0 \to \theta'_0 = \theta_0+\omega_\mu .$$

The *normal hyperbolicity* of Γ_0 under the full family F_μ can be made explicit by using the following change of variables from $\mathbb{C}^2$ to $\mathbb{T}^1\mathrm{x}\mathbb{R}\mathrm{x}\mathbb{C}$:

$$(6)\qquad (z_0,z_1)\to(\theta,x,z),\ \text{with}\ z\sqrt{\varepsilon}=z_1,\ z_0=\rho_0(1+x)e^{2i\pi\theta},\ (\varepsilon\equiv 2\mu_{or}).$$

The map F_μ takes the form :

$$(7)\qquad \begin{cases} \Theta = \theta+\omega_\mu+\varepsilon R(\mu,x,|z|^2)+\hat{R}(\mu,\theta,x,z,\bar z) \\ X = (1-\varepsilon)x+\varepsilon Q(\mu,x,|z|^2)+\hat{Q}(\mu,\theta,x,z,\bar z) \\ Z = z\{\lambda_1(1+\nu).e^{2i\pi\sigma}+\varepsilon S(\mu,x,|z|^2)\}+\hat{S}(\mu,\theta,x,z,\bar z) \end{cases},$$

where

$$(8)\qquad |\hat{R}|+|\hat{Q}|+|\hat{S}| = O(\varepsilon^{(N-2)/2}),$$

and where for convenience we define new parameters ε,ν,σ . The map :

$$(9)\qquad (\mu_0,\mu_1)\to(\varepsilon,\nu,\sigma,\omega_0-\omega_\mu)$$

is a local diffeomorphism and from now on we shall write :

$$(10)\qquad \mu=(\varepsilon,\nu,\sigma,\omega_0-\omega_\mu).$$

In (7) all functions are smooth in their variables, and estimate (8) says at what order the dependency in θ (or the breaking of the rotational invariance) arrives.

We expect to have problems of non normal hyperbolicity near $\varepsilon=0$ and near $\nu=0$. To prove the persistence of a normal hyperbolic invariant closed curve near Γ_0 under the map F_μ, we can use the Ruelle-Takens technique [Ru-Ta] (see also [Lan]). By making the change of variables :

(11) $$x=(\varepsilon)^{\alpha}\hat{x}\ ,\ z=(\varepsilon)^{(1/2)+\alpha}\hat{z},\ \ \alpha=\frac{N}{4}-1>0,$$

we obtain a new form of the map where it is easy to see that the condition :

(12) $$\varepsilon^{N/4}=o(\nu)\ ,$$

gives us the right size for the perturbations with respect to the hyperbolic linear part of (7), where we want the perturbations to have a smaller size than the rate of normal hyperbolicity of Γ_0. Figure 2 shows the region of the parameter plane (ε,ν) corresponding to (12).

2.2. Secondary bifurcation. Normal hyperbolicity region .

On the truncated normal form P_μ (3) the secondary branch is obtained by looking for (ρ_o,ρ_1) solution of the system :

(13) $$\begin{cases} |1+\mu_o+\psi_o(\mu,\rho_o{}^2,\rho_1{}^2)|=1 \\ |1+\mu_1+\psi_1(\mu,\rho_o{}^2,\rho_1{}^2)|=1 \end{cases},$$

where ψ_o and ψ_1 are defined by(3) and

$$z_o=\rho_o e^{2i\pi\theta_o}\quad ,z_1=\rho_1 e^{2i\pi\theta_1}\qquad ,(\theta_o,\theta_1)\in\mathbb{T}^2 .$$

Here we are only interested in what happens near the bifurcation point when ν changes its sign. Hence we now consider the region of the (ε,ν) plane where :

(14) $$\nu=o(\varepsilon),$$

and define x and y by :

$$\begin{cases} z_o=\rho_o.(1+x.\varepsilon^{\alpha})e^{2i\pi\theta_o} \\ z_1=\rho_1.(1+y.\varepsilon^{\alpha})e^{2i\pi\theta_1} \end{cases},$$

then on the map F_μ written in $\mathbb{R}^2\times\mathbb{T}^2$ as $(x,y,\theta_o,\theta_1)\rightarrow(X,Y,\Theta_o,\Theta_1)$ it is easy to show that if $\alpha=\min(2,\frac{N-4}{5})$ the normal hyperbolicity argument gives persisting invariant tori in the region

(15) $$\varepsilon^{(N+1)/5}+\varepsilon^2=o(\nu).$$

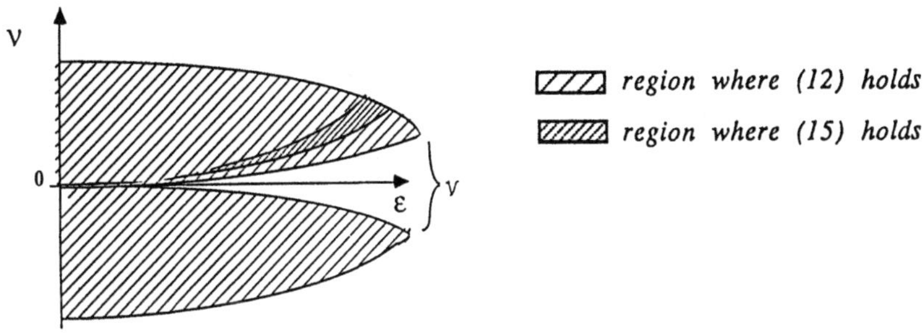

Figure 2. *Persistence of invariant T^1 and of bifurcating invariant T^2 via normal hyperbolicity.*

This condition is compatible with (14), and the corresponding region is indicated at figure 2 .

In §3 and 4 we study what happens in region V not covered by estimates (12) and (15). To find persisting invariant $\mathbb{T}^1$-tori as well as invariant $\mathbb{T}^2$-tori in the region close to the ε-axis, we shall see that we need an additional parameter (in fact σ).

3. Persistence of an invariant curve .

3.1. The rotated-twisted theorem (R-T) .

We consider the family (7) of C^∞ diffeomorphisms F_μ of $\mathbb{T}^1 x(\mathbb{R},0) x (\mathbb{C},0)$. The truncated normal form P_μ defined by $(\hat{R}, \hat{Q}, \hat{S}) \equiv 0$, has a trivial invariant circle:

(16) $$\mathcal{C}_0 = \mathbb{T}^1 x\{0,0\} ,$$

and $P_{\mu|\mathcal{C}_0}$ is simply a rotation of angle ω_μ. We look for a C^∞ invariant closed curve, C^1-close to $\mathcal{C}_0$ for sufficiently small $(\hat{R}, \hat{Q}, \hat{S})$ described by a graph :

(17) $$\mathcal{C} = \{(\theta, \psi(\theta), \tau(\theta)); \theta \in \mathbb{T}^1\},$$

where ψ, τ are C^∞ functions on a neighborhood of the origin in $\mathbb{R}$ and $\mathbb{C}$. The invariance of $\mathcal{C}$ under F_μ is made explicit by the following functional equation :

(18) $$F_\mu(\theta, \psi(\theta), \tau(\theta)) = \{ \Theta(\theta, \psi(\theta), \tau(\theta)), \psi(\Theta), \tau(\Theta)\}.$$

This equation cannot be solved directly in the non normal hyperbolicity region V of the parameter space (see figure 2). We need assumptions on the rotation properties, as in the KAM theorem.

The key result of the paper is a way for defining the *rotation* (λ) and the *twist* (t) parameters the cancellation of which defines the invariant curve we are looking for. Let us define the map :

$$(19)\quad \begin{cases} \mathcal{R}_{\lambda;t}(\theta,x,z) = (\theta+\lambda,\ x,\ e^{2i\pi t}.z)\ ;\ (\theta,x,z)\in \mathbb{T}^1 \times \mathbb{R} \times \mathbb{C} \\ (\lambda;t)\in \mathbb{T}^2 \end{cases},$$

For our purpose it is sufficient to consider the following form of the family F_μ:

$$(20)\quad \begin{cases} F_\mu(\theta,x,z) = (\Theta,X,Z)\ ;\ \mu\in(\mathbb{R}^4,0); \\ \Theta = \theta+\omega_\mu + P^1(\theta,x,z,\bar{z})\ ; \\ X = (1-\varepsilon)x + P^2(\theta,x,z,\bar{z})\ ; \\ Z = e^{2i\pi(\omega_1+\sigma)}(1+\nu)z + P^3(\theta,x,z,\bar{z})\ . \end{cases}$$

Then we have the following :

__R-T Theorem__. Let (ω,Ω) be in $\mathbb{T}^2$ and satisfy the diophantine condition :

$$(21)\ \exists\ C>0,\ \exists\ \beta\geq 0 \text{ such that } \forall \frac{p}{q}\in\mathbb{Q},\ |r\Omega - q\omega - p| \geq \frac{C}{|q|^{1+\beta}},\ r=0,2.$$

If $\Pi=(P^1,P^2,P^3)$ is sufficiently small, there exist $(\lambda;t)\in\mathbb{T}^2$ near $\lambda_o=\omega-\omega_\mu$, $t_o=\Omega-\omega_1-\sigma$, $(a,b)\in\mathbb{R}^2$ near 0, h a C^∞ diffeomorphism of the circle near Id, and a couple of C^∞ maps (ψ,τ), such that the closed curve $\mathcal{C}=\{(\theta,\psi(\theta),\tau(\theta));\theta\in\mathbb{T}^1\}\subset\mathbb{T}^1\times\mathbb{R}\times\mathbb{C}$ *is invariant under* $\mathcal{R}_{\lambda;t}\circ F_\mu$. The first projection ($\pi_1:\mathbb{T}^1\times\mathbb{R}\times\mathbb{C}\to\mathbb{T}^1$) of $(\mathcal{R}_{\lambda;t}\circ F_\mu)_{|\mathcal{C}}$ is C^∞ conjugated by h to the rotation R_ω and the linear part of $(\mathcal{R}_{\lambda;t}\circ F_\mu)_{|\mathcal{C}}$ on the tangent bundle over $\mathcal{C}$ is quasi-conjugated (defined below after (31)) to the constant matrix :

$$\begin{pmatrix} 1-\varepsilon+a & 0 \\ 0 & e^{2i\pi\Omega}(1+\nu+b) \end{pmatrix}.$$

__Idea of the proof__:

We first define the following functional operators :

$$(22)\qquad \Phi(\lambda,(\psi,\tau),P^1)(\theta) = \theta+\omega_\mu+\lambda+P^1(\theta,\psi(\theta),\tau(\theta),\bar{\tau}(\theta)),$$

$$(23)\qquad \mathcal{H}_t((\psi,\tau),\Pi) = \begin{pmatrix} \psi\circ\Phi - (1-\varepsilon)\psi - P^2(.,\psi,\tau,\bar{\tau}) \\ \tau\circ\Phi - e^{2i\pi(\omega_1+\sigma+t)}(1+\nu)\tau - P^3(.,\psi,\tau,\bar{\tau}) \end{pmatrix}.$$

To prove the theorem we have to find $((\lambda;t),(\psi,\tau),h)$ so that :

(24) $$\Phi(\lambda,(\psi,\tau),P^1) = h^{-1} \circ R_\omega \circ h \,,$$

(25) $$\mathcal{H}_t((\psi,\tau),\Pi) = 0 \,.$$

For the truncated normal form, which is reduced here to the linear part, i.e $\Pi \equiv 0$ we have trivially :

(26) $$\begin{cases} \Phi(\lambda_0,(0,0),0) = R_\omega \\ \mathcal{H}_t((0,0),0) = 0 \text{ , for any } t. \end{cases}$$

The proof of the R-T theorem is based on the implicit function theorem on Fréchet spaces in the formulation of R.S.Hamilton [Ham]. There are two main steps :

i) The conjugacy to the rotation R_ω which is a direct consequence of a result by M.Herman (see [Her79],§III.4.2) and the Herman-Yoccoz theorem [Yoc]. This leads to the existence of λ and h as C^1 functionals of (ψ,τ, P^1).

ii) The local inversion of the partial derivative $\dfrac{\partial \mathcal{H}_t}{\partial(t,\psi,\tau)}$ (needed for applying Hamilton's theorem) which means that for any given (C^∞) K and any $((\psi,\tau),\Pi)$ in some neighborhood of 0 the following linear equation :

(27) $$D_1\mathcal{H}_t(\psi,\tau,\Pi)(\Delta t,\Delta\psi,\Delta\tau,\Delta\bar{\tau}) = K \,,$$

has a unique regular solution : $(\Delta t,\Delta\psi,\Delta\tau)$.

By a direct computation we obtain :

(28) $$D_1\mathcal{H}_t(\psi,\tau,\Pi)(\Delta t,\Delta\psi,\Delta\tau,\Delta\bar{\tau}) = \begin{pmatrix} \Delta\psi\circ\Phi + M_{11}\Delta\psi + M_{12}\Delta\tau + M_{13}\Delta\bar{\tau} + L_1\Delta\lambda \\ \Delta\tau\circ\Phi + M_{21}\Delta\psi + M_{22}\Delta\tau + M_{23}\Delta\bar{\tau} + R\Delta t + L_2\Delta\lambda \end{pmatrix}.$$

This matrix is close to the diagonal matrix with eigenvalues :

$(1-\varepsilon \,,\, e^{2i\pi(\omega_1+\sigma+t)}(1+\nu),\, e^{-2i\pi(\omega_1+\sigma+t)}(1+\nu))$.

If we compose equation (27) with h^{-1}, we obtain a system of the form:

(29) $$\begin{cases} T'(\theta+\omega) = M'(\theta).T'(\theta) + L'(\theta)\Delta\lambda + R'(\theta)\Delta t + K'(\theta), \\ \text{where } (T',K') = (T,K)\circ h^{-1}, (M',L',R') = -(M,L,R)\circ h^{-1}, \end{cases}$$

where $\Delta\lambda$ and Δt are linear functionals of $T = (\Delta\psi,\Delta\tau)$ and the matrix M' is in a neighborhood of :

(30) $$M_0 = \begin{pmatrix} 1-\varepsilon & 0 & 0 \\ 0 & e^{2i\pi\Omega}(1+\nu) & 0 \\ 0 & 0 & e^{-2i\pi\Omega}(1+\nu) \end{pmatrix},$$

where $\Omega = \omega_1 + \sigma + t_0$ satisfies condition (21). Then we can apply to the above matrix M' the "twisted" quasi-conjugacy result (which also uses the same implicit function theorem in Fréchet spaces, see [Io-Lo] for the most general result) . There exists an angular dependent matrix H, and matrices D_r, D_φ such that:

$$
(31)\begin{cases} H^{-1}(\theta+\omega)[M'(\theta)-D_\varphi]H(\theta) = M_0 + D_r = M'_0 \text{ , with,} \\ D_r = \begin{pmatrix} a & 0 & 0 \\ 0 & b e^{2i\pi\Omega} & 0 \\ 0 & 0 & b e^{-2i\pi\Omega} \end{pmatrix}, (a,b) \in (\mathbb{R}^2, 0), \\ D_\varphi = \begin{pmatrix} 0 & 0 & 0 \\ 0 & (1+v)(1-e^{2i\pi\varphi})e^{2i\pi\Omega} & 0 \\ 0 & 0 & (1+v)(1-e^{-2i\pi\varphi})e^{-2i\pi\Omega} \end{pmatrix}, \varphi \in (\mathbb{R}, 0), \end{cases}
$$

Definition : we say that a matrix $M(\theta)$ is *quasi-conjugated* to a constant one M'_0 if it satisfies a relation like (31) with $D_\varphi = 0$.

We note that the matrix $H(\theta)$ allows to change $M'(\theta)$ into a constant matrix, *once we modify it by adding a simple twist matrix* D_φ. We remark that as $\Omega=\omega_1+\sigma+t_0$, the quantity $t= t_0+\varphi$, where φ is given by (31) is exactly the parameter we are looking for. We now fix t to this value which depends functionaly on (ψ,τ,Π). The matrix H given by (31) enables us to make the following change of variables:

(32) $\quad T'(\theta) = H(\theta).T''(\theta),$

which transforms (29) into :

$$
(33)\begin{cases} T''(\theta+\omega) = M_0'.T''(\theta) + L''(\theta)\Delta\lambda + R''(\theta)\Delta t + K''(\theta), \\ \text{where } (L'',R'',K'')(\theta) = H^{-1}(\theta+\omega)(M',L',R')(\theta). \end{cases}
$$

This new equation in T" is now solvable, for each Δt, $\Delta\lambda$ due to the fact that M_0' is a constant diagonal matrix whose complex eigenvalue argument satisfies condition (21). Now $\Delta\lambda$ and Δt are in fact known linear funtionals of $(\Delta\psi,\Delta\tau,\overline{\Delta\tau})$ due to the relation of conjugacy and to the twisted quasi-conjugacy. Finally it is not very hard to show that (33) leads to a unique solution $(\Delta\psi,\Delta\tau,\Delta t, \Delta\lambda)$.

3.2. Persistence of C^∞– invariant curves .

To prove the existence of an invariant curve under F_μ, it is sufficient to prove that we have $(\lambda,t) = 0$.

Let us recall that λ belongs to a neighborhood of $\lambda_0 = \omega - \omega_\mu$ and t to a neighborhood of $t_0 = \Omega - \omega_1 - \sigma$. Then $\lambda=0$ and $t=0$ are in these neighborhoods if λ_0 and t_0 are sufficiently small, which implies :

$$
(34)\qquad \begin{cases} |\omega - \omega_0| = O(|\omega_0 - \omega_\mu|), \\ |\Omega - \omega_1| = O(|\sigma|), \end{cases}
$$

But (34) is not always compatible with the diophantine condition (21) since ω_0 is allowed to be rational and ω_1 can be in $\omega\mathbb{Z}$ in order to obtain a generic result. Then we need a more restrictive diophantine condition which has been called "with small constant" in [Los2] :

Definitions . The set $DC_\alpha(\omega_0)$ of diophantine numbers of "order α" close to ω_0 is defined by $\alpha \geq 0$ and :

$$(35)\quad DC_\alpha(\omega_0) = \left\{ \omega \in \mathbb{T}^1 \;/\; \exists C>0,\ \exists \beta \geq 0 \;/\; \forall \tfrac{p}{q} \in \mathbb{Q}/\mathbb{Z},\ \left| \omega - \frac{p}{q} \right| \geq \frac{C|\omega-\omega_0|^\alpha}{|q|^{2+\beta}} \right\}.$$

For a fixed ω in $\mathbb{R}-\mathbb{Q}$ and $\alpha \geq 0$, the set $DC\Omega_\alpha(\omega_1)$ of diophantine numbers of "order α" close to ω_1 is defined by:

$$(36)\qquad DC\Omega_\alpha(\omega_1) = \left\{ \begin{array}{l} \Omega \in \mathbb{T}^1 / \exists C>0,\ \exists \beta \geq 0 / \ \forall \frac{p}{q} \in \mathbb{Q}\ , \\ |2\Omega - q\omega - p| \geq \dfrac{C|\Omega-\omega_1|^\alpha}{|q|^{1+\beta}} \end{array} \right\}.$$

The density of measure of $DC_\alpha(\omega_0)$ goes to 1 when ω goes to ω_0, for every $\alpha \in]0,1]$, and the same holds for $DC\Omega_\alpha(\omega_1)$ when Ω goes to ω_1.

The Rotated-Twisted theorem has been proved for the simplified form of F_μ (20) which is sufficient for the existence part of this theorem, but here we want to obtain the estimates for the true family (7). We therefore define precisely the map $\mathcal{H}_t$ in which we only transform the perturbation Π by :

$$(37)\qquad \Pi = \varepsilon A_\mu + \varepsilon^{(N/2)-1}\, \xi_\mu ,$$

where ξ is bounded for all C^k-topologies. The truncated normal form corresponds to $\xi \equiv 0$. Now, an estimate in particular on matrix $M(\theta)-M_0$ is used with a very careful and systematic use of techniques developped in [Chen] and [Los 2] for obtaining estimates on the derivatives of $\mathcal{H}_t$, since this is necessary to measure the validity of the R.T. theorem.

Denoting by $\delta = \min\,[|\omega-\omega_0|^\alpha, |\Omega-\omega_1|^\alpha]$, then it can be shown that if $\delta = \varepsilon^{1/8}$, and $N \geq 5$, the $\lambda_0$and ω_0-neighborhoods where the R.T. theorem holds are such that

$$(38)\qquad |\lambda-\lambda_0| \leqslant C\,\varepsilon^{5/4}$$

and if we choose ω in a ω_0-neighborhood so that :

$$|\omega-\omega_\mu| \leqslant K\,\varepsilon^{5/4},\ K < C,$$

then the origin lies in the λ_0-neighborhood (38). Moreover it can be proved that generically we have $\frac{\partial\lambda}{\partial\varepsilon}\neq 0$, thus the classical implicit function theorem implies that the equation $\lambda(\mu) = 0$ defines a local smooth manifold of the form :

(39) $$\varepsilon = a\mu_{oi} + b(\omega-\omega_o) + O(\varepsilon^{5/4}),\ \text{for}\ \mu_{oi}= O(\varepsilon).$$

The study of equation $t(\mu) = 0$ is performed in a similar way , with the above choices, the t_0-neighborhood takes the form :

(40) $$|t-t_0| \leqslant C\,\varepsilon^{5/4},$$

then choosing Ω in a ω_1-neighborhood of the form :

$$|\Omega-\omega_1-\sigma| \leqslant K\,\varepsilon^{5/4},\ K<C,$$

insures that the origin lies in the t_0-neighborhood (40), and since we can show that we also have $\frac{\partial t}{\partial\sigma}\neq 0$, this implies the existence of a local smooth manifold on which $t(\mu) = 0$. This manifold is shown to be transversal to the one defined by (39) and takes the form :

(41) $$\sigma = \Omega-\omega_1+ O(\varepsilon^{5/4}).$$

At this point we have proved the main result , namely there exists a family of paths $\mathcal{P}_{\omega,\Omega}$ in the three dimensional parameter space corresponding to a constant μ_{oi} section. Each of these paths is defined by the intersection of the transversal manifolds $\{\mu; \lambda(\mu) = 0\}$ and $\{\mu; t(\mu) = 0\}$ (see figure 3) and is such that the one parameter subfamily $\{F_\mu\}_{\mu\in\mathcal{P}_{\omega,\Omega}}$ admits a family of C^∞ invariant curves. The set of these paths is parametrized by the set of pairs (ω,Ω) satisfying :

(42) $$\begin{cases} (\omega,\Omega)\in DC_{1/8}(\omega_o)\mathrm{x}DC\Omega_{1/8}(\omega_1),\ (\text{see } (35),(36)),\\ |\omega-\omega_\mu| \leqslant K\,\varepsilon^{5/4},\\ |\Omega-\omega_1-\sigma| \leqslant K\,\varepsilon^{5/4}, \end{cases}$$

which is a Cantor set having a positive measure in T^2. The above estimates enable us to be more precise about the dynamics, at least for parameters on the paths $\mathcal{P}_{\omega,\Omega}$. Now let us define the new parameter :

(43) $\chi = \nu+b$ (where b is defined at the R.T. theorem),

which characterizes the weak normal hyperbolicity of the invariant curves on $\mathcal{P}_{\omega,\Omega}$. By the above arguments, the equation $\chi(\mu) = 0$ defines a smooth local manifold of the form :

(44) $$\nu = O(\varepsilon^{5/4}),$$

which is transversal to the two preceding ones (see figure 3) .

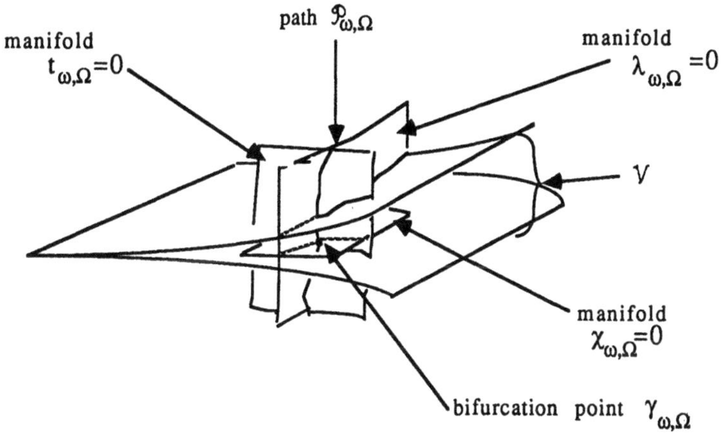

Figure 3: *The transversal manifolds defining the bifurcation point* $\gamma_{\omega,\Omega}$.

The point

$$\gamma_{\omega,\Omega} = \{(\varepsilon,\nu,\sigma)/\ (\lambda,\chi,t)=(0,0,0)\},$$

which belongs to the path $\mathcal{P}_{\omega,\Omega}$ is such that the C^∞ invariant curve is non normaly hyperbolic, so it is the bifurcation point corresponding to the pair (ω,Ω) in T^2. This ends the proof of the following theorem :

Invariant curve theorem . **For each (ω,Ω) in T^2 satisfying (42) there exists a path $\mathcal{P}_{\omega,\Omega}$ in the three dimensional parameter space (for a fixed μ_{0i}) along which the family F_μ (7), satisfying generic assumptions and (2) with $N\geq 5$, admits a C^∞ invariant curve with a rotation number ω. On each of these paths there exists a unique point $\gamma_{\omega,\Omega}$ for which the C^∞ invariant curve is non normaly hyperbolic (bifurcation point).**

Remark . The largest is N, the largest is the Cantor set for (ω,Ω) in T^2, since we can choose a larger α in (35),(36) bounded by a N-dependent constant.

At this point we can now describe the bifurcations of the family F_μ, since along each path $\mathcal{P}_{\omega,\Omega}$ we satisfy the conditions of the Chenciner-Iooss paper [Ch-Io].

4. Bifurcation of invariant T^2 tori. Bubble structure.

At points $\gamma_{\omega,\Omega}$ of the parameter space we precisely arrive to a typical situation from where the analysis of Chenciner and Iooss started in [Ch-Io], except that we have more than one parameter here. In fact let us reconsider the problem by using the result of the R.T.theorem in saying that there exists a curve $\mathcal{C} = \{(\theta,\psi(\theta),\tau(\theta)),\theta\in T^1\}$ which is invariant under $\mathcal{R}_{\lambda,t} o F_\mu$. This implies the relation :

$$(45) \quad \begin{cases} F_\mu(\theta,\psi(\theta),\tau(\theta)) = \mathcal{R}^{-1}_{\lambda,t} \circ (\Phi,\psi(\Phi),\tau(\Phi)) , \\ \text{with } \Phi = h^{-1} o R_\omega o h(\theta), \end{cases}$$

i.e. we have the equality :

$$(46) \qquad F_\mu(\theta,\psi(\theta),\tau(\theta)) = (\Phi - \lambda,\ \psi(\Phi),\ e^{-2i\pi t}\tau(\Phi)) ,$$

which is valid for every μ in a neighborhood of $\gamma_{\omega,\Omega} = \{\mu ;\ \lambda = 0,\ t=0,\ \chi=0\}$, χ being defined by (43). The size of this neighborhood is given by (38),(40), (44). Let us change variables in taking $\mathcal{C}$ as the new origin of radial coordinates. The normalization process derived in [Ch-Io] can be used, provided the following diophantine conditions hold :

$$(47) \quad \begin{cases} \Omega \neq 0,\ 1/2, 1/3,\ 2/3, 1/4,\ 3/4,\ \text{mod } 1 \\ \exists\ C>0,\ \exists \beta \geq 0 \ /\ \forall\ \frac{p}{q} \in \mathbb{Q}\ ,\ |r\Omega - q\omega - p| \geq \dfrac{C\varepsilon^{1/8}}{|q|^{1+\beta}} , \\ \text{for} \quad r=0,1,2,3,4\ . \end{cases}$$

We observe that the coefficient $\varepsilon^{1/8}$ in (47) leads to factors $\varepsilon^{-1/8}$ in the change of variables, but a careful examination shows that this gives no problem on mapping (7) because the angular dependence is contained in small terms.

Now, since after normalization we use again the graph transform technique as in [Ru-Ta], we need that the (weak) hyperbolicity of the linear part be larger than the size of the perturbation. This rate of hyperbolicity is a linear functional of all parameters, so this leads to show persistence of invariant T^1 and bifurcating T^2 when one moves transversaly to an hyperplane defined by an equation of the form $\tilde{\chi} = \chi + 0(\varepsilon^{5/4}) = 0$ passing through $\gamma_{\omega,\Omega}$ (see figure 4).

<u>Bubbles-Theorem</u>. For the generic family of diffeomorphisms F_μ (7), satisfying the non strong resonance condition (2) with N ≥ 5 and generic non degeneracy conditions , there exists a set of two sided paraboloïdal regions tangent at their vertices $\gamma_{\omega,\Omega}$ (see figure 4) getting through the neighborhood $\mathcal{V}$ (defined in §.2, see figure 3). The points $\gamma_{\omega,\Omega}$ are defined, for each (ω,Ω) satisfying (47) and (42), by the invariant curve theorem of §3. In these

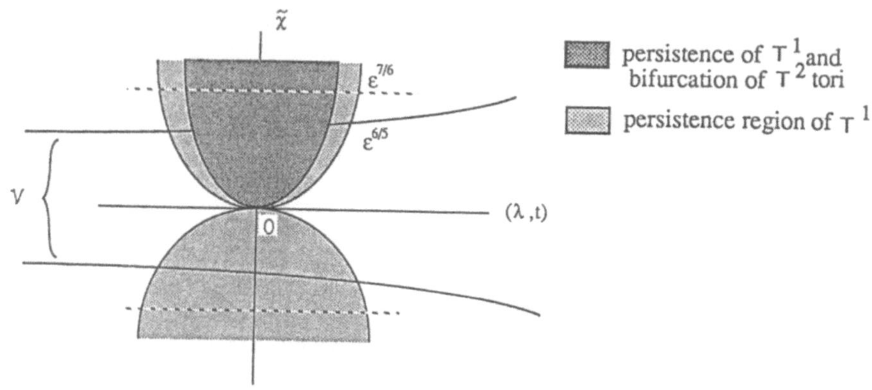

Figure 4 : *Regions of persistence of invariant T^1 and bifurcating invariant T^2.*

regions there is a family of C^k invariant T^1-curves (C^∞ on the special path $\mathcal{P}_{\omega,\Omega}$) loosing their normal hyperbolicity at each vertex of the paraboloïds. Moreover, in each one-sided paraboloïdal region (figure 4), there bifurcates from each vertex a family of C^k normaly hyperbolic invariant T^2-tori.

We finally obtained the bubble structure announced in the introduction as the complementary, in the parameter space, of the paraboloïdal regions (see figure 1).

BIBLIOGRAPHY.

[Bost] J.B.Bost .Tores invariants des systèmes dynamiques hamiltoniens. Séminaire Bourbaki 1984-1985 nº 639.

[B.H.T.] H.W. Broer, G.B. Huitema, F. Takens . Unfoldings of quasi-periodic tori. Preprint (Univ. of Groningen).

[Bra-B] B.L.J.Braaksma, H.W.Broer . On a quasi-periodic Hopf bifurcation . Ann. IHP. Analyse non linéaire, 4,2,p115-168, (1987).

[Ch-Io] A.Chenciner,G.Iooss. Bifurcation de tores invariants. Arch. Rat. Mech. Anal.69,3,109-198,(1979); and Persistance et bifurcation de tores invariants.Arch.Rat.Mech.Anal.71 ,4,301-306,(1979).

[Chen] A.Chenciner . Bifurcation de point fixes elliptiques I, courbes invariantes, Publ Math IHES, 61, 67-127, (1985).

[Go-Be] J.P.Gollub,S.V.Benson. Many routes to turbulent convection .J .Fluid. Mech. 100,3,449,(1980).

[GRS] M.Gorman,L.A.Reith,H.L.Swinney. Modulation patterns, multiple frequencies,and other phenomena in circular Couette flow. Nonlinear dynamics . Ann.N.YAcad.Sci,357,10-21,R.Helleman ed. (1980).

[Gu-Ho] J.Guckenheimer,P.Holmes. Nonlinear oscillations, Dynamical systems, and Bifurcations of vector fields. Appl.Math.Sci 42. Springer Verlag (1983).

[Ham] R.S.Hamilton . The inverse function theorem of Nash-Moser. BAMS, 7, 1,65-222, (1982).

[Her79] M.Herman .Sur la conjugaison differentiable des diffeomorphismes du cercle a des rotations. Pub.Math IHES, 49, 5-233,(1979).

[Her83] M.Herman . Sur les courbes invariantes par les difféomorphismes de l'anneau I. Astérisque, 103-104, (1983) .

[Io-La] G.Iooss,W.F.Langford. Conjectures on the route to turbulence via bifurcations. Nonlinear Dynamics. Ann.N.Y.Acad.Sci, 357, 489-505, R.Helleman ed. (1980).

[Io81] G.Iooss . Bifurcations élémentaires-successions et interactions. Nonlinear phenomena in Chemical dynamics. Vidal- Pacault ed, 71-78,(1981).

[Io-Lo] G.Iooss, J.E.Los. Quasi-genericity of bifurcations to invariant tori for maps. Preprint 1987.

[Jo-Ze] R. Jost, E. Zehnder. A generalization of the Hopf Bifurcation Theorem. Helv. Phys. Acta , 45, 258-276, (1972).

[Lan] O.E.Lanford . Bifurcation of periodic solutions into invariant tori: the work of Ruelle and Takens. Lect.Notes.Math,322, p159-192, Springer Verlag(1973).

[Los1] J.E.Los . Phénomènes de petits diviseurs dans les dédoublements de courbes invariantes. Ann. IHP, Analyse non linéaire (to appear).

[Los2] J.E.Los . Doubling bifurcation for invariant curves : C^∞ invariant curve on C^k cylinder. (Submitted to Nonlinearity). Preprint n° 124 Univ Nice (1986).

[Mos] J.Moser . Convergent series expansion for quasi-periodic motion , Math. Annalen, 169, 136-176 (1967).

[Ru-Ta] D. Ruelle, F. Takens. On the nature of turbulence. Comm. Math. Phys. , 20, 167-192, (1971).

[Russ] H.Rüssmann.Über invariante kurven differenzierbarer abbildungen eines kreisrings.Nach. Akad. Wiss. Göttingen Math. Phys. K1.II, 67-105,(1970).

[Sac] R.Sacker. On invariant surfaces and bifurcation of periodic solutions of ordinary differential equations. Thesis, New-York University, IMM-NYU 333, 1964.

[Sch] J.Scheurle. Bifurcation of quasi-periodic solutions from equilibrium points of reversible dynamical systems. Arch Rat Mech Anal, 97,2,103-139,1987.

[Sell] G.R.Sell . Bifurcation of higher dimensional tori. Arch.Rat.Mech. Anal. 69 ,3,199-230,(1979).

[Yoc] J.C.Yoccoz. Conjugaison différentiable des difféomorphismes du cercle dont le nombre de rotation vérifie une condition diophantienne. Ann.Sci.ENS.17, 333-359 (1984).

Generalized Lieb-Thirring Inequalities and the Dimension of Attractors Associated to the Ginzburg-Landau p.d.e.

J.-M. Ghidaglia

Laboratoire d'Analyse Numérique, CNRS et Université Paris-Sud,
F-91405 Orsay Cédex, France

Abstract. This paper is concerned with the dimension of the global attractor that describes the long time behavior of the solutions to the 2D-Ginzburg-Landau partial differential equation. An upper bound on this dimension is derived via a generalized Lieb-Thirring inequality, while lower bounds follow from the classical (linear) stability analysis. As a matter of fact these bounds are of the same order and therefore optimal.

1. Introduction.

The Ginzburg-Landau partial differential equation

$$\frac{\partial A}{\partial t} - (1+iD)\Delta A + (1+iN) \mid A \mid^2 A - A = 0, \tag{1}$$

describes the amplitude evolution of instability waves in a very large variety of dissipative systems in fluid mechanics which are close to criticality e.g. in Taylor-Couette flow, Bénard convection or plane Poiseuille flow... Although various boundary conditions can be considered for the functions A in (1), we shall restrict ourselves to the case where A is defined on the whole plane $\mathcal{R}^2$ and is L-periodic in each space direction :

$$A(x_1 + L, x_2, t) = A(x_1, x_2 + L, t) = A(x_1, x_2, t), \forall x_1, x_2, \forall t \geq 0. \tag{2}$$

As reported by MOON *et al.* [10,11], KEEFE [8,9] and HUERRE [7], for certain values of the real parameters D, N and L, the solutions to the one-dimensional analogue of (1)(2) do have chaotic behavior and strange attractors occur. Our joint work with B. HERON ([3]), on which we report here, was motivated by these facts and the following questions

(i) Does the permanent regime of (1)-(2) actually depends on a finite number of degrees of freedom ?
(ii) If so, is it possible to estimate accurately this number in terms of the parameters (D, N and L) of the problem ?

Since (1)-(2) is a mild nonlinear dissipative parabolic equation, the existence of a finite dimensional global attractor is now classical (see [12] and the references therein). This gives a positive answer to (i). Concerning the question (ii), we will outline the main steps of the

THEOREM 1 : *There exists two positive universal constants* c_0 *and* c_0' *such that the fractal dimension* $d_F(\mathcal{A})$ *of the global attractor* $\mathcal{A}$ *satisfies*

$$d_F(\mathcal{A}) \leq c_0(1 + L^2(1 + 2\omega)) \tag{3}$$

where $\omega = max\left(\sqrt{1+N^2} - 2, 0\right)$,

$$d_F(\mathcal{A}) \geq -1 + c_0' L^2, \tag{4}$$

and finally for $1 + ND < 0$,

$$d_F(\mathcal{A}) \geq c_0' L^2 \frac{| 1 + ND |}{1 + D^2}. \tag{5}$$

We recall that the fractal dimension of a subset $\mathcal{E}$ of a metric space H is defined as

$$d_F(\mathcal{E}) = \lim_{\epsilon \to 0} \frac{log N_\epsilon(\mathcal{E})}{log 1/\epsilon}$$

where $N_\epsilon(\mathcal{E})$ denotes the minimal number of balls of radius ϵ which are needed to cover $\mathcal{E}$.

Finally, let us mention that a more general study of (1) is intended in [4], in particular we consider the Ginzburg-Landau equation on an arbitrary bounded set of $\mathcal{R}^n, n = 1$ or 2 with general boundary conditions. The bounds on the dimension of the global attractor are identical to those obtained in Theorem 1 provided L^2 is replaced by the $n-$dimensional volume of Ω. A similar point of view in the one-dimensional case has been independently given by C.R. DOERING *et al.* [3]. However our bounds ([4]) are sharper, at least in the case $D \neq 0$, $\omega \neq 0$, since we use the generalized Lieb-Thirring inequalities.

2. Generalized Lieb-Thirring inequalities.

Some of these inequalities, that we describe subsequently (see (8)), are useful in proving the upper bound (3). Since they are interesting in their own, we begin by giving them in a rather large generality, refering to [5] for a more complete exposition.

We consider on an open set Ω in $\mathcal{R}^n$, a finite family $\{v_j\}_{j=1}^N$ of vector valued functions which belong to $H^m(\Omega)^k$, where $H^m(\Omega)$ is the usual Sobolev space of order $m \geq 1$ on $L^2(\Omega)$ ($| \alpha | = \alpha_1 + ... + \alpha_n$) :

$$H^m(\Omega) = \left\{ v \in L^2(\Omega), \frac{\partial^{|\alpha|} v}{\partial x_1^{\alpha_1} ... \partial x_n^{\alpha_n}} \in L^2(\Omega), \ \forall \alpha, | \alpha | \leq m \right\}.$$

We say that the family $\{v_j\}_{j=1}^N$ is *suborthonormal* in $L^2(\Omega)^k$ if

$$\sum_{i,j=1}^N \xi_i \xi_j \int_\Omega v_i \cdot v_j dx \leq \sum_{p=1}^N \xi_p^2, \forall \xi \in \mathcal{R}^N. \tag{6}$$

Let p be any real number with

$$\max(1, n/2m) < p \leq 1 + (n/2m), \tag{7}$$

the generalized Lieb-Thirring inequalities relate the $L^{2/(p-1)}-$norm of the v_j to their H^m-norm :

THEOREM 2. *Let Ω be a smooth bounded open set in $\mathcal{R}^n$. For every p satisfying (7), there exist two constants κ and χ such that for every finite family $\{v_1, ..., v_n\}$ in $H^m(\Omega)^k$ which is suborthonormal in $L^2(\Omega)^k$, we have*

$$\int_\Omega \left(\left(\sum_{j=1}^{N} | v_j(x) |^2 \right)^{p/(p-1)} dx \right)^{2m(p-1)/n}$$

$$\leq \sum_{j=1}^{N} \int_\Omega \left(\kappa \sum_{|\alpha|=m} | D^\alpha v_j |^2 + \frac{\chi}{\delta(\Omega)^{2m}} | v_j |^2 \right) dx, \tag{8}$$

where $\delta(\Omega)$ denotes the diameter of Ω.

A few comments.

1. The constants κ and χ depend on m, k, p and Ω. However they can be chosen so that they remain unchanged if we translate, rotate or apply a homothety to Ω.

2. When $N = 1$, (8) reduces to the classical Sobolev-Gagliardo-Nirenberg inequality·

$$\left(\int_\Omega | v |^{2p/(p-1)} dx \right)^{((2mp/n)-1)} \leq$$

$$\leq C(\Omega) \left(\int_\Omega | v |^2 dx \right)^{((2mp/n)-1)} \int_\Omega \left\{ \frac{1}{\delta(\Omega)^{2m}} | v |^2 + \sum_{|\alpha|=m} | D^\alpha v |^2 \right\} dx. \tag{9}$$

Applying (9) to each of the v_j and taking (6) into account leads to (8) with κ and χ replaced by $C(\Omega)N^{2m/n}$ which tends to infinity as $N \to \infty$, while κ and χ are independent of N!

3. If one considers families of functions vanishing up to the order m on the boundary of Ω, then (8) is true with $\chi \equiv 0$ and moreover Ω can be ***any*** subset in $\mathcal{R}^n$, bounded or unbounded, smooth or not and κ is a universal constant.

4. In fact (8) is true for a very general class of unbounded domains, see [5] for a precise statement.

5. Let us finally mention that the original inequality, which is due to Lieb and Thirring, was given in the particular case $\Omega = \mathcal{R}^n, k = 1, m = 1$ and under the very restrictive hypothesis that the $\{v_j\}$ are orthonormal in $L^2(\Omega)$:

$$\int_\Omega v_i v_j dx = \delta_{ij}, \forall i, j = 1, ..., N. \tag{10}$$

Although the application of (8) to Ginzburg-Landau equations that we describe later uses functions that satisfy (10), in fact (6) seems more natural when one considers partly parabolic equations (see e.g. [6]).

6. There is also an analogue of Theorem 2 for Ω an open set in an n–dimensional oriented riemannian manifold (see [5] where an application to geophysical flows is given).

7. Other inequalities of the same type that concern functions which are orthonormal in other scalar products than that of L^2 have been proven by P. CONSTANTIN [1].

3. Construction of the global attractor.

The basic Hilbert space for (1)-(2) is $L^2(\Omega)^2$, $\Omega =]0, L[^2$ (i.e. the complex valued functions on Ω whose modulus belongs to $L^2(\Omega)$). If we are given $A_0 \in H$, then (1)-(2) admits a unique solution A which is continuous from $[0, +\infty[$ into H and satisfy the initial condition $A(x,0) = A_0(x)$ for a.e. $x \in \Omega$. This allows us to define a semi-group on H by

$$S(t) : A_0 \to A(\cdot, t), t \geq 0.$$

Moreover for every $t > 0, S(t)$ is a compact operator on H (i.e. if B is bounded in H then the closure of $S(t)B$ in H : $cl(S(t)B)$ is compact in H). In fact $S(t)$ possesses a ***bounded absorbing*** set in H, indeed according to (1)-(2) we have

$$\frac{1}{2}\frac{d}{dt}\int_\Omega |A|^2\,dx + \int_\Omega |\nabla A|^2\,dx + \int_\Omega |A|^4\,dx = \int_\Omega |A|^2\,dx. \tag{11}$$

Hence using that $|A|^4 - 2|A|^2 \geq -1$, we have

$$\frac{d}{dt}\int_\Omega |A|^2\,dx + \int_\Omega |A|^2\,dx \leq 2L^2,$$

therefore

$$\int_\Omega |S(t)A_0|^2\,dx \leq \left(\int_\Omega |A_0(x)|^2\,dx\right)e^{-t} + 2L^2(1 - e^{-t}). \tag{12}$$

This shows that each trajectory enters in a finite time in the ball

$$B_a = \{v \in L^2(\Omega)^2, \int_\Omega |v|^2\,dx \leq 3L^2\},$$

which is, by definition, a bounded absorbing set.

It follows then (all the details can be found in [4] and references therein) that the omega-limit set of B_a is the global attractor for $S(t)$ in H, i.e. :

$$\mathcal{A} = \bigcap_{s>0} \mathrm{cl}\left(\bigcup_{t\geq s} S(t)B_a\right) \tag{13}$$

satisfies

$$\mathcal{A} \text{ is compact, non empty (in } H) \tag{14}$$

$$\mathcal{A} \text{ is invariant } (S(t)\mathcal{A} = \mathcal{A},\ \forall t \geq 0) \tag{15}$$

$$\mathcal{A} \text{ is attractive } : \forall B \text{ bounded in } H, d(S(t)B, \mathcal{A}) \to 0, \text{ as } t \to +\infty. \tag{16}$$

4. <u>An upperbound on the dimension of $\mathcal{A}$.</u>

The $\mathcal{R}$-differential of the semi-group $S(t)$ at $A_0 \in H$ is the mapping $v_0 \to v(t)$ where v is the solution of the linearized equation

$$\frac{\partial v}{\partial t} - (1+iD)\Delta v + (1+iN)[\mid A \mid^2 v + 2Re(\bar{A}v)A] - v = 0, \tag{17}$$

$$v(0,t) = v_0(x),$$

v satisfying the analogue of (2) and $A = A(t) = S(t)A_0$.

As was proven in CONSTANTIN *et al.* [2], the fractal dimension of $\mathcal{A}$ is related to the global Lyapunov exponents on this invariant set. These are related to the time average of the Nth-trace of the linear Schrödinger operator that appears in (17) :

$$\sigma_N(t) = \sum_{j=1}^{N} \frac{1}{t} \int_0^t Re \int_\Omega \Gamma(A) v_j \bar{v}_j dx \tag{18}$$

where (17) is written $\frac{\partial v}{\partial t} + \Gamma(A)v = 0$, and $\{v_j\}_{j=1}^N$ is an arbitrary family of functions in $H^1(\Omega)^2$ satisfying (2) and orthonormal in $L^2(\Omega)^2$.

In order to evaluate (18), we take $A_0 \in \mathcal{A}$ and consider

$$Re \int_\Omega \Gamma(A) v\bar{v} dx = \int_\Omega \mid \nabla v \mid^2 dx + \int_\Omega (\mid A \mid^2 - 1) \mid v \mid^2 dx + 2N \int_\Omega Re(\bar{A}v) Im(\bar{A}v) dx,$$

$$Re \int_\Omega \Gamma(A) v\bar{v} dx \geq \int_\Omega \mid \nabla v \mid^2 dx - \int_\Omega \mid (1 + \omega \mid A \mid^2) \mid v \mid^2 dx \tag{19}$$

where ω is given in Theorem 1. Then we set

$$\rho(x) = \sum_{j=1}^{N} \mid v_j(x) \mid^2 \ \ for\ a.e.\ x \in \Omega; \tag{20}$$

and deduce from (18) and (19) :

$$\sigma_N(t) \geq \sum_{j=1}^{N} \int_\Omega \mid \nabla v \mid^2 dx - \frac{1}{t} \int_0^t \int_\Omega (1 + \omega \mid A \mid^2) \rho dx \tag{21}$$

Now we apply Theorem 2 with $k = 2, m = 1, n = 2$, $\Omega =]0, L[^2$, $\delta(\Omega) = \sqrt{2}L$, $p = 1 + n/2M = 2$: there exists κ and χ such that

$$\int_\Omega \rho^2 dx \leq \kappa \sum_{j=1}^{N} \int_\Omega \mid \nabla v_j \mid^2 dx + \frac{\chi}{L^2} \int_\Omega \rho dx. \tag{22}$$

Therefore (21) implies

$$\sigma_N(t) \geq \frac{1}{\kappa}\int_\Omega \rho^2 dx - \frac{1}{t}\int_0^t \int_\Omega \left(1 + \frac{\chi}{\kappa L^2} + \omega \mid A \mid^2\right)\rho dx. \tag{23}$$

But since the $\{v_j\}_{j=1}^N$ are orthonormal in $L^2(\Omega)^k$,

$$N = \int_\Omega \rho dx \leq \left(\int_\Omega \rho^2 dx\right)^{1/2}\left(\int_\Omega 1 dx\right)^{1/2}$$

hence

$$\int_\Omega \rho^2 dx \geq \frac{N^2}{L^2}. \tag{24}$$

On the other hand

$$\int_\Omega^t \mid A \mid^2 \rho dx \leq \left(\int_\Omega \mid A \mid^4 dx\right)^{1/2}\left(\int_\Omega \rho^2 dx\right)^{1/2}$$

which leads together with (24) in (23) to

$$\sigma_N(t) \geq \left(\int_\Omega \rho^2 dx\right)^{1/2}\left[\frac{1}{\kappa}\frac{N}{L} - \left(1 + \frac{\chi}{\kappa L^2}\right)L - \omega\left(\frac{1}{t}\int_0^t \int_\Omega \mid A \mid^4 dx\right)^{1/2}\right] \tag{25}$$

Using the fact that $A(t) \in B_a, \forall t \geq 0$ and (11) we find

$$\frac{1}{t}\int_0^t \int_\Omega \mid A \mid^4 dx \leq \frac{3L^2}{2t} + 3L^2.$$

Returning to (25), we have

$$\sigma_N(t) \geq \left(\int_\Omega \rho^2 dx\right)^{1/2}\left[\frac{1}{\kappa}\frac{N}{L} - \left(1 + \frac{\chi}{\kappa L^2}\right)L - 2\omega L\right],\ t \geq \frac{2}{3}.$$

It is obvious on this last expression that when

$$N \geq N_0 = 1 + \text{ integer part of } (\chi + \kappa(1 + 2\omega)L^2) \tag{26}$$

we have

$$\sigma_N(t) \geq k > 0, t \geq \frac{2}{3}$$

where k is independent of $A_0 \in \mathcal{A}, t$ and the $\{v_j\}$. It follows then from a general result of [2] that $d_F(\mathcal{A}) \leq cN_0$ and (2) is proven.

5. Lower bounds on the dimension of $\mathcal{A}$.

The lower bounds (4) and (5) are based on the study of the linear stability of two homogeneous solutions, namely the trivial solution $u(x,t) = 0$ and the Stokes solution $u^S(x,t) = u^S(t) = \exp(-iNt)$. Indeed the linearized flow around $u \equiv 0$ reads

$$\frac{\partial v}{\partial t} - (1 + iD)\Delta v - v = 0$$

and since v is $]0, L[^2$-space periodic, the number of unstable modes is greater than $-1 + c_0' L^2$, where c_0' is a universal constant. This leads to (4) (see [4] for the details).

In order to study the stability of the Stokes solution, it is convenient to set

$$A(x,t) = u^S(t)(1 + \tilde{A}(x,t)),$$

then

$$\frac{\partial \tilde{A}}{\partial t} - (1 + iD)\tilde{A} + (1 + iN)(1 + \tilde{A})(|\tilde{A}|^2 + 2Re\tilde{A}) = 0.$$

As the previous case, the study of the linear stability of the trivial solution $\tilde{A} \equiv 0$ leads to (5).

REFERENCES

1. P. Constantin, Indiana University Math. J., to appear.
2. P. Constantin, C. Foias and R. Temam, Mem. Amer. Math. Soc., *53*, 314 (1985), see also J. Fluid Mech., *150*, 427-440 (1985).
3. C.R. Doering, J.D. Gibbon, D.D. Holm and B. Nicolaenko, Los Alamos preprint LA-UR; 87-1546, to be published.
4. J.M. Ghidaglia and B. Héron, Physica *28* D, 282-304 (1987).
5. J.M. Ghidaglia, M. Marion and R. Temam, Differential and Integral Equations, *1*, 1-21 (1988).
6. J.M. Ghidaglia and R. Temam, Asymptotic Analysis, 1,23-49 (1988).
7. P. Huerre, in ***Equations aux Dérivées Partielles non Linéaires et Systèmes Dynamiques***, J.M. Ghidaglia and J.C. Saut Ed., Travaux en cours, Hermann, Paris, 1988.
8. L. Keefe, Phys. Fluids, *29*, 3135-3141 (1986).
9. L. Keefe, Stud. in Appl. Math., *73*, 91-153 (1985).
10. H.T. Moon, P. Huerre and L.G. Redekopp, Phys. Rev. Lett.,*49*, 458-460 (1982).
11. H.T. Moon, P. Huerre and L.G. Redekopp, Physica 7D, 135-150 (1983).
12. R. Temam, ***Infinite Dimensional Dynamical Systems in Mechanics and Physics***, Springer, Berlin, to appear.

Low Order Spectral Models in Hydrodynamics

A. Doelman and J. Grasman

Departments of Mathematics, University of Utrecht,
P.O. Box 80 010, NL-3508 TA Utrecht, The Netherlands

1. Introduction

In this contribution we study low order spectral models of two different partial differential equations: the Ginzburg-Landau equation and the barotropic vorticity equation. Both equations govern a two dimensional hydrodynamic flow: the Ginzburg-Landau equation arises in systems such as plane Poiseuille flow, the vorticity equation governs the atmospheric flow in a beta-plane. For both equations we use a spectral expansion of their solutions to derive an infinite dimensional system of coupled nonlinear ordinary differential equations. Truncation of this system to a finite, n-dimensional system leads, for n small, to a low order spectral model.

We are interested in the effects of this truncation, i.e. we want to investigate two problems:

I: is the truncated system a (qualitative) approximation of the full untruncated system?

II: if truncation is permitted: determine, if possible, the minimal value of n necessary to mimic the dynamics of the full system.

Since the two equations we study are based on the Navier-Stokes equations, the results of Foias and Temam [7] are important, in relation to problem I. They proved that for n sufficiently large, and n depending on the control parameter R, solutions of the n-dimensional Galerkin approximation of the Navier-Stokes equations have the same asymptotic behaviour for $t \to \infty$ as solutions of the full Navier-Stokes equations. There are no results known, other than numerical experiments, which deal with problem II.

We study, mainly numerically, for both spectral models the properties of the flow, for different values of n. We compare our results with results obtained from the untruncated system. Or, in the case of the vorticity equation, for which no such results exist, we compare the properties of the truncated system with the real physical process which is described by our model, i.e. the atmospheric circulation.

In the following we will split up each section in an A and a B part. The A parts correspond to the Ginzburg-Landau equation. In this case we observe that the complex four dimensional system already exhibits exactly the same complex behaviour, such as two frequency motion, complete period doubling sequences and two chaotic regions, in a certain parameter range, as was observed by Keefe [9] in his study of the dynamics of space-periodic solutions of the Ginzburg-Landau equation. The three dimensional system differs from the four dimensional system only in the subtle process of transition to chaos. The B parts correspond to the vorticity equation. In that case the three dimensional system shows already some characteristics of the atmospheric circulation. However, the system only has stable equilibrium solutions corresponding to fixed weather patterns, which is unrealistic. The ten-dimensional system has a chaotic solution with a more realistic behaviour: it irregularly visits points in the phase space that correspond with various circulation patterns. We also study the three-dimensional system with small white noise forcing. Solutions of this randomly perturbed system exhibit behaviour similar to the ten-dimensional system. More details on the A respectively B parts can be found in respectively Doelman [4] and De Swart and Grasman [3].

2. Two Partial Differential Equations and Spectral Expansions of their Solutions.

A The Ginzburg-Landau Equation

The Ginzburg-Landau equation arises in many hydrodynamical problems with a control parameter R (for instance: plane Poiseille flow, R = Reynoldsnumber,

Rayleigh-Bénard convection, R = Rayleigh number, Taylor-Couette flow, R = Taylor number) and also in many other areas (such as reaction-diffusion problems). In the hydrodynamical problems the Ginzburg-Landau equation describes the slow evolution of a modulated wave $\epsilon\Phi(z,t)e^{i(k_cX+w_cT)}$ for $R = R_c+O(\epsilon^2)$, with R_c the critical value of R at which the basic flow becomes unstable and the wave solution $e^{i(k_cX+w_cT)}$ begins to develop:

$$(2.1)\qquad \frac{\partial\Phi}{\partial t} = (\sigma + \beta|\Phi|^2)\Phi + \lambda\,\frac{\partial^2\Phi}{\partial z^2}.$$

The complex coefficients σ,β and λ are determined by the physics of the problem. This equation has been derived by many authors, for instance Stewardson and Steward [11] and, more recently by Eckhaus and Iooss [5].

We are mainly interested in space-periodic, even solutions, hence we set

$$(2.2)\qquad \Phi(z,t) = \sum_{n=-\infty}^{\infty} Z_n(t)e^{inqz}.$$

with $Z_n(t) = Z_{-n}(t)$ for all n. The period q will be used as a parameter.

B The Vorticity Equation

We consider a large scale barotropic flow over a slowly varying topography in a midlatitude beta-plane. The potential vorticity equation for the atmospheric circulation model reads, in nondimensional form:

$$(2.3)\qquad \frac{\partial}{\partial t}\nabla^2\Psi + J(\Psi,\nabla^2\Psi) + \gamma J(\Psi,h) + \bar{\beta}\,\frac{\partial\Psi}{\partial x} + c\nabla^2(\Psi-\Psi^*) = 0$$

where Ψ is the stream function, h the position of earth's surface and Ψ^* a forcing stream function. Furthermore

$$J(a,b) = \frac{\partial a}{\partial x}\frac{\partial b}{\partial y} - \frac{\partial a}{\partial y}\frac{\partial b}{\partial x}$$

and γ, $\bar{\beta}$ are coefficients related with earth rotation; c is a coefficient related with dissipation of energy.

The boundary conditions suggest an expansion in, again space-periodic functions

$$(2.4)\qquad \Psi(x,y,t) = \sum_{n=1}^{\infty} x_n(t)\phi_n(x,y).$$

With $(\phi_i)_{i=1}^{\infty}$ an orthonormal set of eigenfunctions of the Laplace operator for the domain of the channel (the rescaled beta-plane), i.e.

$$\phi_1(x,y) = \sqrt{2}\cos(y/b)$$

$$\phi(x,y) = 2\cos x \sin(y/b) \text{ etc.},$$

where $2\pi b$ is the width of the channel; the rescaled length of the channel is 2π.

3. The Low Order Spectral Models.

A The Ginzburg-Landau Equation.

The Ginzburg-Landau equation reads, after rescaling and selecting the appropriate signs of Re β and Re λ in (2.1)

$$\frac{\partial \Psi}{\partial t} = (1 - (1+ib)|\Psi|^2)\Psi + (1+ia)\frac{\partial^2 \Psi}{\partial z^2} \tag{3.1}$$

with a and b real coefficients. One obtains, after substitution of the spectral expansion (2.2) into (3.1) and truncation of the infinite dimensional system

$$\dot{Z}_n = (1-n^2q^2(1+ia))Z_n - (1+ib)\sum_{\substack{k+l+m=n \\ |k|,|l|,|m|\leq N}} Z_k Z_l Z_m^* \tag{3.2}$$

for $n = 0,1,\ldots,N$ and with $Z_j = Z_{-j}$ for all j, Z_m^* the complex conjungated of Z_m. This is a complex N+1 dimensional system, hence a 2N+2 dimensional real system.

First we list some analytical results obtained for the truncated system.

a. Stability of the Stokes Solution.

The solution $Z(t) = (Z_0(t),\ldots,Z_N(t))$ with $Z_0(t) = e^{-ibt}$, $Z_j(t) = 0$ for $j = 1,\ldots,N$, which corresponds to the Stokes solution of the Ginzburg-Landau equation, $\Psi(z,t) = e^{-ibt}$, is stable for

$$q > q_0(a,b) = \sqrt{-2\,\frac{1+ab}{1+a^2}}.$$

This solution remains stable for all q if $1 + ab > o$.

b. Symmetries.

There are two transformations, T_ϕ and S_0 which carry over solutions of (3.2) to other solutions of this system

$$T_\phi(Z_0,\ldots,Z_N) = (e^{i\phi}Z_0,\ldots,e^{i\phi}Z_N),\ \phi \in [0,2\pi) \tag{3.3}$$

$$S_\sigma(Z_0,\ldots,Z_N) = ((-1)^{0+\sigma}Z_0,\ldots,(-1)^{N+\sigma}Z_N),\ \sigma = 0,1. \tag{3.4}$$

Using symmetry T_ϕ one derives that an isolated periodic solution of (3.2) has to be of the form $Z(t) = Ae^{iwt}$, with $A \in \mathbb{C}^{N+1}$, a complex vector. Due to symmetry T_ϕ the N+1 dimensional complex system can be reduced to a 2N+1 dimensional real system.

c. Decay.

For any $\alpha > 0$ there exists a $L = L(\alpha)$, L independent of N, such that for every N

$$(3.5) \qquad \sup_{t>T_0} |Z_k(t)| < \frac{L}{k^\alpha} ,$$

for T_0 large enough, $1 \le k \le N$. Thus individual modes Z_k decay faster than any power of k.

Moon et. al [10] investigated (3.2) for N = 63. They checked that choosing another, large enough value of N did not alter the properties of the flow. We consider the paper by Keefe [9] as a study of the untruncated model, although Keefe investigated (3.2) with N = 31. We compare our results with those obtained from the untruncated system. Computations of the energy spectrum showed that the relative energy content of a mode Z_n, P_{nq}, is a rapidly, exponentially decaying function of n; and for all $q \in [0.6, 1.33]$: $P_{nq} < 10^{-4}$ for $n \ge 4$. This suggests that higher order modes do not have a significant influence on the flow. Hence we investigated, numerically, the truncated system (3.2) for N = 1,2,3 with a = 4 and b = -4 and the bifurcation parameter $q \in [0.6, 1.33]$, the same values as Keefe took in his paper. Remark that $q_0(4,-4)$, the value at which the Stokes solution becomes unstable, is equal to 1.328.. . We study the flow by its Poincaré map: by plotting the projection in the $Re(Z_0)$, $Re(Z_1)$ plane of the Poincaré section of a solution Z(t) of (3.2) at $Im\ Z_0 = 0$ with $\frac{d}{dt}(Im\ Z_0) > 0$. The set of points, for a certain value of q, in the $Re(Z_0)$, $Re(Z_1)$ plane is called ω_q.

Summary of the Results

a. N = 3.

We observe no qualitative or quantitative differences between our Poincaré sections and those induced by the untruncated flow: both systems exhibit the same, complex behaviour: chaos, period-doubling sequences etc. (see fig. 1.b). This means that for $q \in [0.6, 1.33]$ the spectral system can be truncated to a complex four-simensional system, which is, due to symmetry T_ϕ, essentially a seven-dimensional real system.

b. N = 2.

Only a slightly different behaviour, compared to N = 3, can be observed only for values of q close to a chaotic region. By definition, ω_q is a projection of a section of the ω-limit set of a solution Z(t) of system (3.2). Consequently $S_0\omega_q$ is the projection of the Poincaré section of $S_0Z(t)$, $S_0\omega_q$ is, due to symmetry property S_0, the mirror image of ω_q found by reflection with respect to the $Re(Z_0)$ axis. There is an interplay of merging and splitting between ω_q and $S_0\omega_q$, which is the cause of the differences between the N = 2 and N = 3 cases. The ω_q's are, for N = 2 and N = 3, idential and symmetrical, i.e. $\omega_q = S_0\omega_q$, at q = 0.900. For increasing q we observe (see fig. 1.a and 1.b):

N = 2: ω_q becomes asymmetrical ($S_0\omega_q$ splits off) - ω_q and $S_0\omega_q$ merge again - a sequence of "loop-doublings" - chaos.

N = 3: ω_q becomes asymmetrical - a sequence of "loop-doublings" - chaos, ω_q still asymmetrical - ω_q and $S_0\omega_q$ merge.

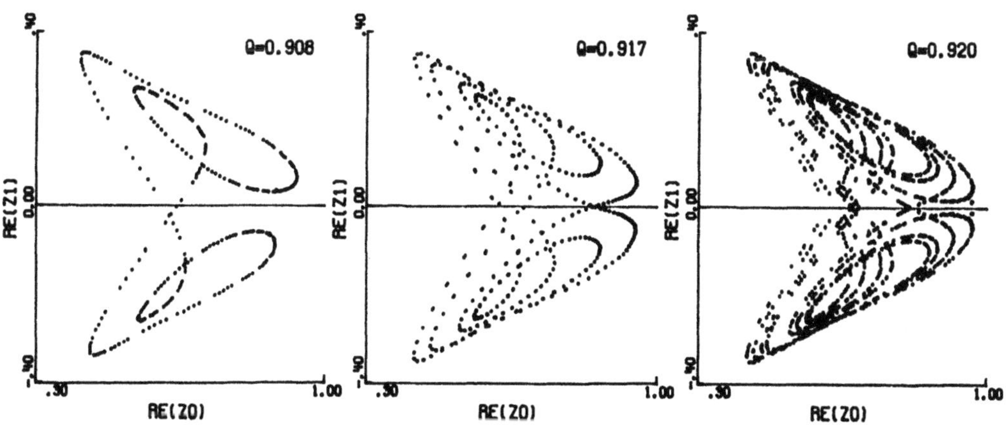

fig 1.a Transition to chaos for N = 2

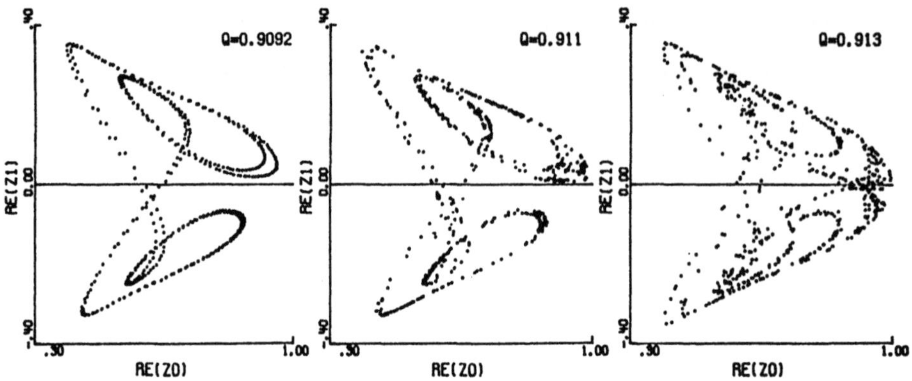

fig 1.b Transition to chaos for N = 3

This switching of the "loop doubling sequence - merging" order occurs two times, in two small regions of the parameter q. N = 2 behaves as N = 3 outside these regions.

c. N = 1.

The Lyapunov dimension of the chaotic attractor of the untruncated system exceeds 3 for certain values of q (the Lyapunov dimension attains its maximum, 3.05, at q = 0.950). Hence the N = 1 system has to be different from the other cases: it is essentially a three dimensional system. Still a chaotic attractor, of dimension less than three, can be found. Remarkable similarities between the N = 1 and the untruncated case can be observed for values of q outside the chaotic regions of Lyapunov dimension greater than 3. It should be remarked that the rôle of the higher order modes, Z_4, Z_5 etc., will become more important for $q < 0.6$: this was already observed in the numerical work of Moon et al [10].

Two theoretical results confirm this numerical results:

- Ghidaglia and Heron [8] derived a (sharp) upperbound on the dimension of the attractor of the Ginzburg-Landau equation. This upperbound tends to ∞ as $q \to 0$.
- We derived estimate (3.5) for the truncated system. The constant L depends on q: L tends to ∞ as $q \to 0$.

B. The Vorticity Equation.

We study the truncated spectral model for n = 3, 6 and 10. Values of n inbetween are less interesting from physical point of view.

B.a n = 3.

By substitution of the spectral expansion (2.4) into the vorticity equation (2.3) and truncation we obtain the three dimensional system

$$(3.5)\qquad \begin{cases} \dfrac{dx_1}{dt} = f_1(x) = bx_3 - C(x_1-x_1^*) \\ \dfrac{dx_2}{dt} = f_2(x) = -ab(x_1-\frac{1}{2}\beta)x_3 - Cx_2 \\ \dfrac{dx_3}{dt} = f_3(x) = ab(x_1-\frac{1}{2}\beta)x_2 - \frac{1}{2}ax_1 - Cx_3 \end{cases}$$

with constants

$$a = \frac{2b}{1+b^2}, \quad \beta = 2.55, \quad C = 0.2$$

defined by the physics of the problem. The coefficient x_1^* is due to the forcing stream function Ψ^*, x_1^* will be used as bifurcation parameter.

In figure 2 the first component of the equilibrium $\bar{x}$ is given as a function of x_1^*. For certain values of x_1^*, for instance $x_1^* = 10$, three equilibria, E_1, E_2 and E_3 (E_1, E_3 stable, E_2 unstable) exist. The atmospheric circulation patterns corresponding to the stable equilibria E_1 and E_3 can be drawn. These patterns resemble, according to Charney and Devore [1], respectively the zonal - and the blocked preference state of the

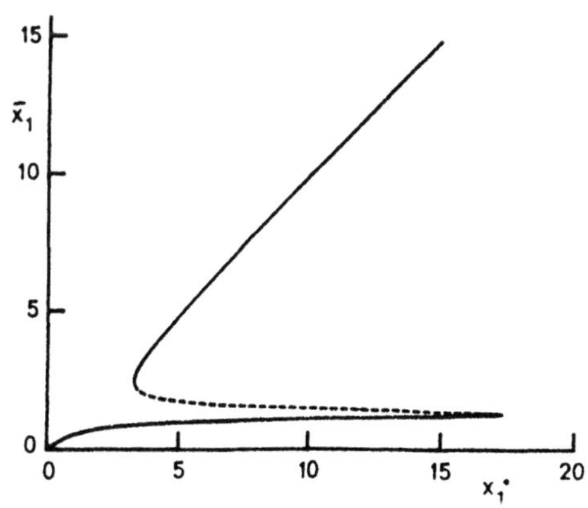

fig. 2: Equilibrium solution $\bar{x}_1$ as a function of x_1^* for b = 1.

atmosphere. However, the model behaviour is unrealistic in the sense that the solutions of (3.5) always end in an equilibrium state. For a system describing the atmospheric behaviour more accurately we expect frequent transitions between these stationary states.

Before we study the spectral models of higher dimension we try a different approach: we force each term of (3.5) by white noise of intensity ϵ. This stochastic input should compensate the absence of the higher order terms. We are now mainly interested in the behaviour of T(x): for a solution of the randomly perturbed system starting at t = 0 in x, T(x) is the expected first passage time of arriving at the separatrix Γ, Γ separates the attraction domains of E_1 and E_3, Ω_1 and Ω_3, in the deterministic (unperturbed) system, $E_2 \in \Gamma$. T(x) now satisfies Dynkin's equation

$$(3.6) \qquad \begin{cases} \frac{1}{2}\epsilon^2 \Delta T + f(x).\nabla T = -1 \text{ in } \Omega_i, \ i = 1,3, \\ T = 0 \text{ at } \Gamma. \end{cases}$$

Eq. (3.6) is an elliptic singular perturbation problem, it can be studied using asymptotic analysis. It is concluded that most of the time the system is in an ϵ-neighbourhood of one of the two stable equilibria. The attraction domain of such an equilibrium is most likely left through the separatrix Γ in an ϵ-neighbourhood of the unstable equilibrium E_2. The expected residence times in the attraction domains of E_1 and E_3, Ω_i i = 1,3 satisfy

$$T_i \approx C_i \, e^{K_i \epsilon^2} \qquad i = 1,3$$

where C_i, K_i can be computed explicitly. The system remains a time of order

$$T_2 \approx \frac{1}{\lambda} \ln \frac{1}{\epsilon}$$

near the unstable equilibrium E_2, with λ the largest eigenvalue of the deterministic system linearized at E_2. Estimates of ϵ for atmospheric models are given by Egger and Shilling [6], they found $\epsilon^2 = 0.2$. The characteristic residence times T_i, i = 1,2,3, found at the basis of this ϵ-value are too large compared to observational data.

B.b n = 6,10.

In higher dimensional spectral models the system exhibits irregular dynamics from itself. No stochastic forcing is needed to obtain vacillation between states with a zonal flow of different intensities. In this section we summarize some results of De Swart [2]. The purpose is to formulate a spectral model with the lowest dimension that still has chaotic behaviour, and that has a zonal component (x_1). In figure 3 the bifurcation

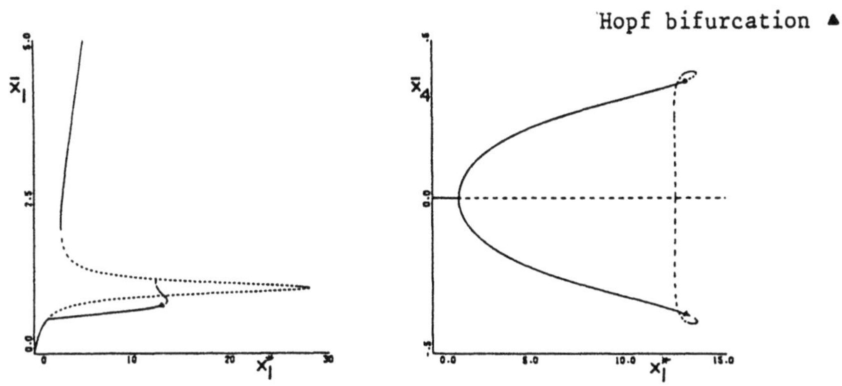

(a) six dimensional spectral model

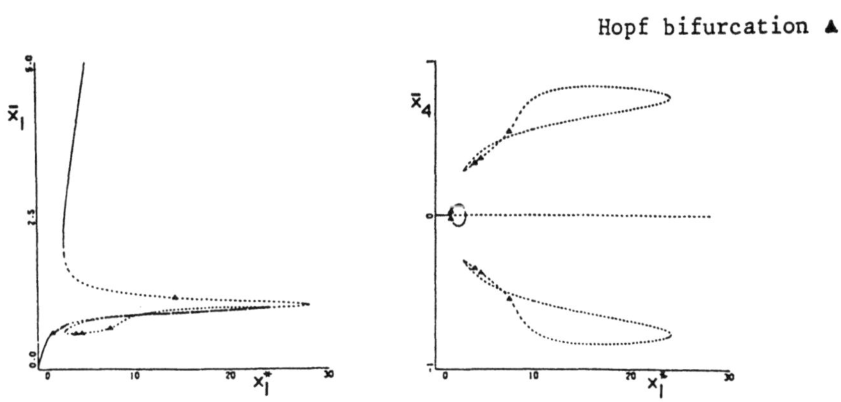

(b) ten dimensional model

fig. 3. Bifurcation diagrams for higher dimensional spectral models. Solid (dotted) lines represent stable (unstable) stationary solutions. Δ: Hopf bifurcation.

diagrams conencted to the equilibria of the three dimensional model are given for the six and ten dimensional model. One of the stable equilibria undergoes a pitchfork bifurcation with the stable branches turning unstable through a Hopf bifurcation. However, the other equilibrium always remains stable. This is due to the symmetry in the forcing stream function. Therefore we take $x_4^* \neq 0$. It is verified that only in the ten dimensional model all equilibria get unstable for some x_4^*. Using physical arguments it is understood that ten dimensions has to be the minimum, as only then energy exchange between wave triades is possible. It is possible to find solution of which the largest Lyapunov exponent has a positive value, which indicates the presence of a strange attractor. Examining the course of the trajectory of a chaotic solution, we observe that it remains from time to time close to the three different periodic orbits. The behaviour of a chaotic solution of the ten dimensional system strongly resembles the behaviour of solutions of the stochastically forced three dimensional problem: there is an irregular transition between preferent regions in state space.

References.

1. J.G. Charney, J.G. Devore: Multiple flow equilibria in the atmosphere and blocking, J. Atmos. Sci. 36 (1979) p. 1205-1216
2. H.E. De Swart: Studies on low order spectral models of the atmospheric circulation: chaotic motion, predictability and vacillation behaviour, Thesis to appear (1988)
3. H.E. De Swart, J. Grasman: Effects of stochastic perturbations on a low-order spectral model of the atmospheric circulation, Tellus 39A (1987) p. 10-24
4. A. Doelman: Low dimensional truncated models of the Ginzburg-Landau equation, Preprint to appear (1988)

5. W. Eckhaus, G. Iooss: Non-classical bifurcations I: strong pattern selection or rejection, Preprint 477, R.U.U., Utrecht (1987)

6. J. Egger, H.D. Schilling: Stochastic forcing of planetary scale flow, J. Atmos. Sci. 41 (1984) p. 779-788

7. C. Foias, R. Temam: Asymptotic numerical analysis for the Navier-Stokes equations, in Nonlinear dynamics and turbulence, ed. by G.I. Barenblatt, G. Iooss, D.D. Joseph (Pitman, Boston, 1983) p. 139-155

8. J.M. Ghidaglia, B. Heron: Dimension of the attractors associated to the Ginzburg-Landau equation, Phys. D. 28 (1987)

9. L.R. Keefe: Dynamics of perturbed wavetrain solutions to the Ginzburg-Landau equation, Stud. in Appl. Math. 73 (1985) p. 91-153

10. H.T. Moon, P. Huerre, L.G. Redekopp: Transitions to chaos in the Ginzburg-Landau equation, Phys. D. 7 (1983) p. 135-150

11. K. Stewardson, J.T. Stuart: A non-linear instability theory for a wave system in plane Poisseuille flow, J. Fluid. Mech. 48 (1971)

Asymptotical Behaviour of Solutions to Quasi-Linear Schrödinger Equations

H. Lange

Mathematisches Institut, Universität Köln, Weyertal 86–90,
D-5000 Köln 41, Fed. Rep. of Germany

1. Introduction.

Brüll and Lange (1986 a, 1986 b) and Hadeler and Lange (1987) introduced the following class of quasilinear Schrödinger equations in one space dimension

$$iu_t = -u_{xx} + W(u)\cdot u \tag{1}$$

with an amplitude dependent nonlinear potential of the form

$$W(u) = f(|u|^2) + \beta h(|u|^2)_{xx}\cdot h'(|u|^2) \tag{2}$$

with some real functions $f \in C(R^+)$, $h \in C^2(R^+)$, and a real constant β; (1) has many applications in physics (see e.g. Lange (1986, 1987) and the references cited therein). In this note we consider the asymptotical behaviour of the wave function $u(x,t)$ as a solution of (1) for $t \to \infty$ for the special model case when

$$f(s) = \lambda s^p, \quad h(s) = s^q$$

(with real constants λ, p, q; $p, q > 0$); in this situation (1) has applications in the dynamics of superfluids ($q=1$) or in nonlinear quantum mechanics ($q=\frac{1}{2}$) (see the references cited in Lange (1987)). Here we write (1) in the form

$$iu_t = -u_{xx} + \lambda|u|^{2p}\cdot u + \beta(|u|^{2q})_{xx}\cdot|u|^{2(q-1)}\cdot u \tag{3}$$

with some real constant β. We are interested in determining the decay rate $\delta > 0$ in an estimate of the type

$$\max_{x\in R} |u(x,t)| \le \frac{C}{t^\delta} \qquad \text{(for all } t > 0) \tag{4}$$

with a certain constant C>0. The main result of this note will be that in the defocusing or repulsive case $\lambda>0$, $\beta<0$ we have

$$\delta = \frac{1}{2} \tag{5}$$

for $p\geq 2$ and any $q>0$; i.e. the decay rate is as in the linear Schrödinger equation corresponding to (3); also in the case $p=1+\varepsilon$ ($0\leq\varepsilon<1$) and small q ($2q\leq\varepsilon$) one has the decay rate (5), whereas in the situation where $p=1+\varepsilon$, $0\leq|\varepsilon|<1$ and $2q>|\varepsilon|$ one has a decay rate which is smaller then $\frac{1}{2}$, e.g. for p=1 and $q \geq \frac{1}{2}$ one has $\delta = \frac{1}{3}$.
For equations of type (1), (2) with n=0 the asymptotical behaviour of solutions has been treated by Lin and Strauss (1978) Ginibre and Velo (1979), Tsutsumi and Yajima (1984), Tsutsumi (1984), Guang-Chang Dong and Shujie Li (1981), Dias and Figueira (1982), Cazenave, Dias, Figueira (1982), Cazenave (1986), Hayashi and Tsutsumi (1986).

We will use the following notation and assumptions:
By L^p ($1\leq p\leq\infty$) and H^1 we denote the usual $L^p(\mathbb{R})$, $H^1(\mathbb{R})$ spaces resp. with their natural norms; the L^p-norm is written as $\|u\|_p$; by H we mean the space of all $u\in H^1$ such that $x\cdot u\in L^2$ (with natural norm induced by H^1 and L^2). We always assume $f\in C(\mathbb{R}^+)$, $h\in C^2(\mathbb{R}^+)$, and define for $s\geq 0$

$$g(s) := \int_0^s f(t)dt; \qquad G(s) := 2(3g(s)-sf(s)),$$

$$H(s) := s\, h'(s)^2;$$

$$F(u) = F(u(x,t)) := G(|u|^2)+2H'(|u|^2)(|u|_x^2)^2;$$

here and in the following we consider any derivative of a function u(x,t) to exist at least in the distributional sense. Furthermore we will need a set of assumptions for the wave function u(x,t): For given T>0, p>0 let

$$(A)_T : u\in C([0,T],H),\ 2g(|u|^2) - \beta[h(|u|^2)_x]^2 \in C([0,T],L^1);$$

$$(B)_T : t\cdot F(u(x,t)) \in C([0,T], L^{2(p+1)});$$

$$(C)_T : u \in C([0,T], H \cap L^{2(p+1)}).$$

Sometimes, to avoid lengthy formulas we will write simply s instead of $|u(x,t)|^2$; also we will use the symbol C for any positive constant appearing in a variety of estimates.

2. Conservation laws.

The following conservation laws may be proved in a standard way (see e.g. Ginibre and Velo (1979) or Lange (1987)):

Lemma 1 Let u(x,t) *be any distributional solution of* (1), (2) *of class* $(A)_T$ *(finite energy solution). Then we have for any* $t\in[0,T]$

$$\int_{\mathbb{R}} |u(x,t)|^2\,dx = \text{const.}; \tag{6}$$

$$\operatorname{Im}\int_{\mathbb{R}} u(x,t)\,\overline{u_x(x,t)}\,dx = \text{const.}; \tag{7}$$

$$\int_{\mathbb{R}} \{|u_x(x,t)|^2 + g(|u(x,t)|^2) - \frac{\beta}{2}[h(|u(x,t)|^2)_x]^2\}dx=\text{const.} \tag{8}$$

A generalization of the quasi-conformal conservation law (for the case $\beta=0$) is the content of the following lemma (see Ginibre and Velo (1979)) for the case $\beta=0$):

Lemma 2 For any distributional solution of (1), (2) *of class* $(A)_T$ *and* $(B)_T$ *we have for all* $t\in[0,T]$

$$\int_{\mathbb{R}} \{|(x+2it\,\partial_x)u(x,t)|^2 + 2t^2[2g(s) - \beta(h(s)_x)^2]\}\,dx$$

$$= \int_{\mathbb{R}} |xu(x,0)^2\,dx + 2\int_0^t \tau\int_{\mathbb{R}} \{2[3g(s_\tau) - s_\tau f(s_\tau)] +$$

$$+ \beta[h'(s_\tau)^2 + 2s_\tau h'(s_\tau)h''(s_\tau)]s_{\tau,x}^2\}\,dxd\tau$$

$$(\textit{with } s = |u(x,t)|^2,\ s_\tau = |u(x,\tau)|^2). \tag{9}$$

The following lemma is due to Cazenave, Dias and Figueira (1982); for the sake of completeness we give a proof which is a bit different from theirs.

Lemma 3 For any $u\in H^1(\mathbb{R}^n)$ *with* $x\cdot u\in L^2(\mathbb{R}^n)$ *and any* $t\in\mathbb{R}^+$ *we have*

$$2t\|\nabla|u|\|_{L^2(\mathbb{R}^n)} \le \|(x+2it\nabla)u\|_{L^2(\mathbb{R}^n)}.$$

<u>Proof</u> We can approximate any $u\in H^1$ with $x\cdot u\in L^2$ in the norm of the space $H=\{w\in H^1\,|\,x\cdot w\in L^2\}$ by functions of the type $v(x)=\rho(x)\exp(i\varphi(x))$ with real functions ρ,φ such that $\rho\in H$, $\rho\cdot\nabla\varphi\in L^2$. Thus we may assume that u itself belongs to that class. Then we have

$$\nabla u = (\nabla\rho + i\rho\nabla\varphi)e^{i\varphi},\quad |\nabla u|^2 = |\nabla\varphi|^2 + \rho^2|\nabla\varphi|^2$$

and

$$\|(x+2it\nabla)u\|_2^2 = \int_{\mathbb{R}^n} x^2|u|^2dx+4t^2\int_{\mathbb{R}^n} |\nabla u|^2dx-4t\operatorname{Im}\int_{\mathbb{R}^n} x\cdot\nabla u\cdot\bar{u}\,dx$$

$$= \int_{\mathbb{R}^n} x^2\rho^2 dx + 4t^2 \int_{\mathbb{R}^n} |\nabla\rho|^2 dx + 4t^2 \int_{\mathbb{R}^n} \rho^2 |\nabla\varphi|^2 dx - 4t \int_{\mathbb{R}^n} x\rho^2\nabla\varphi dx.$$

Since $4t \int_{\mathbb{R}^n} x\rho^2\nabla\varphi dx \le \int_{\mathbb{R}^n} x^2\rho^2 dx + 4t^2 \int_{\mathbb{R}^n} \rho^2|\nabla\varphi|^2 dx$

we arrive at

$$\|(x+2it\nabla)u\|_2^2 \ge 4t^2 \int_{\mathbb{R}^n} |\nabla\rho|^2 dx = 4t^2 \int_{\mathbb{R}^n} (\nabla|u|)^2 dx.$$

QED.

3. Asymptotical behaviour.

Let us first consider the general case (1), (2). Here we need the following assumptions for the nonlinearities:

$$g(s) \ge 0 \qquad \text{(for all } s \ge 0\text{);} \tag{10}$$

There exists a $c_0 \ge 0$: $3g(s)-sf(s) \le c_0\, g(s)$,

$$h'(s)^2 + 2sh'(s)h''(s) \ge -c_0\, h'(s)^2 \qquad \text{(for all } s \ge 0\text{).} \tag{11}$$

Let us remark that assumption (10) is made only to ensure the uniform boundedness of the Dirichlet integral $\int_{\mathbb{R}} |u_x|^2 dx$ for any finite energy solution of (1), (2); this condition may be weakened considerably as a growing condition for $g^-(s)$ (see e.g. Ghidaglia (1987)).

Proposition 1 *Let* $u(x,t)$ *be any distributional solution of* (1), (2) *on* $[0,\infty)$ *of class* $(A)_T$ *and* $(B)_T$ *for any* $T>0$, *and let the assumptions* (11) *and* $\beta<0$ *be valid. Then for* $t \ge t_0 > 0$ *we have*

$$\int_{\mathbb{R}} [2g(|u|^2) + |\beta|(h(|u|^2)_x{}^2]dx \le \frac{K_1}{t^{2-c_0}} \tag{12}$$

with a constant $K_1 = K_1(t_0) > 0$. *If* $c \le 2$ (12) *is valid for any* $t>0$ *with a constant* $K_1>0$.

<u>Proof</u> Let $t \ge t_0 > 0$. By (9) of *Lemma 2* and by assumption (11) we have (using the notation $s = |u(x,t)|^2$, $s_0 = |u(x,t_0)|^2$, $s_\tau = |u(x,\tau)|^2$, $C(t_0) = \int_{\mathbb{R}} |(x+2it_0\partial_x)u(x,t_0)|^2 dx$):

$$2t^2 \int_{\mathbb{R}} [2g(s)+|\beta|h'(s)^2 s_x{}^2]dx - 2t_0{}^2 \int_{\mathbb{R}} [2g(s_0)+|\beta|h'(s_0)^2 s_{0,x}{}^2]dx \le$$

$$2\int_{t_0}^{t} \tau\{2\int_{\mathbb{R}} [4g(s_\tau)-s_\tau f(s_\tau)]dx$$

$$-|\beta| \int_{\mathbb{R}} [h'(s_\tau)^2+2s_\tau h(s_\tau)h''(s_\tau)s_{\tau,x}{}^2]dx\}d\tau + C(t_0)$$

$$\le 2c_0 \int_{t_0}^{t} \tau \int_{\mathbb{R}} [2g(s_\tau)+|\beta|h'(s_\tau)^2 s_{\tau,x}{}^2]dxd\tau + C(t_0).$$

Setting $\varphi(t) := 2t^2 \int_{\mathbb{R}} [2g(s)+|\beta|(h(s)_x)^2 dx$ the last inequality means

$$\varphi(t) \le \tilde{\varphi}(t_0) + c_0 \int_{t_0}^{t} \frac{\varphi(\tau)}{\tau} d\tau \qquad (\tilde{\varphi}(t_0) = \varphi(t_0) + C(t_0))$$

which implies by using the Gronwall-Lemma

$$\varphi(t) \le \varphi(t_0) \left\{1 + \int_{t_0}^{t} \frac{c_0}{\tau} \exp\left(c_0 \int_{\tau}^{t} \frac{dy}{y}\right) d\tau\right\}$$

$$= \varphi(t_0) \left(\frac{t}{t_0}\right)^{c_0} = K_0 t^{c_0}.$$

From this we get (12). If we have $c_0 \le 2$ then we may assume that

$$t^{2-c_0} \int_{\mathbb{R}} [2g(s) + |\beta| (h(s)_x{}^2] dx$$

is bounded in $[0,t_0]$ (u belongs to class $(A)_T$ for any T>0!); together with the inequality just proved for $t \ge t_0$ this implies that (12) is true for any t>0 with a universal constant $K_1>0$.

QED.

For the special case we have in mind, namely $f(s)=\lambda \cdot s^p, h(s)=s^q$ ($\lambda>0$; $p,q>0$) we see that (10) is satisfied; the first inequality of (11) is true if and only if $2-p \le c_0$, whereas the second is valid if and only if $c_0 \ge 1-2q$.
Thus we may choose

$$c_0 = \max(0,2-p, 1-2q) \tag{13}$$

(which is ≤ 2). Applying *Proposition 1* this gives

Proposition 2 For any distributional solution u(x,t) *of* (3) *on* $[0,\infty)$ *of class* $(C)_T$ *for any* T>0 *we have for all* t>0

$$\int_{\mathbb{R}} |u(x,t)|^{2(p+1)} dx \le \frac{K_2}{t^\gamma}, \tag{14}$$

$$\int_{\mathbb{R}} [\frac{\partial}{\partial x}|u(x,t)|^{2q}]^2 dx \le \frac{K_3}{t^\gamma} \tag{15}$$

with certain constants $K_2, K_3>0$ *and*

$$\gamma = \min(2,p,2q+1).$$

<u>Proof</u> We only have to remark that

$$2-\max(0,2-q,1-2q) = \min(2,p,2q+1).$$

Also any distributional solution of class $(C)_T$ belongs to class $(A)_T$ and $(B)_T$ because of the Sobolev-type inequality

$$\|u\|_\infty \le C\|u_x\|_2^{\frac{2}{r+2}} \cdot \|u\|_r^{\frac{r}{r+2}} \quad (1 \le r \le \infty) \tag{16}$$

if we choose $r=2(p+1)$.

Now we can formulate our main result:

Theorem. Let $u(x,t)$ *be any distributional solution of* (3) *with* $\lambda>0$, $\beta<0$ *on* $[0,\infty)$ *which belongs to class* $(C)_T$ *for any* $T>0$. *Then there is a constant* $C>0$ *such that for any* $t>0$ *we have*

$$\max_{x\in R}|u(x,t)| \le \frac{C}{t^\delta} \tag{17}$$

where $\delta := \begin{cases} \frac{1}{2} & \textit{for } p \ge 2 \textit{ or } 2q+1 \le p < 2 \\ \frac{p}{p+\alpha} & \textit{otherwise (for } 0 < p < 2) \end{cases}$

with $\alpha := \min(2,2q+1)$.

<u>Proof</u> Again, let us note that any solution of class $(C)_T$ is of class $(A)_T$ and $(B)_T$ (by using (16)); furthermore we may assume $u\in C(R\times R^+)$, also assumptions (10) and (11) are true in our case. Thus we may apply *Proposition 2*. By *Lemma 1* and (10) we get

$$\int_R |u_x(x,t)|^2 dx \le C$$

for any $t\in[0,\infty)$. Thus by using the inequalities (14) and (16) with $r=2(p+1)$ we see that

$$\|u\|_\infty \le C\|u_x\|_2^{\frac{1}{p+2}} \cdot \|u\|_{2(p+1)}^{\frac{p+1}{p+2}} \le C\left(\int_R |u|^{2(p+1)}dx\right)^{\frac{1}{2(p+2)}}$$

$$\le \frac{C}{t^{\frac{\gamma}{2(p+2)}}} . \tag{18}$$

Furthermore by (15) and (16) we see that for $\tilde{u} := |u|^{2q}$ we have

$$\|\tilde{u}\|_\infty \le C\|\tilde{u}_x\|_2^{\frac{2}{r+2}} \|\tilde{u}\|_r^{\frac{r}{r+2}}$$

which implies

$$\|u\|_\infty^{2q} \le \frac{C}{t^{\frac{\gamma}{r+2}}} \left(\int_{\mathbb{R}} |u|^{2qr} dx \right)^{\frac{1}{r+2}} . \tag{19}$$

If we choose $r \ge 1$, $r \ge \dfrac{p+1}{q}$ and set $\rho := qr-(p+1)$ then by (14) and (19) we get:

$$\|u\|_\infty^{2q} \le \frac{C}{t^{\frac{\gamma}{r+2}}} \|u\|_\infty^{\frac{2\rho}{r+2}} \left(\int_{\mathbb{R}} |u|^{2(p+1)} dx \right)^{\frac{1}{r+2}} ,$$

$$\|u\|_\infty^{2(q-\frac{\rho}{r+2})} \le \frac{C}{t^{2\frac{\gamma}{r+2}}} ,$$

$$\|u\|_\infty \le \frac{C}{t^{\frac{\gamma}{2q+p+1}}} .$$

Together with (18) this implies

$$\|u\|_\infty \le \frac{C}{t^{\alpha_0 \gamma}} \qquad \left(\alpha_0 := \max \left(\frac{1}{2(p+2)} , \frac{1}{2q+p+1} \right) \right) . \tag{20}$$

By applying (6) of *Lemma 1* and (20) we arrive at

$$\int_{\mathbb{R}} |u|^{2(p+1)} dx \le \|u\|_\infty^{2p} \int_{\mathbb{R}} |u|^2 dx \le \frac{C}{t^{2p\gamma\alpha_0}} , \tag{21}$$

which gives something new only if $2q+1<p$ (for only then we have $2\alpha_0 p>1$). With the same procedure as before (using the same notation for r and ρ too) we then get by applying (21)

$$\|u\|_\infty^{2(q-\frac{\rho}{r+2})} \le \frac{C}{t^{\frac{\gamma}{r+2} \cdot (1+2\alpha_0 p)}}$$

or $$\|u\|_\infty \le \frac{C}{t^{\alpha_1 \cdot \gamma}} \qquad \left(\alpha_1 := \frac{1}{2} \frac{1+2\alpha_0 p}{2q+p+1} \right) .$$

By setting

$$\alpha_{n+1} := \frac{1}{2} \frac{1+2p\alpha_n}{2q+p+1} = \frac{\alpha_0}{2} (1+2p\alpha_n)$$

this procedure can be iterated and gives

$$\|u\|_\infty \le \frac{C}{t^{\alpha_n \cdot \gamma}} \qquad (n = 0,1,\ldots) \tag{22}$$

with a constant C>1 depending not on n. If we define $\beta_n := \frac{1}{2(2q+1)} - \alpha_n$ we get

$$\beta_{n+1} = \frac{p}{2q+p+1}\beta_n; \ \beta_n = \left(\frac{p}{2q+p+1}\right)^n \beta_0,$$

$$\alpha_n = \frac{1}{2(2q+1)}\left\{1 + \frac{2q+1-p}{2q+1+p}\left(\frac{p}{2q+p+1}\right)^n\right\}.$$

One sees that (α_n) is monotone increasing if 2q+1<p; one has $\lim_{n\to\infty} \alpha_n = \frac{C}{2(2q+1)}$. Letting $n\to\infty$ in (22) we get

$$\|u\|_\infty \le \frac{C}{t^{\frac{\gamma}{2(2q+1)}}}. \tag{23}$$

By (9) of *Lemma 2* and (11) we see that for any $t\ge 0$ we have

$$\|(x+2it\partial_x)u\|_2^2 \le 2c_0 \int_0^t \tau \int_{\mathbb{R}} [2g(s_\tau)+|\beta|(h(s_\tau)_x)^2]dxd\tau+C$$

$$\le 2c_0 K_1 \int_0^t \tau^{c_0-1} d\tau + C \le C\, t^{c_0}.$$

Thus by applying *Lemma 3* (for n=1) and (16) for |u| instead of u one gets for t>0

$$\|u\|_\infty \le C\, \|u\|_2^{\frac{1}{2}} \, \||u|_x\|_2^{\frac{1}{2}} \le C\, \||u|_x\|_2^{\frac{1}{2}} \le$$

$$\le \frac{C}{t^{\frac{1}{2}(1-\frac{c_0}{2})}} = \frac{C}{t^{\frac{\gamma}{4}}}. \tag{24}$$

Furthermore again by using (16) (with |u| instead of u) one arrives at

$$\|u\|_\infty \le \||u|_x\|_2^{\frac{2}{r+2}} \, \|u\|_r^{\frac{r}{r+2}} \le \frac{C}{t^{2\frac{\gamma}{r+2}}} = \frac{C}{t^{\frac{\gamma}{p+2}}} \tag{25}$$

for r=2(p+1).

Collecting all the results on estimates of $\|u\|_\infty$ (see (20), (23), (24), (25)) one gets finally

$$\|u\|_\infty \le \frac{C}{t^{\sigma\cdot\gamma}} \tag{26}$$

with $\sigma := \max\left(\frac{1}{4}, \frac{1}{p+2}, \frac{1}{2(2q+1)}, \frac{1}{2q+p+1}\right)$. A short calculation shows that (26) is equivalent to (17) i.e. $\sigma\cdot\gamma=\delta$. QED.

Remark Our proof shows that the decay estimate (15) of *Proposition 2* can be refined to

$$\int_{\mathbb{R}} \left(\frac{\partial}{\partial x} |u|^{2q}\right)^2 dx \le \frac{C}{t^{2(2q-1)\sigma\cdot\gamma}}$$

for 2q>p+3 using the analogue procedure as in deriving (21).

References.

[1] Brüll, L. and Lange, H. (1986 a). Stationary, Oscillatory and solitary wave type solutions of singular nonlinear Schrödinger equations. Math. Methods Appl. Sci. 8: 559-575.

[2] Brüll, L. and Lange, H. (1986 b). Solitary waves for quasilinear Schrödinger equations. Expos. Math. 4: 279-288.

[3] Cazenave, T. (1986). Some remarks on the asymptotic behaviour of solutions to nonlinear Schrödinger equations. Preprint, Zentrum für interdisziplinäre Forschung, Universität Bielefeld.

[4] Cazenave, T., Dias, J.P. and Figueira, M. (1982). A remark on the decay of the solutions of some Schrödinger equations. Rend. Sem. Mat. Univ. Politec. Torino 40: 129-137.

[5] Dias, J.P. and Figueira, M. (1982). On the decay of the solution of some nonlinear Schrödinger equation. Portugalia Matematica 41: 33-41.

[6] Ghidaglia, J.-M. (1987). Finite dimensional behaviour for weakly damped driven Schrödinger equations. Preprint, Laboratoire d'Analyse Numérique, C.N.R.S., Université Paris-Sud.

[7] Ginibre, J. and Velo, E. (1979). On a class of nonlinear Schrödinger equations I & II. J. Functional Analysis 32: 1-71.

[8] Guang-Chang Dong and Shujie Li (1981). On the initial value problem for a nonlinear Schrödinger equation. J. Differential Equations 42: 353-365.

[9] Hadeler, K.P. and Lange, H. (1987). Two comments on nonlinear Schrödinger Equations. J. Math. Phys. 28: 1091-1093.

[10] Hayashi, N. and Tsutsumi, M. (1986). $L^{\infty}(\mathbb{R}^n)$-decay of classical solutions for nonlinear Schrödinger equations. Proc. Roy. Soc. Edinburgh 104A: 309-327.

[11] Lange, H. (1986). Nonexistence of solutions to singular nonlinear Schrödinger equations. Mathematische Modellierung (A. Kempf, F. Wille, eds.) McGraw-Hill, Hamburg, pp. 114-124.

[12] Lange, H. (1987). Blow up of solutions to singular nonlinear Schrödinger equations and colliding solitary waves. Preprint, Mathematisches Institut. Univ. Köln (submitted).

[13] Lin, J.E. and Strauss, W.A. (1978). Decay and scattering of solutions of a nonlinear Schrödinger equation. J. Functional Analysis 30: 245-263.

[14] Tsutsumi, Y. (1984). Global existence and asymptotic behaviour of solutions for nonlinear Schrödinger equations. Preprint, Dept. Pure and Applied Sciences, Univ. of Tokyo.

[15] Tsutsumi, Y. and Yajima, K. (1984). The asymptotic behaviour of nonlinear Schrödinger equations. Bull. Amer. Math. Soc. 11: 186-188.

Lyapounov Stability of Electro-Magneto-Elastic Systems

J.B. Alblas

Eindhoven University of Technology,
Department of Mathematics and Computing Science,
P.O. Box 513, NL-5600 MB Eindhoven, The Netherlands

1. Introduction

In this paper we are concerned with the Lyapounov stability of an electric or magnetic body, placed in an electric or magnetic field under mechanical forces, cf. [1]. Because there are not many bodies known, which can be electrified *and* magnetized, we shall consider the electric and magnetic cases separately. The mechanical forces are potential forces, in fact we only consider here dead loads, and the electro-magnetic fields are *conservative* fields, a concept that will be defined. These fields have their origin in sources: a condensor or a coil, placed on a large, but finite, distance from the body under consideration. The work by the condensor is

$$P_{el} = \dot{Q}\,\phi_0\,, \tag{1.1}$$

in which Q is the charge on the plates and ϕ_0 is the potential, while the work by the coil may be written as

$$P_{magn} = \int_{\Omega_1} \dot{A}_K J_K \, dX = J \frac{d\phi_{magn}}{dt}\,, \tag{2.1}$$

where A_K is the vector potential, J_K the density of current in the loading body with volume Ω_1 and ϕ_{magn} the magnetic flux. We call the fields *conservative* if ϕ_0, J and J_K are constants. The importance of this concept will become evident in the following.

2. Lyapounov stability

We consider three Cartesian coordinate systems: the $\underline{X}$, $\underline{x}$ and $\underline{y}$-systems, which may coincide. In the $\underline{X}$-system the coordinates of the particles of the undeformed body are measured, in the $\underline{x}$- or intermediate system we measure them in an equilibrium state under investigation. In the $\underline{y}$-system the perturbations of the equilibrium state are measured. This state is time-dependent. The volumes of the body in the states are denoted by Ω_0, Ω and Ω^*, respectively. The spaces outside the body are $\overline{\Omega}_0$, $\overline{\Omega}$ and $\overline{\Omega}^*$ and the infinite space is R.

If we want to compare two different solutions $x(t)$ and $y(t)$ of the system, satisfying the same equations and boundary conditions, but different initial conditions at $t = t_0$, we introduce the metric ρ_t by

$$\rho_t = \rho_t(x(t), y(t))\,. \tag{2.1}$$

We define the solution $x(t)$ to be Lyapounov stable in $T \equiv (t_0, t_1(>t_0))$, if, cf. [2]

$$\rho_0(x(t_0), y(t_0)) < \delta \Rightarrow \rho = \sup_{t\in T} \rho_t(x(t), y(t)) < \varepsilon\,. \tag{2.2}$$

We note that ρ_0 and ρ_t may be different (and will be different here too). They contain kinetic and potential parts. We introduce the class K of monotonous, strictly increasing functions, zero at the origin and $\overline{V}_t(x(t), y(t))$, the Lyapounov functional, satisfying

$$\frac{d\,\overline{V}_t(x,y)}{dt} \leq 0 . \tag{2.3}$$

If there exists a $\overline{V}_t$ such that

$$\delta > \mu(\rho_0) \geq \overline{V}_{t_0} \geq V_t \geq \lambda(\rho) , \tag{2.4}$$

we may conclude that

$$\rho_0 < \delta \Rightarrow \rho < \varepsilon , \tag{2.5}$$

Lyapounov stability. In (2.4), $\lambda, \mu \in K$ and V_t is the potential energy at time t.

Note that the problem consists of three parts:

a) the construction of $\overline{V}_t$, often closely related to the energy functional,

b) the proof of existence of a lower bound for V_t

$$V_t \geq \lambda(\rho) , \tag{2.6}$$

(c) and of an upper bound with ρ_0

$$\delta > \mu(\rho_0) \geq V_t . \tag{2.7}$$

3. The physical problem

a) Notations. The quantities referring to the $\underline{y}$-state are given an upper mark *, while those, referring to the $\underline{X}$-state are denoted by a lower index $_0$. Differentiation with respect to $\underline{X}$ is denoted by a bar, with respect to $\underline{x}$ (or $\underline{y}$) by a comma: e.g.

$$x_{k/\alpha} = \frac{\partial x_k}{\partial X_\alpha} ; \qquad y_{k,l} = \frac{\partial y_k}{\partial x_l} . \tag{3.1}$$

b) In this paper we limit the discussion to problem 2a) of the former section, applied to an electro-elastic system. The Maxwell equations in Minkowski's form are for the intermediate state

$$\begin{aligned} \operatorname{curl} \underline{E} &= -\frac{\partial \underline{B}}{\partial t} , \\ \operatorname{curl} \underline{H} &= \frac{\partial \underline{D}}{\partial t} + \underline{J} , \\ \operatorname{div} \underline{B} &= 0 , \\ \operatorname{div} \underline{D} &= \rho_{el} = 0 , \end{aligned} \tag{3.2}$$

with

$$\underline{D} = \varepsilon_0 \underline{E} + \rho \underline{P} ; \qquad \underline{B} = \mu_0(\underline{H} + \rho \underline{M}) . \tag{3.3}$$

In (3.2) and (3.3), $\underline{E}$ and $\underline{H}$ are the electric and magnetic field intensities, $\underline{B}$ is the induction, $\underline{D}$ the dielectric displacement, $\underline{J}$ is the current density, ρ_{el} the free electric charge in the body, which will be taken equal to zero in Ω, $\underline{P}$ is the polarization and $\underline{M}$ the magnetization, both per unit of mass. The quantities ε_0 and μ_0 are the permittivity and permeability of vacuum respectively, satisfying

$$\varepsilon_0 \mu_0 = \frac{1}{c^2} \quad (\approx 0) \, , \tag{3.4}$$

in which c is the velocity of light. In the non-relativistic theory we neglect terms of order $1/c^2$.
We limit the discussion to the case of *quasi-electrostatics*.

4. Quasi-electrostatics

In quasi-electrostatics we have

$$\underline{J} = \underline{M} = 0 \, , \tag{4.1}$$

and there is no external magnetic field. It follows from (3.2) and (3.3) that

$$\underline{P} = O(\varepsilon_0); \quad \underline{D} = O(\varepsilon_0); \quad \underline{B} = O(\varepsilon_0 \mu_0) = O\left[\frac{1}{c^2}\right] \approx 0 \, , \tag{4.2}$$

thus the quasi-electrostatic equations become

$$\operatorname{curl} \underline{E} = 0 \, , \tag{4.3}$$

$$\operatorname{div}(\varepsilon_0 \underline{E}) = -\operatorname{div}(\rho \underline{P}) \, ,$$

$$\operatorname{curl} \underline{H} = \frac{\partial \underline{D}}{\partial t} \, .$$

For the convected quantities we have

$$\underline{E}' = \underline{E} + (\underline{v} \times \underline{B}) \approx \underline{E} \, , \tag{4.4}$$

$$\underline{D}' = \underline{D} \, ,$$

$$\underline{H}' = \underline{H} - (\underline{v} \times \underline{D}) \, .$$

We introduce the material representations

$$\hat{E}_j = E_i \, x_{i/j} \quad ; \quad E_i = \hat{E}_j X_{j,i} \, , \tag{4.5}$$

$$\hat{D}_j = J \, D_i X_{j,i} \quad ; \quad D_i = \frac{1}{J} \, D_j \, x_{i/j} \, , \tag{4.6}$$

$$\hat{P}_j = P_i X_{j,i} \quad ; \quad P_i = \hat{P}_j \, x_{i/j} \, , \tag{4.7}$$

where

$$J = \rho_0 / \rho \, , \tag{4.8}$$

with ρ_0 and ρ the densities in the $\underline{X}$-state and $\underline{x}$-state respectively.

In mechanics we consider the transformation

$$\underline{x} = \underline{x}(\underline{X}) \, , \tag{4.9}$$

where $\underline{x}$ and $\underline{X}$ refer to the same material point. The Maxwell equations are not invariant with respect to the transformation (4.9). Thus if we have

$$D_{k,k} = 0\ , \tag{4.10}$$

we find

$$D_{k/k} \neq 0\ . \tag{4.11}$$

However, with (4.5), (4.6) and (4.8), we easily may prove that

$$\mathrm{curl}_{\underline{x}}\, \underline{E} = 0 \Rightarrow \mathrm{curl}_{\underline{X}}\, \hat{\underline{E}} = 0\ , \tag{4.12}$$

and

$$\mathrm{div}_{\underline{x}}\, \underline{D} = 0 \Rightarrow \mathrm{div}_{\underline{X}}\, \hat{\underline{D}} = 0\ . \tag{4.13}$$

However, the relationship (3.3^1) is not invariant, but becomes

$$\hat{D}_k = \varepsilon_0 J\, X_{k,p}\, X_{l,p}\, \hat{E}_l + \rho_0 \hat{P}_k\ , \tag{4.14}$$

which means that the material representations are no real intensities. In the constitutive equation for G, the enthalpy we have

$$G = G(x_{k/\alpha}, T, E_k)\ , \tag{4.15}$$

which becomes because of frame-indifference

$$G = G(C_{\alpha\beta}, T, \hat{E}_\alpha)\ , \tag{4.16}$$

with $C_{\alpha\beta}$ the Green deformation tensor and T the temperature.

In the theory of stability we introduce still a third representation for the electric quantities:

$$\tilde{E}_k(y) = E_l(x)\, x_{l,k} = E_l\, \frac{\partial x_l}{\partial y_k}\ , \tag{4.17}$$

$$\tilde{D}_k(y) = \frac{\rho^*}{\rho}\, D_l(x)\, y_{k,l} = \frac{\rho^*}{\rho}\, D_l\, \frac{\partial y_k}{\partial x_l}\ , \tag{4.18}$$

$$\tilde{P}_k(y) = P_l\, y_{k,l} = P_l\, \frac{\partial y_k}{\partial x_l}\ . \tag{4.19}$$

We also find

$$\mathrm{curl}_{\underline{x}}\, \underline{E} = 0 \Rightarrow \mathrm{curl}_{\underline{y}}\, \tilde{\underline{E}} = 0\ , \tag{4.20}$$

$$\mathrm{div}_{\underline{x}}\, \underline{D} = 0 \Rightarrow \mathrm{div}_{\underline{y}}\, \tilde{\underline{D}} = 0\ , \tag{4.21}$$

and the relation between $\tilde{\underline{D}}$ and $\tilde{\underline{E}}$ becomes

$$\tilde{D}_k = \varepsilon_0\, \frac{\rho^+}{\rho}\, y_{k,p}\, y_{l,p}\, \tilde{E}_l + \rho^* \tilde{P}_k\ . \tag{4.22}$$

There are two important identities:

$$E_k P_k = \hat{E}_k \hat{P}_k = \tilde{E}_k \tilde{P}_k\ , \tag{4.23}$$

$$\hat{E}_k(\underline{X}) = E_l(x)\, \frac{\partial x_l}{\partial X_k} = \tilde{E}_p(y)\, \frac{\partial y_p}{\partial x_l} \cdot \frac{\partial x_l}{\partial X_k} = \tilde{E}_l\, \frac{\partial y_l}{\partial X_k}\ . \tag{4.24}$$

The basic equation for the conservation of energy may be written as

$$\frac{d}{dt}\left[\int_{\Omega_0} \rho_0 U \, dX + K + W\right] = P + P_{el} + \phi , \tag{4.25}$$

in which Ω_0 is the volume of the body in the undeformed state, U is the internal energy per unit of mass, K the kinetic energy

$$K = \frac{1}{2}\int_{\Omega_0} \rho_0 \dot{x}_k \dot{x}_k \, dX , \tag{4.26}$$

P the rate of external mechanical energy

$$P = \int_{\Omega_0} \rho_0 b_k \dot{x}_k \, dX + \int_{\partial\Omega_0} T_k \dot{x}_k \, dA , \tag{4.27}$$

P_{el} is given by (1.1) and ϕ is the added heat per sec.

$$\phi = \int_{\Omega_0} \rho_0 r \, dX - \int_{\partial\Omega_0} H_\alpha N_\alpha dA . \tag{4.28}$$

In (4.27) and (4.28) b_k is the body force, T_k the Piola-Kirchhoff stress vector, r the heat supply per unit time and mass, H_α the outgoing heat flux per sec. and N_α the external normal to $\partial\Omega_0$.

The total potential energy in (4.25) is $\int_{\Omega_0} \rho_0 U \, dX + W$. A unique separation of U and W is impossible, but if we choose for W

$$W = \frac{1}{2}\int_{R} \varepsilon_0 E_k E_k \, dX , \tag{4.29}$$

it may be proved (cf. [1]), that U then satisfies

$$\rho\dot{U} = E_k \rho \dot{P}_k + (E_k J_k) + T_{kl} v_{k,l} + \rho r - h_{k,k} , \tag{4.30}$$

with T_{kl} the Cauchy stress tensor and h_k the outgoing heat flux in the intermediate state. We shall also introduce the free energy F per unit of mass by

$$F = U - TS , \tag{4.31}$$

with S the entropy, and the enthalpy G by

$$G = F - P_k E_k . \tag{4.32}$$

From (4.30), (4.31) and (4.32) we obtain

$$E_k = \frac{\partial U}{\partial P_k} = \frac{\partial F}{\partial P_k} ; \qquad P_k = -\frac{\partial G}{\partial E_k} ; \tag{4.33}$$

$$S = -\frac{\partial F}{\partial T} = -\frac{\partial G}{\partial T} .$$

Transformation of F

We write (4.25) in the form

$$\int_{\Omega_0} \rho_0 \dot{F} \, dX + \int_{\Omega_0} \rho_0 T\dot{S} \, dX + \int_{\Omega_0} \rho_0 \dot{T} S \, dX + \dot{K} + \dot{W} = P + P_{el} + \phi . \tag{5.1}$$

This equation holds in any state and it is written here for the intermediate state. A corresponding equation for the y state has the same form with the quantities written with stars. Subtracting the two equations results in

$$\int_{\Omega_0} \rho_0(\dot{F}^* - \dot{F})dX + \int_{\Omega_0} \rho_0(\dot{T}^* S^* + T^* \dot{S}^* - \dot{T}S - T\dot{S})dX + \tag{5.2}$$

$$+ (\dot{K}^* - \dot{K}) + (\dot{W}^* - \dot{W}) - (P^* - P) - (P^*_{\mathrm{el}} - P_{\mathrm{el}}) - (\phi^* - \phi) = 0 .$$

We have

$$F^* = F^*(y_{k/\alpha}, T^*, P_k^*) \tag{5.3}$$

and introduce

$$\bar{F}^* = F^*(y_{k/\alpha}, T, P_k^*) , \tag{5.4}$$

the free energy of the $\underline{y}$-state with the temperature of the $\underline{x}$-state. We apply the mean value theorem to transform (5.3) into (5.4) and use the second law of thermodynamics in the form

$$\phi^* \le \int_{\Omega_0} \rho_0 T^* \dot{S}^* dX - \int_{\Omega_0} \frac{H^*_\alpha T^*_{,\alpha}}{T^*} dX . \tag{5.5}$$

Then we find for the *isothermal equilibrium* intermediate $\underline{x}$-state with, (cf. [1], [3]),

$$\dot{T} = T_{,\alpha} = \dot{S} = \phi = r = H_\alpha = K = P = \dot{F} = P_{\mathrm{el}} = \dot{W} = 0 , \tag{5.6}$$

and

$$r^* = 0 , \tag{5.7}$$

$$\int_{\Omega_0} \rho_0(\dot{\bar{F}}^* - \dot{F})dX + \frac{d}{dt} \int_{\Omega_0} \rho_0 \frac{\hat{c}}{\hat{T}} (T - T^*)^2 dX + \tag{5.8}$$

$$+ \dot{K}^* + \dot{W}^* - P^* - P^*_{\mathrm{el}} \le 0 ,$$

where

$$\frac{\hat{c}}{\hat{T}} = - \left(\frac{\partial^2 F^*}{\partial T^2} \right)_{\hat{T}} > 0 ,$$

and $\hat{T}$ is between T and T^*, while c is the specific heat and $\hat{c}$ the value at $\hat{T}$.

6. Transformation of G

The enthalpy G in the intermediate state $\underline{x}$ is defined by

$$G = F - P_k E_k = G(x_{k/\alpha}, T, E_k) . \tag{6.1}$$

Correspondingly, we have

$$G^* = G^*(y_{k/\alpha}, T^*, E_k^*) . \tag{6.2}$$

We now define

$$\bar{G}^* = G^*(y_{k/\alpha}, T, E_k^*) \tag{6.3}$$

and

$$\overset{=*}{G} = G^*(y_{k/\alpha}, T, \tilde{E}_k) . \tag{6.4}$$

With the mean value theorem we have

$$\overset{=*}{G} = \overline{G}^* - P_k^*(\tilde{E}_k - E_k^*) - \frac{1}{2}\hat{c}_{kl}(\tilde{E}_k - E_k^*)(\tilde{E}_l - E_l^*) , \tag{6.5}$$

where the permittivity c_{kl} is defined by

$$c_{kl} = - \frac{\partial^2 G}{\partial E_k \, \partial E_l} , \tag{6.6}$$

and $\hat{c}_{kl}$ is an intermediate value for E_k between $\tilde{E}_k$ and E_k^*. We always have the fundamental property

$$c_{kl} \, E_k \, E_l > 0 . \tag{6.7}$$

From (6.3) and (6.5) follows

$$\overline{F}^* = \overset{=*}{G} + P_k^* \tilde{E}_k + \frac{1}{2}\hat{c}_{kl}(\tilde{E}_k - E_k^*)(\tilde{E}_l - E_l^*) . \tag{6.8}$$

With the abbreviation

$$H = \frac{1}{2} \int_{\Omega_0} \rho_0 \, dX \left[\frac{\hat{c}}{\hat{T}} (T - T^*)^2 + \hat{c}_{kl}(\tilde{E}_k - E_k^*)(\tilde{E}_l - E_l^*) \right] > 0 , \tag{6.9}$$

we may write (5.8) in the form

$$\int_{\Omega_0} \rho_0 [\dot{\overset{=*}{G}} - \dot{G}] \, dX + \dot{H} + \frac{d}{dt} \int_{\Omega_0} \rho_0 P_k^* \tilde{E}_k \, dX + \dot{K} + \dot{W} - P^* - P_{\text{el}}^* \leq 0 . \tag{6.10}$$

7. The effective internal energy $\overline{U}$

We write W^* according to (4.29) in the following form

$$W^* = W_1 - A - \int_R \varepsilon_0 E_k^* \tilde{E}_k \, dy , \tag{7.1}$$

with

$$W_1 = \frac{1}{2} \int_R \varepsilon_0 (\tilde{E}_k - E_k^*)(\tilde{E}_k - E_k^*) dy , \qquad (> 0) , \tag{7.2}$$

and

$$A = \frac{1}{2} \int_R \varepsilon_0 \tilde{E}_k \tilde{E}_k \, dy . \tag{7.3}$$

Because we have

$$D_k^* = \varepsilon_0 E_k^* + \rho^* P_k^* , \quad \text{in } \Omega^* , \tag{7.4}$$

$$D_k^* = \varepsilon_0 E_k^* \quad , \quad \text{in } \overline{\Omega}^* ,$$

we may take some terms in (6.10) and (7.1) together, to find

$$\int_{\Omega^*} \rho^*(\dot{\overset{=}{G}}{}^* - \dot{G})dy + \dot{H} + \dot{K}^* - P^* + \dot{W}_1 - \dot{A} - P^*_{\text{el}} + \frac{d}{dt}\int_R D^*_k \tilde{E}_k\, dy \leq 0 . \tag{7.5}$$

First we calculate the last integral in (7.5). As a consequence of (4.20) we have

$$\tilde{E}_k = -\frac{\partial \Phi(y)}{\partial y_k} , \tag{7.6}$$

with

$$\Phi(y) = \Phi(y(x)) = \phi(x) , \tag{7.7}$$

where

$$E_k = -\frac{\partial \phi(x)}{\partial x_k} . \tag{7.8}$$

The displacement D^*_k satisfies

$$\frac{\partial D^*_k(y)}{\partial y_k} = 0 , \quad \text{outside } \Omega_1 , \tag{7.9}$$

$$D^*_{k,k} = q^* \quad , \quad \text{inside } \Omega_1 ,$$

with q^* the density of charge in Ω_1. Because D^*_k is continuous on $\partial\Omega^*$, we find

$$\int_R D^*_k \tilde{E}_k\, dy = \int_{\Omega_1} q^* \Phi\, dy = Q^* \phi_0 , \tag{7.10}$$

with Q^* the total charge on the condensor and ϕ_0 its constant potential. From (7.10) it follows

$$\frac{d}{dt}\int_R D^*_k \tilde{E}_k\, dy = \dot{Q}^* \phi_0 = P^*_{\text{el}} , \tag{7.11}$$

thus, with this result, the inequality (7.5) simplifies to

$$\int_{\Omega^*} \rho^*(\dot{\overset{=}{G}}{}^* - \dot{G})dy + \dot{H} + \dot{K}^* + \dot{W}_1 - P^* - \dot{A} \leq 0 . \tag{7.12}$$

For the calculation of A we split the integral into two parts

$$A = \tfrac{1}{2}\varepsilon_0 \int_{\Omega^*} \tilde{E}_k \tilde{E}_k\, dy + \tfrac{1}{2}\varepsilon_0 \int_{\overline{\Omega}^*} \tilde{E}_k \tilde{E}_k\, dy . \tag{7.13}$$

The second integral in (7.13) is constant. For we have

$$\tfrac{1}{2}\varepsilon_0 \int_{\overline{\Omega}^*} \tilde{E}_k \tilde{E}_k\, dy = \tfrac{1}{2}\int_{\overline{\Omega}^*} \tilde{D}_k \tilde{E}_k\, dy = \tfrac{1}{2}\int_R \tilde{D}_k \tilde{E}_k\, dy + \tag{7.14}$$

$$-\tfrac{1}{2}\int_{\Omega^*} \tilde{D}_k \tilde{E}_k\, dy = -\tfrac{1}{2}\int_R (\tilde{D}_k \Phi)_{,k}\, dy + \tfrac{1}{2}\int_{\Omega_1} \tilde{D}_{k,k} \Phi\, dy +$$

$$-\tfrac{1}{2}\int_{\Omega^*} \tilde{D}_k \tilde{E}_k\, dy = \tfrac{1}{2} Q\, \phi_0 - \tfrac{1}{2}\varepsilon_0 \int_{\Omega^*} \frac{\rho^*}{\rho} y_{k,p}\, y_{l,p}\, \tilde{E}_k \tilde{E}_l\, dy +$$

$$-\tfrac{1}{2}\int_{\Omega^*} \rho^* \tilde{P}_k \tilde{E}_k\, dy = \tfrac{1}{2} Q\, \phi_0 - \tfrac{1}{2}\int_{\Omega} \rho P_k E_k\, dx - \tfrac{1}{2}\varepsilon_0 \int_{\Omega} E_k E_k\, dx = \text{const.}$$

Thus we have

$$\dot{A} = \frac{d}{dt}\,\frac{1}{2}\varepsilon_0 \int_{\Omega^*} \tilde{E}_k \tilde{E}_k \, dy = \frac{d}{dt}\,\frac{1}{2}\varepsilon_0 \int_{\Omega} S_{kl} E_k E_l \, dx \,, \tag{7.15}$$

with

$$S_{kl} = \rho/\rho^* \, x_{k,p} \, x_{l,p} \,. \tag{7.16}$$

We take this term together with the first one in (7.12) and introduce

$$\rho^* \overline{U} = \rho^* \overset{=*}{G} - \frac{1}{2}\varepsilon_0 \tilde{E}_k \tilde{E}_k \,. \tag{7.17}$$

Now the basic inequality (7.12) becomes

$$\int_{\Omega^*} \rho^* \dot{\overline{U}} \, dy + \dot{W}_1 + \dot{H} + \dot{K}^* - P^* \le 0 \tag{7.18}$$

or written in the $\underline{x}$-coordinates

$$\int_{\Omega} \rho \dot{\overline{U}} \, dx + \dot{W}_1 + \dot{H} + \dot{K}^* - P^* \le 0 \,. \tag{7.19}$$

The effective internal energy is

$$\int_{\Omega} \rho \overline{U} \, dx = \int_{\Omega} \rho \, dx \left[\overset{=*}{G} - \frac{1}{2}\varepsilon_0 \frac{1}{\rho^*} \tilde{E}_k \tilde{E}_k \right] = \tag{7.20}$$

$$= \int_{\Omega} \rho \, dx \left[\overset{=*}{G} - \frac{1}{2} \frac{\varepsilon_0}{\rho^*} G^*_{\alpha\beta} \hat{E}_\alpha \hat{E}_\beta \right],$$

with $G_{\alpha\beta}$ and $G^*_{\alpha\beta}$ defined by

$$G_{\alpha\beta} = X_{\alpha,k} X_{\beta,k} = C^{-1}_{\alpha\beta} \,; \qquad G^*_{\alpha\beta} = \frac{\partial X_\alpha}{\partial y_k} \frac{\partial X_\beta}{\partial y_k} \,. \tag{7.21}$$

8. The Lyapounov formulation

We write (7.19) again with $\dot{U}_0 = 0$.

$$\int_{\Omega} \rho(\dot{\overline{U}} - \dot{U}_0) dx + \dot{W}_1 + \dot{H} + \dot{K}^* - P^* \le 0 \,. \tag{8.1}$$

Using (4.27) we have

$$P^* = \int_{\Omega_0} \rho_0 b_k^* \dot{y}_k \, dX + \int_{\partial\Omega_0} T_k^* \dot{y}_k \, dA = \int_{\Omega_0} \rho_0 b_k^* \dot{u}_k \, dX + \int_{\partial\Omega_0} T_k^* \dot{u}_k \, dA \,, \tag{8.2}$$

where we have introduced the displacement u_k by

$$u_k = y_k - x_k \,; \qquad \dot{u}_k = \dot{y}_k \,. \tag{8.3}$$

For dead loads we have

$$P^* = -\frac{\partial \chi_1}{\partial t} - \frac{\partial \chi_2}{\partial t} \,, \tag{8.4}$$

with the potentials χ_1 and χ_2 defined by

$$\chi_1 = -\int_{\Omega_0} \rho_0 b_k^* u_k \, dX \; ; \qquad \chi_2 = -\int_{\partial\Omega_0} T_k^* u_k \, dA \; . \tag{8.5}$$

With (8.4), (8.5) we find for the Lyapounov function $\overline{V}_t$

$$\overline{V}_t = \int_{\Omega} \rho(\overline{U} - U_0) dx + W_1 + H + K^* + \chi_1 + \chi_2 \, , \tag{8.6}$$

that satisfies

$$\frac{d\overline{V}_t}{dt} \leq 0 \, . \tag{8.7}$$

Because

$$H > 0, \quad W_1 > 0, \quad K^* > 0, \tag{8.8}$$

we find for the potential energy V_t, defined by

$$V_t = \int_{\Omega} \rho(\overline{U} - U_0) dx + \chi_1 + \chi_2 \, , \tag{8.9}$$

$$V_t \leq \overline{V}_t \, , \qquad t \geq t_0 \, . \tag{8.10}$$

To conclude stability we must find the metric functions ρ_0 and ρ, that satisfy (2.6) and (2.7). For the chain inequality, (2.4) implies, that if, at $t = t_1$ ($> t_0$)

$$\rho(t_1) < \varepsilon ; \qquad K^*(t_1) < \varepsilon \, , \tag{8.11}$$

we must have

$$V_{t_0} + K^*(t_0^+) < \delta \, . \tag{8.12}$$

In general, it is easy to prove (2.7), but it seems that the proof of (2.6) can only be given for a system with internal dissipation, e.g. heat conduction.

9. Some remarks

(a) A corresponding analysis as has been presented in the Sections 4 to 7 may be given for the case of *quasi-magneto-statics*, with zero current. This theory can be extended to include conductive material, without an external electric field, because in this case the entropy production σ^* in the $\underline{y}$-state, which satisfies the inequality

$$\sigma^* > 0 \, , \tag{9.1}$$

is augmented by a term

$$\sigma_2^* = \int_{\Omega^*} J_k^* E_k^* \, dy = \int_{\Omega^*} g \, E_k^* E_k^* \, dy \quad (> 0) \, , \tag{9.2}$$

where g is the (positive) conductivity. In fact, we have now split the entropy production into two parts

$$\sigma^* = \sigma_1^* + \sigma_2^* \, , \tag{9.3}$$

with σ_1^* due to the heat conduction. It is σ^*, which leads to to the inequality (5.8).

(b) The analysis is directed on *stability* and not on *bifurcation*. However, if for a certain configuration Ω, stability cannot be proved, or even instability can be proved, it might be possible to prove stability for another configuration e.g. Ω'. Because Ω and Ω' depend on $\underline{\hat{E}}$, a critical $\underline{\hat{E}}$ may be found.

References

[1] J.B. Alblas: General Theory of Electro- and Magneto Elasticity, Ch. 11, in Electro Magnetic Interactions in Elastic Solids, ed. H. Parkus, Springer Verlag, Wien etc., 1979.

[2] R.J. Knops and E.W. Wilkes: Theory of elastic Stability in Handbook der Physik, ed. S. Flügge, Vol. VI a/3, pag. 133, Springer Verlag, Berlin etc., 1973.

[3] J.L. Ericksen: A thermo-kinetic view of elastic stability theory, Int. J. Sol. Struct., 2, 573, 1966; J.L. Ericksen, Thermo-elastic Stability, Proc. 5th U.S. Nat. Congress of Appl. Mech. ASME, pag. 187, 1966.

Criticality of an Exothermic Reaction-Diffusion System of nth-Order

D. Meinköhn

Fakultät für Physik, Universität Bielefeld, Postfach 8640,
D-4800 Bielefeld 1, Fed. Rep. of Germany

1. Physical Background

A porous fuel pellet immersed in an oxidizing atmosphere or a porous catalyst particle surrounded by gaseous reactants is giving rise in time to a conversion process during which reactants are consumed and heat is liberated. Such an exothermic chemical reaction is termed heterogeneous, as it is supposed to proceed via an intermediate adsorption step due to which the forward reaction is overwhelmingly predominant and thus causing the overall reaction to become irreversible :

$$\text{Reactants} \rightarrow \text{Products} + q \tag{1}$$

Here, q designates the heat liberated in the course of the overall reaction.

A porous body D of high porosity may be conceived as pseudo-homogeneous which implies a nonvanishing concentration of gaseous reactants throughout D. Consequently, for an irreversible exothermic reaction of nth-order, a source density of heat Q exists in D which is supposed to depend in the following way on the temperature T and the concentration c of one of the gaseous reactants to be converted :

$$Q \sim qc^n \cdot \exp(-E/RT) \tag{2}$$

Here, E = activation energy, R = universal gas constant. Due to the assumption of pseudo-homogeneity, $Q = Q(x)$ with $x \in D$, independent of the geometry of the pores.

The heat Q liberated in the course of the process may serve to maintain the system in a stationary state far from thermodynamic equilibrium, which depends in a sensitive way on the interaction of chemical reaction and transport processes governing the supply of gaseous reactants and the removal of heat. Due to the exponential dependence of Q on the temperature T, two characteristic stationary regimes of the reaction process exist which are defined in terms of the temperature : for low temperatures, the rate of conversion (and therefore the concomitant generation of heat) is much lower than the rate of transport. Thus, the reaction is slow and inefficient, i.e., "switched-off", so to speak. For high temperatures, the rate of conversion is much larger than the rate of transport and thus the reaction is effectively "switched-on".

For such a reaction-diffusion system, the attainment of a particular stationary state may be influenced by choosing appropriate values for the so-called control parameters which represent in effect the independent variables of the problem : temperature and composition of the surrounding atmosphere, transport and transfer coefficients, and parameters characterizing the geometry of the system. In consequence of the assumption of pseudo-homogeneity, effective transport coefficients l, k are defined in order to describe heat and mass transport within the porous body D . Because of the strong nonlinear dependence of Q on c and T (cf.(2)), regions are found to exist in the space of the control parameters representing parameter combinations to which correspond more than one stationary state. These multiple stationary states may be classified according to their stability. While a unique stationary state is necessarily stable [1], non-uniqueness leads to the occurrence of stable as well as unstable stationary states. The various stable stationary states are accessible (or physically connected) through the expedient of a quasi-stationary process, a concept well established in the context of similar phenomena, e.g., phase transitions. For a unique designation of a particular stationary state, an appropriate number of state variables of the system (i.e., mathematically speaking, dependent variables) have to be added to the control parameters. Consequently, the state space of the system comprises the control parameters in a proper subspace.

The boundary C of the regions of non-uniqueness in control parameter space is necessarily associated with discontinuous behaviour of the process (if considerations are restricted exclusively to the set of stationary states), e.g., jumps from one branch of stable stationary solutions to another. In the present case, jumps between the different stationary regimes of the process are termed ignition or extinction, depending on the direction of the associated temperature change. The boundary C comprises the critical states of the process and therefore ignition and extinction criteria may be derived from C .

The boundary C in control parameter space necessarily possesses a characteristic cusp singularity which would correspond to a critical point with respect to separation into distinct phases in the analogous case of a thermodynamic system in equilibrium. A critical point is indicative of a system's incapacity for staying homogeneous. Here, the cusp singularity of C for a stationary system far from equilibrium characterizes the capacity of the system for ignition/extinction jumps.

In the present context, in contrast to the theory of phase transitions, a boundary value problem of a fundamental differential equation exists (cf.(3)), from which the state surface in an appropriate state space can be derived, with the state surface comprising all of the stationary solutions of the problem. A projection of the state surface into the proper subspace of the control parameters leads to the appearance of the boundary C which represents therefore a certain contour of the state surface, with its characteristic cusp resulting from projecting in an asymptotic direction.

2. Mathematical Representation

In order to just preserve the essence of the phenomenon, boundary conditions for the reaction-diffusion system are chosen such that the mathematical problem can be reduced to the boundary value problem of a single differential equation (of ODE or PDE form). Then the following formulation is obtained by employing nondimensional variables :

$$\left.\begin{aligned} \Delta y + \lambda(1 - \xi y)^n \cdot \exp\big(y/(1 + \beta y)\big) &= o \qquad \text{in } D \\ y &= o \qquad \text{on } \partial D \end{aligned}\right\} \tag{3}$$

with

$$y = (T - T_o)E/RT_o^2 \qquad \text{(dimensionless temperature)}$$

$$\beta = RT_o/E \qquad \text{(activation energy parameter)}$$

$$\lambda = \frac{Q}{T}\frac{E}{RT_o^2} c_o^n \cdot \exp(-E/RT_o) \qquad \text{(Frank-Kamenetzki parameter)}$$

$$\xi = 1RT_o^2/kEQc_o \qquad \text{(Lewis number)}$$

Here, T_o = temperature of the surroundings, c_o = concentration in the surroundings of the selected gaseous reactant, Δ = Laplacian, ∂D = boundary of D .

Equation (3) represents the boundary value problem of the nonlinear Fourier equation, with the nonlinearity :

$$w(y) = (1 - \xi y)^n \cdot \exp\big(y/(1 + \beta y)\big) > o \quad , \quad y < 1/\xi \tag{4}$$

Due to their definition, the parameters λ and ξ are nonnegative. In order to include functions w(y) with $(1 \pm \xi y)^n$, positive and negative values are formally admitted for the parameter ξ :

$$\xi \gtreqless o \tag{5}$$

As (3) represents a quasi-linear elliptic differential equation with

$$\Delta y \leqq o \qquad \text{in D} \quad (\text{with } \lambda \geqq o) \tag{6}$$

on account of (4), the maximum principle [2] is found to hold whereby a nonconstant y can have no minimum in D. Consequently, there exists a unique maximum y_m of y in D. Assuming the existence of a center $x_o \in D$ of symmetry of (3) implies of necessity :

$$y_m = \max y(x) = y(x_o) \tag{7}$$

Allowing for y_m to be an independent parameter is equivalent to trying to find a solution of (3) in parametric representation :

$$\left(\lambda(y_m)\ ,\ y(x;y_m)\right) \tag{8}$$

Due to providing for y_m to be an independent parameter of the problem, the Frank-Kamenetzki parameter λ becomes dependent upon the choice of y_m : $\lambda = \lambda(y_m)$, with $\lambda(y_m)$ assuming the role of a characteristic function. The set of all stationary solutions of (3) decomposes into branches of stable and branches of unstable solutions, with separate branches joining in branching points. Those particular branching points which connect a stable with an unstable solution branch are designated critical solutions of (3) because they represent the very points from which the system jumps off into an unstationary evolution. Determining the critical conditions of ignition and extinction is thus equivalent to finding the critical solutions of (3) in the set of all its solutions.

It can be shown [3], that some part of the structure of $\lambda(y_m)$ is related to the stability properties of interest. For stable solutions of (3), $\lambda(y_m)$ is of necessity strictly increasing and a critical point of (3) is of necessity related to a stationary point of $\lambda(y_m)$.

In the example of Fig.1, $\lambda(y_m)$ contains two stable branches and two critical points P_1 and P_2, where P_1 is associated with an ignition jump and P_2 with an extinction jump.

For the boundary value problem (3), the nonlinearity is exclusively contained in $w(y)$ (cf.(4)), which depends on the parameters β, ξ, n controlling the strength of the nonlinearity. By continuously changing the value of the parameter β, for instance, the strength of the nonlinearity of $w(y)$ may be altered which entails a continuous process of deformation of $\lambda(y_m)$, such that the maximum P_1 and the minimum P_2 are set to approach each other. Upon collision of P_1 with P_2, the maximum-minimum pair disappears, leaving $\lambda(y_m)$ to be strictly increasing for all positive values of y_m.

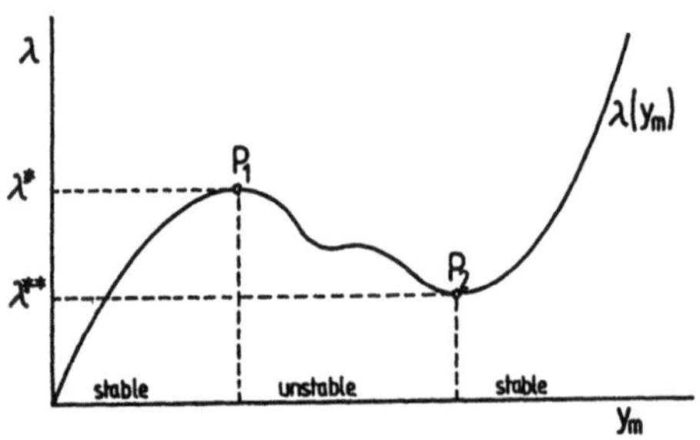

Figure 1 : Characteristic Function $\lambda(y_m)$

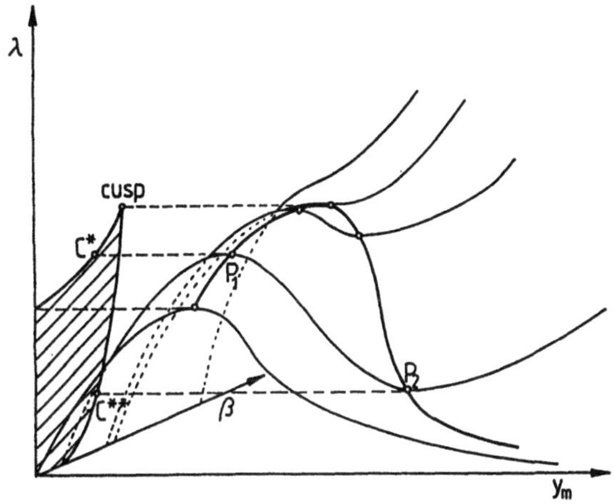

Figure 2 : State Surface in State Space $\{\lambda,\beta,y_m\}$, with Contour $C = C^* \cup C^{**}$ of the State Surface with Respect to Projection into Subspace $\{\lambda,\beta\}$ of Control Parameters

The very collision of P_1 with P_2 corresponds to a degenerate critical solution of (3) which is associated with a stationary point P^+ of $\lambda(y_m)$ joining two branches of stable solutions of (3), i.e., a point of inflection of $\lambda(y_m)$ with a horizontal tangent.
Projection of the state surface in $\{\lambda,\beta,y_m\}$-state space into the subspace $\{\lambda,\beta\}$ of control parameters entails a cusp in the resulting contour C at P^+ because the projection at P^+ is in the direction of an asymptotic tangent which is of an order of tangency larger than 1 . The cusp divides the curve C into segments C^* and C^{**}, with C^* comprising the ignition conditions (cf. e.g., point P_1) and C^{**} comprising the extinction conditions (cf. point P_2). $C = C^* \cup C^{**}$ represents the boundary of the hatched area of Fig.2, which is multiply covered by projecting the state surface into the $\{\lambda,\beta\}$-plane along the direction of the y_m-axis.

3. Critical Conditions in Terms of Lower Bounds

In order for critical phenomena to appear, i.e., temperature jumps corresponding to ignition or extinction, the nonlinearity (4) of the problem (3) needs to be strong enough. This prerequisite may be formulated in terms of a generalized concavity concept whereby a function w(y) is termed concave in the generalized sense at $y = y_m$, if

$$w(y_m) - y_m w'(y_m) \geqq 0 \tag{9}$$

Geometrically speaking, w(y) is of generalized concavity at $y=y_m$, if the tangent of w at $y=y_m$ does not intercept the positive y-axis. If a function w(y) is concave in the generalized sense over an interval, for instance for y_m with $0 \leqq y_m \leqq \bar{y}_m$,

then it is easily seen that a certain weakly convex comportment of w(y) is allowed for w(y) still to be designated as concave in the generalized sense. If, in particular, w(o) > o , then there always exists a nontrivial interval $o \leqq y_m \leqq \overline{y}_m$ where concavity in the generalized sense holds for any function w(y).
The requirement of a sufficiently strong nonlinearity for the appearance of critical phenomena may now be given in terms of generalized concavity [3] : for a function w(y) which is concave in the generalized sense for $o \leqq y \leqq \overline{y}_m$, no stationary point of $\lambda(y_m)$ will occur for y_m with $o \leqq y_m \leqq \overline{y}_m$. Consequently, no critical solution of (3) exists with $o \leqq y_m \leqq \overline{y}_m$. Particularly, if $\overline{y}_m = \infty$, then the system has no capacity for critical phenomena at all.

The convexity of w(y) is governed by the control parameters β, ξ (cf.(4)) and possibly n. For any value of n, continuous changes in β and ξ can bring about a reduction in the strength of the nonlinearity of w(y) to such an extent, that w(y) becomes concave in the generalized sense for all nonnegative values of y, so that the system loses completely its capacity for critical behaviour. This changeover may be investigated in terms of lower bounds of $\lambda(y_m)$ derived with the help of shifted linear majorants of w(y) [4, 5]. A particularly simple representation is obtained by employing shifted constant majorants of w(y), from which a continuous curve $\lambda_M(y_m)$ of lower bounds results (cf. Fig.3) with :

$$\lambda_M(y_m) \leqq \lambda(y_m) \tag{1o}$$

The first maximum of $\lambda_M(y_m)$ is attained at $y_m = \overline{y}_m$, with $\overline{y}_m$ the right hand end point of the concavity interval of w(y), i.e., w(y) is concave in the generalized sense for any y_m with $o \leqq y_m \leqq \overline{y}_m$. Consequently, $\lambda(y_m)$ will have no maximum over this interval and thus, with (1o) :

$$\lambda_M^* \leqq \lambda^* \; , \; \lambda_M^{**} \leqq \lambda^{**} \tag{11}$$

Here, λ^* = first stationary point of $\lambda(y_m)$ to occur beyond $y_m = \overline{y}_m$, λ^{**} = any minimum of $\lambda(y_m)$, $\lambda_M^* = \lambda_M(\overline{y}_m)$ = maximum, $\lambda_M^{**} = \lambda_M(\overline{\overline{y}}_m)$ minimum of $\lambda_M(y_m)$.

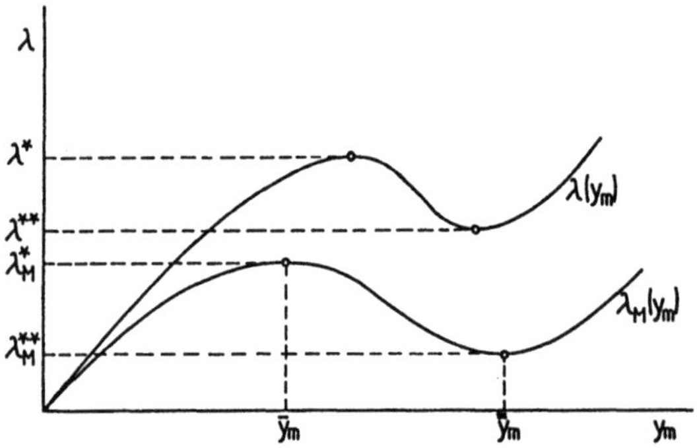

Figure 3: Curve $\lambda_M(y_m)$ of Lower Bounds of the Characteristic Curve $\lambda(y_m)$

A continuous change in the control parameters β, ξ will effect a change in the shape of $w(y)$ such that it becomes concave in the generalized sense for all nonnegative values of y. This change-over is represented in terms of lower bounds $\lambda_M(y_m)$ of $\lambda(y_m)$ by a collision of the maximum λ_M^* with the minimum λ_M^{**}, with susequent disappearance of this maximum/minimum pair. The collision proper is represented for $\lambda_M(y_m)$ by a point of inflection λ_M^+ with a horizontal tangent at $y_m^+ = \overline{y}_m = \overline{\overline{y}}_m$. Consequently, the projection of the lower bounds $\lambda_M(y_m)$ into the space $\{\lambda,\beta\}$ of control parameters will reproduce a continuous curve of lower bounds of the curve C of interest, and, in particular, will produce a lower bound replica of the cusp of C .

In order to provide an example, D is assumed to be spherical and ξ is held constant ($\xi = o$, for instance). Then the change-over for $w(y)$ to become concave in the generalized sense can be neatly displayed in the (β,y_m)-plane. The roots of (9) define a curve $\beta_M(y_m)$ which intersects at y_m^+ the curve β_{WP} representing the locus of the points of inflection of $w(y)$ (cf.Fig.4).

β designates a control parameter and may thus be manipulated independently. From Fig.4, it appears that for β approaching β_o from below, $\overline{y}_m$ and $\overline{\overline{y}}_m$ are closing in. According to Fig.3, this is equivalent to a process of deformation of $\lambda_M(y_m)$, whereby the corresponding extrema are closing in. For $\beta = \beta_o$, the two extrema merge in a point of inflection of $\lambda_M(y_m)$ with a horizontal tangent. Because of (1o), $\lambda_M(y_m;\beta)$ represents a surface in $\{\lambda,\beta,y_m\}$ - state space which is below the surface $\lambda(y_m;\beta)$ everywhere. A projection of the surface $\lambda_M(y_m;\beta)$ into $\{\lambda,\beta\}$ - control parameter space along the y_m-axis produces a contour which is given by the following parametric representation [4] :

$$\left\{\lambda_M(y_m) \ ; \ \beta_M(y_m)\right\} \tag{12}$$

with

$$\lambda_M = \frac{y_m}{w(y_m)\,\overline{G}} \quad , \quad \overline{G} = \int_D G(x_o,x')\,dx' > o \tag{13}$$

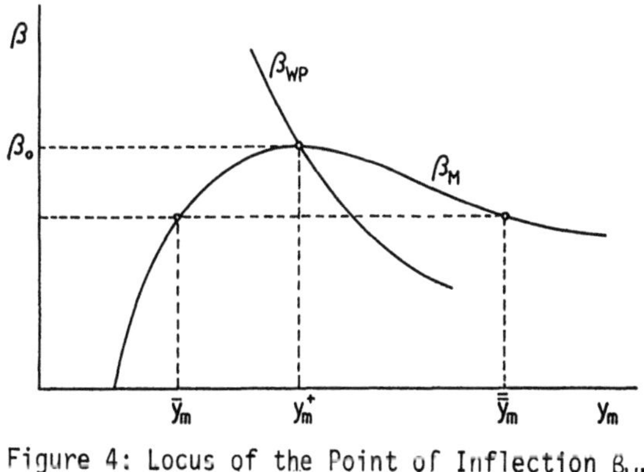

Figure 4: Locus of the Point of Inflection β_{WP} and of the Concavity Condition β_M for $w(y)$ in the (β,y_m) -Plane

In the parametric representation (12), y_m assumes the role of the parameter, and $G(x',x)$ represents Green's function of (3). According to (9), β_M is implicitly given as the root of the following equation :

$$\beta_M = \beta_M(y_m) \quad \longleftrightarrow \quad w(y_m;\beta) - y_m \frac{\partial w(y_m;\beta)}{\partial y_m} = 0 \tag{14}$$

It can be easily shown that the contour $\{\lambda_M(y_m) , \beta_M(y_m)\}$ contains a cusp which provides a lower bound due to (1o) of the cusp of $\{\lambda,\beta\}$.

It is demonstrated in [5] that through use of secants as linear majorants improved lower bounds result which again reproduce the characteristic cusp. This is displayed in Fig.5 which for comparison includes some upper bounds and the exact results derived numerically by [6] .

If beyond β the second control parameter ξ is allowed to vary, the cusps of the previous discussion trace the edge of a cusp ridge in the 3-dimensional $\{\lambda,\beta,\xi\}$ - control parameter space. Consequently, a family of cusp ridges is obtained for various values of the parameter n (which designates the reaction order). This result will be presented in a forthcoming publication.

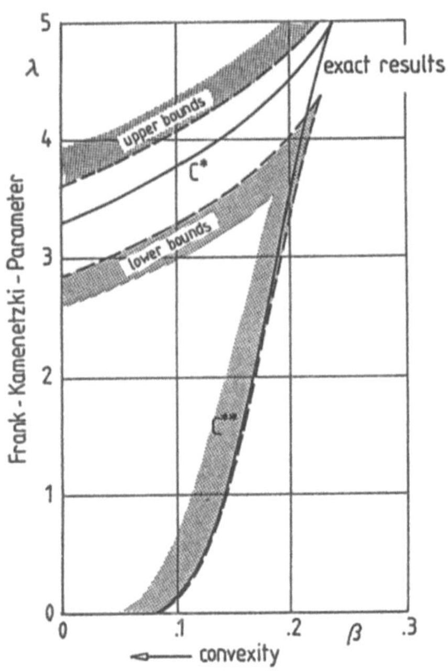

Figure 5: Exact Results and Upper and Lower Bounds Displaying Cusp

4. Literature

1. D.H.Sattinger : Indiana Univ.Math.J. 21, 979 (1972)
2. M.H.Protter, H.F.Weinberger : Maximum Principles in Differential Equations, Prentice-Hall, Englewood Cliffs 1967
3. D.Meinköhn : J.Chem.Phys. 74, 36o3 (1981)
4. D.Meinköhn : SIAM J.Appl.Math. 48 (1988), in press
5. D.Meinköhn : SIAM J.Appl.Math. 48 (1988), in press
6. W.Kordylewski : Combust.Flame 34, 1o9 (1979)

Flow Induced Bifurcations to Three-Dimensional Motions of Tubes with an Elastic Support

A. Steindl and *H. Troger*
Technische Universität Wien, A-1040 Vienna, Austria

1 Introduction

The fascinating behavior of tubes conveying fluid has attracted great interest among engineers as well as mathematicians. A nice and rather complete review of the relevant literature is given in [1]. In a fundamental paper Bajaj and Sethna ([2]) studied the Hopf bifurcation (flutter instability) of the downhanging equilibrium position of a cantilever tube. The mathematical interest in this problem stems from the fact that due to the rotational and reflectional symmetry (O(2)-symmetry) each eigenvalue of the linearized problem, associated with the loss of stability, occurs with multiplicity two. The consequence of this double multiplicity is that the system of bifurcation equations (amplitude equations of the critical modes) is rather high dimensional and hence has a richer solution set. However, on the other hand the bifurcation equations must obey the same symmetry properties as the equations of motion. Therefore the number of terms which are consistent with the symmetry requirements is pretty small. Thus the complication resulting from the high dimension is to some extend compensated. In [2] it is shown that depending on the mass ratio β between tube and fluid (see eq. (14) below) two distinct oscillatory motions occur after loss of stability. Both follow from symmetry breaking bifurcations. One is a planar oscillation which breaks the rotational symmetry and the other is a rotating motion which breaks the reflectional symmetry. Both solutions can be easily observed in experiments.

In this paper we want to extend the work presented in [2] in that respect that by introducing a rotationally symmetric elastic support (in Fig. 1 only some of the springs are drawn) more

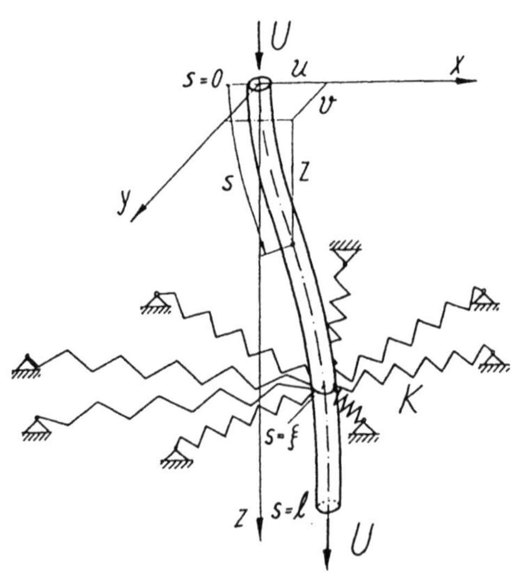

Fig. 1: Mechanical model of the tube conveying fluid (flow rate U), clamped at $s = 0$, free at $s = \ell$ and with a rotationally symmetric elastic support at $s = \xi$.

complicated cases of loss of stability can be studied if the stiffness K of the support is not kept at a fixed value but considered to be a parameter. Hence it will be possible to investigate more degenerate cases of loss of stability as it is done in [2]. In [3] it is shown that external and internal damping has an important influence on the structure of the critical eigenvalues at loss of stability, hence also external and internal damping is included in the mechanical model. The investigation of more degenerate cases is not only of academic interest but is of great importance in applications. Though in technical systems always slight perturbations of such degenerate cases occur and hence the complicated case breaks up in several simple cases, these are located closely together. This leads to secondary bifurcations yielding a transition sequence from simple to more complicated states by a small variation of a distinguished parameter.

The analysis proceeds in several steps. Firstly we supplement the equations of motion used in [2] which are a set of two nonlinear partial differential equations. Then for a quasistatic variation of the flow rate U and for fixed values of the stiffness K an eigenvalue problem is solved to calculate the stability boundary in parameter space spanned by U and K respectively ρ and k in nondimensional quantities. In the next step the reduction to bifurcation equations is performed by means of Center Manifold Theory ([4]), yielding bifurcation equations of dimension two, four and six. These bifurcation equations are simplified by means of symmetry and Normal Form theory. Finally the unfolded bifurcation equations (one and two parameter families occur) are solved and the results are interpreted physically.

2 Mechanical model and equations of motion

The derivation of the equations of motion which are used in [2] is given in ([5]). All nonlinearities follow from geometry. The material is assumed to be linearly viscoelastic. All assumptions made in [5] are valid also for the equations given below. However, in addition to [5] the effects of external and internal damping, gravitiy and of the rotationally symmetric support will be included in the equations. Why, in general, it is necessary to include both types of damping is explained in [3]. The external damping force is assumed to be proportional to the velocity of the axis of the tube. The internal damping terms are derived from the Kelvin Voigt law of viscoelasticity. As the assumptions of Bernoulli-Euler beam theory are made, the bending moment $\mathbf{M}$ for an elastic material is given by $\mathbf{M} = EJ(\mathbf{t} \times \mathbf{t}')$. Hence for a viscoelastic material the term in the moment following from the speed of deformation is given by $\alpha EJ(\mathbf{t} \times \dot{\mathbf{t}}')$ and the bending moment is given by

$$\mathbf{M} = EJ[(\mathbf{t} \times \mathbf{t}') + \alpha(\mathbf{t} \times \dot{\mathbf{t}}')]. \tag{1}$$

$\mathbf{t}$ is the unit tangent vector

$$\mathbf{t} = \mathbf{r}' = u'\mathbf{i} + v'\mathbf{j} + z'\mathbf{k}. \tag{2}$$

( )' is the derivative with respect to arc length s and $(\ )^{\cdot}$ is the derivative with respect to time. EJ is the bending stiffness and α is the viscosity coefficient. Furthermore the curvature κ is given by

$$\kappa^2 = |(\mathbf{t} \times \mathbf{t}')|^2 = \mathbf{t}' \cdot \mathbf{t}' = (u'')^2 + (v'')^2 + (z'')^2. \tag{3}$$

The equations of motion read

$$\begin{aligned} G_1(u,v,\lambda) &= m_T\ddot{u} + m_F(\ddot{u} + 2U\dot{u}' + U^2u'') + \mu\dot{u} + EJ(u^{IV} + \alpha\dot{u}^{IV}) \\ &\quad -[(N - P - EJ\kappa^2)u' - \alpha EJ(\kappa^2\dot{u}' + (\mathbf{t}\cdot\dot{\mathbf{t}}''))u']' = 0 \\ G_2(u,v,\lambda) &= m_T\ddot{v} + m_F(\ddot{v} + 2U\dot{v}' + U^2v'') + \mu\dot{v} + EJ(v^{IV} + \alpha\dot{v}^{IV}) \\ &\quad -[(N - P - EJ\kappa^2)v' - \alpha EJ(\kappa^2\dot{v}' + (\mathbf{t}\cdot\dot{\mathbf{t}}''))v']' = 0. \end{aligned} \tag{4}$$

m_T and m_F are the specific masses of the tube and the fluid respectively. U is the constant flow rate, μ the coefficient of external damping, N is the axial force in the tube, P the pressure in the fluid times the area of the cross-section, and λ the parameter vector which is given in nondimensional form in (19). $(N-P)$ can be calculated from the following expressions. For

$$
\begin{aligned}
\xi_+ \le s \le \ell: \quad \left(N-P+\tfrac{1}{2}EJ\kappa^2\right)\Big|_s^{\ell} &= \int_s^{\ell}[\mathbf{t}\cdot\mathbf{Q}' - \alpha EJ(-\kappa^2(\mathbf{t}\cdot\dot{\mathbf{t}}') + (\mathbf{t}'\cdot\dot{\mathbf{t}}''))]ds \\
0 \le s \le \xi_-: \quad \left(N-P+\tfrac{1}{2}EJ\kappa^2\right)\Big|_s^{\xi_-} &= \int_s^{\xi_-}[\mathbf{t}\cdot\mathbf{Q}' - \alpha EJ(-\kappa^2(\mathbf{t}\cdot\dot{\mathbf{t}}') + (\mathbf{t}'\cdot\dot{\mathbf{t}}''))]ds
\end{aligned}
\tag{5}
$$

where $\mathbf{t}\cdot\mathbf{Q}'$ is given by

$$
\begin{aligned}
\mathbf{t}\cdot\mathbf{Q}' &= (m_T+m_F)(u'\ddot{u}+v'\ddot{v}+z'\ddot{z})+ \\
&\quad \mu(u'\dot{u}+v'\dot{v}+z'\dot{z}) - (m_T+m_F)gz'.
\end{aligned}
\tag{6}
$$

For a tube which is clamped at $s=0$, free at $s=\ell$ and where the spring is acting at $s=\xi$ the following boundary and intermediate conditions hold:

$$
s=0: \quad u=u'=0, \quad v=v'=0 \tag{7}
$$

$$
\begin{aligned}
s=\ell: \quad & u''+\alpha\dot{u}''=0, \quad v''+\alpha\dot{v}''=0 \\
& u'''+\alpha\dot{u}'''=0, \quad v'''+\alpha\dot{v}'''=0
\end{aligned}
\tag{8}
$$

$s=\xi$: There is a discontinuity for the axial force N and transversal force R. P and κ, however, are continuous. Hence the following relations must hold

$$
\begin{aligned}
R(\xi_-)-R(\xi_+) &= \mathbf{S}\cdot\mathbf{m} \\
N(\xi_-)-N(\xi_+) &= \mathbf{S}\cdot\mathbf{t} \\
P(\xi_-)-P(\xi_+) &= 0 \\
\kappa(\xi_-)-\kappa(\xi_+) &= 0
\end{aligned}
\tag{9}
$$

where the force S of the support and $\mathbf{m}$ the unit vector orthogonal to $\mathbf{t}$ in the direction to the z axis are given by

$$
\mathbf{S}=-K(u\mathbf{i}+v\mathbf{j}), \qquad \mathbf{m}=\frac{\mathbf{S}-(\mathbf{S}\cdot\mathbf{t})\mathbf{t}}{\|\mathbf{S}-(\mathbf{S}\cdot\mathbf{t})\mathbf{t}\|}. \tag{10}
$$

3 Symmetry properties of the equations of motion

As already mentioned in the Introduction the equations (4) to (10) possess certain symmetry properties, which express themselves by the fact that the equations of motion are equivariant under group operations. Mathematically the equivariance of a system of governing equations $\mathbf{G}(\mathbf{w})=\mathbf{0}$ under the elements of a symmetry group $\tilde{G}$ is expressed by

$$
\mathbf{S}_g\mathbf{G}(\mathbf{w})=\mathbf{G}(\mathbf{S}_g\mathbf{w}) \tag{11}
$$

where $\mathbf{S}_g$ is a representation of $\tilde{G}$ ([6]). For the tube problem $\mathbf{S}_g$ is given by the following two matrices $\mathbf{S}_R$ and $\mathbf{S}_S$

$$
\mathbf{S}_R=\begin{pmatrix}\cos\vartheta & \sin\vartheta \\ -\sin\vartheta & \cos\vartheta\end{pmatrix}, \qquad \mathbf{S}_S=\begin{pmatrix}0 & 1 \\ 1 & 0\end{pmatrix} \tag{12}
$$

denoting a rotation by an angle ϑ about the z axis and a reflection with respect to a plane including the z axis respectively. Relation (11) is easy to check for (4) using (12). A system possessing the symmetries expressed by (12) is called O(2) symmetric ([6]). Now we change to nondimensionalized variables by setting

$$
s \to \frac{s}{\ell}, \qquad t \to \sqrt{\frac{EJ}{m_T+m_F}}\frac{t}{\ell^2}, \qquad u \to \frac{u}{\ell}, \qquad v \to \frac{v}{\ell}. \tag{13}
$$

The following coefficients obtained from the nondimensionalisation procedure will be important for the further investigation

$$\begin{array}{lll} \rho = \sqrt{\frac{m_F}{EJ}}U\ell, & \beta = \frac{m_F}{m_F+m_T}, & \alpha^* = \sqrt{\frac{EJ}{m_T+m_F}}\frac{\alpha}{\ell^2} \\ \mu^* = \frac{\mu\ell^2}{\sqrt{EJ(m_T+m_F)}}, & \gamma = \frac{m_T+m_F}{EJ}\ell^2 g, & k = \frac{K\ell^3}{EJ}. \end{array} \tag{14}$$

To transform (4) into a system of equations of first order new variables

$$w_1 = u, \qquad w_2 = \dot{u}, \qquad w_3 = v, \qquad w_4 = \dot{v} \tag{15}$$

are introduced, yielding

$$\dot{\mathbf{w}} = \mathbf{L}(\boldsymbol{\lambda})\mathbf{w} + \mathbf{N}(\mathbf{w}, \boldsymbol{\lambda}). \tag{16}$$

$\mathbf{w}(s,t) = (w_1, w_2, w_3, w_4)^T$, the matrix $\mathbf{L} = \mathrm{diag}(\mathbf{L}_1, \mathbf{L}_1)$ contains the terms obtained by linearization about the trivial downhanging position

$$\mathbf{L} = \begin{pmatrix} \mathbf{L}_1 & 0 \\ 0 & \mathbf{L}_1 \end{pmatrix} \qquad \mathbf{L}_1 = \begin{pmatrix} 0 & 1 \\ -C & -B \end{pmatrix}. \tag{17}$$

The differential operators C and B are defined by

$$\begin{array}{rcl} Cw_1 & = & w_1^{IV} + \rho^2 w_1'' - \gamma[(1-s)w_1']' \\ Bw_2 & = & \mu w_2 + 2\sqrt{\beta}\rho w_2' + \alpha^* w_2^{IV}. \end{array} \tag{18}$$

The block-diagonal structure of $\mathbf{L}$ is a consequence of the symmetry property. The components of the vector $\mathbf{N} = (0, N_2, 0, N_4)^T$ of the nonlinear terms up to third order are easily obtained from (4). The parametervector $\boldsymbol{\lambda}$ is given by

$$\boldsymbol{\lambda} = (\rho, k, \beta, \alpha^*)^T \in R^4. \tag{19}$$

ρ and k are the main parameters and are according to (14) proportional to the flow rate U and the stiffness K of the support. β is the mass ratio and α^* is the nondimensionalized internal damping coefficient.

4 Stability boundary in parameter space and eigenvalue problem

We assume the tube to be in a stable downhanging equilibrium position which for appropriately chosen parameter values in (19) (small flow rate ρ) will be certainly possible. Now the parameter values are determined for which this stable position loses its stability. This requires to solve a linear stability problem leading to a linear eigenvalue problem for either ρ or k, according to (14). Because of the block diagonal form (17) only half of the linearized problem of (16) must be studied. Hence the linearized problem reads

$$\dot{\mathbf{w}}_I = \mathbf{L}_1(\boldsymbol{\lambda})\mathbf{w}_I \tag{20}$$

where $\mathbf{L}_1$ is given by (17) and (18), $\boldsymbol{\lambda}$ by (19) and $\mathbf{w}_I = (w_1, w_2)^T$. The corresponding boundary and intermediate conditions are

$$\begin{array}{ll} s = 0: & w_1 = 0, \quad w_1' = 0 \\ s = \xi: & w_1'''(\xi_+) + \alpha^* w_2'''(\xi_+) - w_1'''(\xi_-) - \alpha^* w_2'''(\xi_-) = -k w_1(\xi) \\ & w_1, \quad w_1', \quad w_1'' + \alpha^* w_2'' \qquad \text{must be continuous} \\ s = 1: & w_1'' + \alpha^* w_2'' = 0, \quad w_1''' + \alpha^* w_2''' = 0. \end{array} \tag{21}$$

For a purely imaginary eigenvalue $\sigma = \pm i\omega$ follows from (20) with (17), and the use of complex variables $\mathbf{w}_I = e^{\sigma t}\mathbf{z}(s)$

$$i\omega \begin{pmatrix} z_1 \\ z_2 \end{pmatrix} = \begin{pmatrix} 0 & 1 \\ -C & -B \end{pmatrix} \begin{pmatrix} z_1 \\ z_2 \end{pmatrix} \tag{22}$$

or

$$\begin{aligned} z_2 &= i\omega z_1 \\ -Cz_1 - Bz_2 &= i\omega z_2. \end{aligned} \tag{23}$$

From (23) follows with $z_i = x_i + iy_i$

$$\begin{aligned} x_2 + iy_2 &= i\omega x_1 - \omega y_1 \\ -Cx_1 - Bx_2 &= -\omega y_2 \\ -Cy_1 - By_2 &= \omega x_2. \end{aligned} \tag{24}$$

By eliminating y_1 and y_2 from the second and third equation (24) by means of the first equation

$$\begin{aligned} (C - \omega^2)x_1 + Bx_2 &= 0 \\ -\omega^2 Bx_1 + (C - \omega^2)x_2 &= 0 \end{aligned} \tag{25}$$

is obtained with C and B defined by (18). The boundary and intermediate conditions for x_1 and x_2 are for

$$\begin{aligned} s = 0: \quad & x_1 = 0, \quad x_1' = 0, \quad x_2 = 0, \quad x_2' = 0 \\ s = \xi: \quad & x_1'''(\xi_+) + \alpha^* x_2'''(\xi_+) - x_1'''(\xi_-) - \alpha^* x_2'''(\xi_-) = -kx_1(\xi) \\ & -x_2'''(\xi_+) + \alpha^*\omega^2 x_1'''(\xi_+) + x_2'''(\xi_-) - \alpha^*\omega^2 x_1'''(\xi_-) = kx_2(\xi) \\ & x_1, \quad x_2, \quad x_1', \quad x_2', \quad x_1'' + \alpha^* x_2'', \quad \alpha^*\omega^2 x_1'' - x_2'' \qquad \text{continuous} \\ s = 1: \quad & x_1'' + \alpha^* x_2'' = 0, \quad \alpha^*\omega^2 x_1'' - x_2'' = 0 \\ & x_1''' + \alpha^* x_2''' = 0, \quad \alpha^*\omega^2 x_1''' - x_2''' = 0. \end{aligned} \tag{26}$$

The solutions $x_1(s)$ and $x_2(s)$ of the eigenvalue problem (25), (26) yield the real part of the eigenfunctions. The imaginary parts follow from the first equation (24).

Setting $\omega = 0$ in (25) results in the eigenvalue problem for the zero eigenvalue

$$\begin{aligned} C_1 x_1 + Bx_2 &= 0 \\ C_2 x_2 &= 0 \end{aligned} \tag{27}$$

with x_2 as eigenvector and x_1 as first principal vector. The solution of the eigenvalue problem is given numerically by means of the boundary value problem solver BNDSCO ([7]).

The results are shown in Fig. 2 for the support located at the midspan of the tube ($\xi = 0.5\ell$), for two different values of the damping parameters α^* and μ^*. This data is the same as used in [8] where qualitatively similar results are obtained and good agreement with experiments is shown. All eigenvalues obtained must be doubled because of the block diagonal structure of $\mathbf{L}$ in (16). Another argument for this double multiplicity is given in [9]. From Fig. 2 it is clearly visible that three qualitatively different cases of loss of stability can occur:

(a) For a stiff support (k big) two zero roots are found.

(b) For a soft support (k small) two identical pairs of purely imaginary eigenvalues $\pm i\omega_0, \pm i\omega_0$ are given.

(c) There exists a special value $k = k_c$ where a qualitative change in the type of loss of stability takes place. This is a more degenerate case than (a) and (b) because six eigenvalues with zero real part are given.

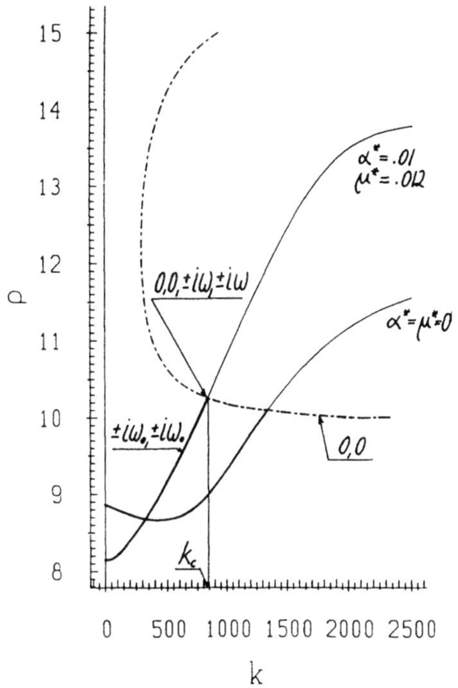

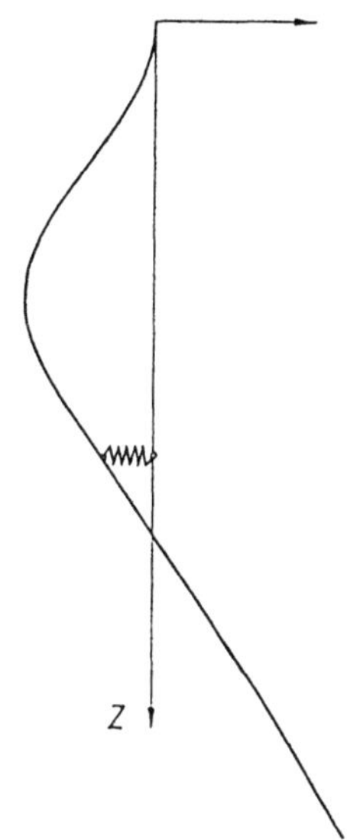

Fig. 2: Stability boundary with the corresponding eigenvalues in the ρ, k – parameter plane for $\beta = 0.25$ and $\xi = 0.5\ell$.

Fig. 3: Eigenmode of the statically buckled tube (SB).

Case (a) is called divergence bifurcation or static instability because the system buckles statically. Case (b) is called Hopf bifurcation or flutter instability because after loss of stability of the equilibrium position an oscillatory motion sets in. Case (c) is a coupling of the two simple cases (a) and (b) and gives rise to more complicated motions as will be seen below.

It must be pointed out, however, that the behavior of the system after loss of stability cannot be assessed from a linear but only by a nonlinear analysis which will be given in the next sections.

5 Reduction to bifurcation equations

A decisive step in solving nonlinear stability problems is the possibility of a reduction of the original system of partial differential equations (4) or (16) to a finite dimensional system of ordinary differential equations called bifurcation equations. The dimension n of this system of bifurcation equations is equal to the number of critical eigenvalues. I. e. $n = 2, 4$ and 6 for cases (a), (b) and (c) respectively. These bifurcation equations possess locally in the neighborhood of the bifurcation point the same solutions as the original system. This concept of locality is sometimes expressed by saying that only small solutions are obtained and will be made more precise in the next section where the results are discussed. The reduction can be performed by means of Center Manifold theory ([4]). For a reader not familiar with Center Manifold theory we assume it appropriate to note that up to terms of third order in the bifurcation equations Center Manifold theory and the Ritz-Galerkin method yield exactly the same bifurcation equations.

A detailed description of the reduction process with all assumptions is given in [4] p. 117. We only give the idea, which is a splitting of the unknown function $\mathbf{w}(s,t)$ and of equation (16) into two parts. Firstly

$$\mathbf{w}(s,t) = \mathbf{q}(s,t) + \mathbf{r}(s,t) \tag{28}$$

where $\mathbf{q}(s,t) \in Q$ is spanned by the eigenfunctions corresponding to eigenvalues with zero real part and $\mathbf{r}(s,t) \in R$ is in a closed space of functions R complementary to Q. Secondly a projection P on Q along R is defined. Let

$$\mathbf{L}_c = P\mathbf{L} \quad \text{and} \quad \mathbf{L}_s = (I-P)\mathbf{L} \tag{29}$$

and

$$\mathbf{N}_c = P\mathbf{N}(\mathbf{q}+\mathbf{r}) \quad \text{and} \quad \mathbf{N}_s = (I-P)\mathbf{N}(\mathbf{q}+\mathbf{r}). \tag{30}$$

Then (16) can be written

$$\begin{aligned} \dot{\mathbf{q}} &= \mathbf{L}_c\mathbf{q} + \mathbf{N}_c(\mathbf{q},\mathbf{r}) \\ \dot{\mathbf{r}} &= \mathbf{L}_s\mathbf{r} + \mathbf{N}_s(\mathbf{q},\mathbf{r}). \end{aligned} \tag{31}$$

The first of equations (31) are the bifurcation equations from which still $\mathbf{r}$ must be eliminated. This can be done with the help of (31_2) because $\mathbf{L}_s$ is regular and hence there exists locally a Center Manifold $\mathbf{h}$ given by

$$\mathbf{r} = \mathbf{h}(\mathbf{q}) = O(\|\mathbf{q}\|^2). \tag{32}$$

Introducing (32) into (31_1) the equation on the Center Manifold

$$\dot{\mathbf{q}} = \mathbf{L}_c\mathbf{q} + \mathbf{N}_c(\mathbf{q},\mathbf{h}(\mathbf{q})) \tag{33}$$

is obtained. It is important to note that if only third order terms are retained in (33) the contribution from (32) can be neglected because it is at least of fourth order. Hence (33) reduces to

$$\dot{\mathbf{q}} = \mathbf{L}_c\mathbf{q} + \mathbf{N}_c(\mathbf{q},0) \tag{34}$$

which are the same equations as are obtained from the Ritz-Galerkin method.

As $\mathbf{L}_1$ according to (17) is not self adjoint for the projection operator P the eigenfunctions of the operator $\mathbf{L}_1^*$ adjoint to $\mathbf{L}_1$ must be known. $\mathbf{L}_1^*$ is defined by

$$\mathbf{L}_1^* = \begin{pmatrix} 0 & C^* \\ -1 & B^* \end{pmatrix} \tag{35}$$

where C^* and B^* are differential operators adjoint to (18). Furthermore appropriate boundary conditions must be formulated which are

$$\begin{aligned} s = 0: &\quad w_2 = 0, \quad w_2' = 0 \\ s = \xi: &\quad w_2'''(\xi_+) - w_2'''(\xi_-) - \alpha^*[\dot{w}_2'''(\xi_+) - \dot{w}_2'''(\xi_-)] = -kw_2(\xi) \\ s = \ell: &\quad w_2'' + \rho^2 w_2 - \alpha^*\dot{w}_2'' = 0, \quad w_2''' + \rho^2 w_2' + 2\sqrt{\beta}\rho\dot{w}_2 - \alpha^*\dot{w}_2''' = 0. \end{aligned} \tag{36}$$

The result of the reduction process is a system of first order nonlinear ordinary differential equations in the amplitudes and phases of the critical modes. The dimension of these systems is either two, four or six.

6 Simplification of the bifurcation equations by Symmetry and Normal Form Theory and Unfolding

We study now the three different cases (a), (b) and (c) of section (4). The application of the symmetry properties and Normal Form Theory is given in detail in [10,3,9]. Hence it is sufficient to give the results.

(a) One zero root with multiplicity two

The two dimensional system of bifurcation equations ($x = q_1, y = q_2$)

$$\dot{x} = \varepsilon x + ax(x^2 + y^2), \quad \dot{y} = \varepsilon y + ay(x^2 + y^2) \tag{37}$$

has the steady state solutions: (i) $x = y = 0$. This is the unbuckled state. It is stable for $\varepsilon < 0$ and unstable for $\varepsilon > 0$. (ii) $x = 0, y = \pm\sqrt{\frac{-\varepsilon}{a}}$ or $x = \pm\sqrt{\frac{-\varepsilon}{a}}, y = 0$. Only one number a must be calculated in the reduction process. If $a > 0$ or $a < 0$ a subcritical (unstable) or supercritical (stable) solution is obtained. (iii) $x = y = \pm\sqrt{\frac{-\varepsilon}{2a}}$. This solution is identical to (ii). Physically a planar statically buckled configuration of the tube is obtained (Fig. 3), where the angle ϑ locating the plane is arbitrary.

(b) A pair of purely imaginary eigenvalues with multiplicity two

Now the bifurcation system is four-dimensional. Finally after introducing polar coordinates the two amplitude equations read

$$\dot{r}_1 = \varepsilon r_1 + r_1(a_1 r_1^2 + a_2 r_2^2), \quad \dot{r}_2 = \varepsilon r_2 + r_2(a_2 r_1^2 + a_1 r_2^2). \tag{38}$$

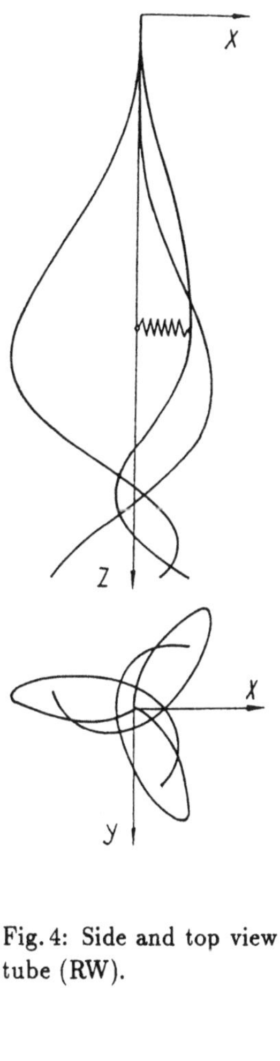

Fig. 4: Side and top view of the rotating tube (RW).

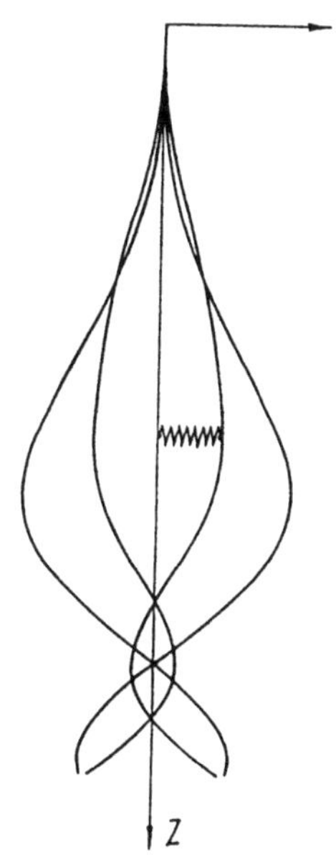

Fig. 5: Side view of the planar oscillations of the tube (SW).

The two equations for the polar angles (phases) are decoupled from (38) and not relevant. Again three steady state solutions are obtained: (i) $r_1 = r_2 = 0$, being the trivial state. (ii) $r_1 = 0, r_2 \neq 0$ or $r_1 \neq 0, r_2 = 0$. This solution is a rotating tube (Fig. 4). (iii) $r_1 = r_2$. The physical interpretation of this solution is a planar oscillation (Fig. 5). Both types of solutions have been found in [2].

The stability of these solutions depends only on two parameters a_1, a_2 and is shown in the bifurcation diagram of Fig. 3 in [11].

(c) One zero root and a pair of purely imaginary eigenvalues with multiplicity two

Now the system of bifurcation equations is six dimensional and its derivation is given in [10,3]. After introducing polar coordinates one obtains the following set of equations ([3])

$$\begin{aligned}
\dot{r}_1 &= (\varepsilon_1 + a_1 r_1^2 + a_2 r_2^2 + a_3 r_3^2) r_1 + r_2 r_3^2 (a_4 \cos\psi + b_4 \sin\psi) \\
\dot{r}_2 &= (\varepsilon_1 + a_2 r_1^2 + a_1 r_2^2 + a_3 r_3^2) r_2 + r_1 r_3^2 (a_4 \cos\psi - b_4 \sin\psi) \\
\dot{r}_3 &= (\varepsilon_2 + a_5 (r_1^2 + r_1^2) + a_6 r_3^2 + a_7 r_1 r_2 \cos\psi) r_3 \\
\dot{\psi} &= (b_1 - b_2 - 2b_5)(r_1^2 - r_2^2) + r_3^2 b_4 (\tfrac{r_2}{r_1} - \tfrac{r_1}{r_2}) \cos\psi \\
&\quad - r_3^2 a_4 (\tfrac{r_1^2}{r_2} + \tfrac{r_2^2}{r_1}) \sin\psi - 2 a_7 r_1 r_2 \sin\psi .
\end{aligned} \tag{39}$$

In (37–39) ε respectively ε_1 and ε_2 are the unfolding parameters. The relationship of ε_1 and ε_2 to the physical parameters ρ and k is, for example, given in Fig. 6. We found the following steady state solutions of (39)

(I)		$r_1 = r_2 = r_3 = 0:$	Trivial state (TS), downhanging tube
(II)		$r_1 = r_2 = 0, r_3 \neq 0:$	Statically buckled tube (SB)
(III)		$r_1 = r_2 \neq 0, r_3 = 0:$	Planar oscillating tube about TS (SW)
(IV)		$r_1 = r_2 \neq 0, r_3 \neq 0,$ $\sin\psi = 0:$	Oscillation of the tube about a statically buckled state. There exist two different solutions: One in the plane (SWB0) and one out of the plane (SWBπ).
(V)	(i)	$r_1 \neq 0, r_2 = r_3 = 0:$	Rotating tube (RW)
	(ii)	$r_1 = 0, r_2 \neq 0, r_3 = 0:$	RW in opposite direction
(VI)		$r_1 \neq 0, r_2 \neq 0, r_3 = 0:$	Rotating and oscillating tube (MW)
(VII)		$r_1 \neq 0, r_2 = O(\lvert r_1 r_3^2 \rvert),$ $r_3 \neq 0:$	Statically buckled and rotating tube (SRW).

In Fig. 6 the bifurcation diagram is given which is a stratification of the parameter plane $\varepsilon_1, \varepsilon_2$ or ρ, k in the neighborhood of the critical point $\varepsilon_1 = \varepsilon_2 = 0$. The domains of existence of certain solutions are shown using the notation above. The stability of the solutions is indicated by the signs of the real part of the eigenvalues of a linear stability analysis. All signs must be negative for a stable solution. In Fig. 7 an application of Fig. 6 for a slightly perturbed system according to the dotted line in Fig. 6 is given. We start in the sector where the trivial state is stable. By increasing the flow rate ρ a sequence of bifurcations is obtained leading via statically buckled (SB) and oscillating states (SWB0 and SWBπ) about the buckled configuration finally to a planar standing wave solution SW. In between there is a small gap where degenerate Hopf bifurcations occur the analysis of which would require terms of at least fifth order.

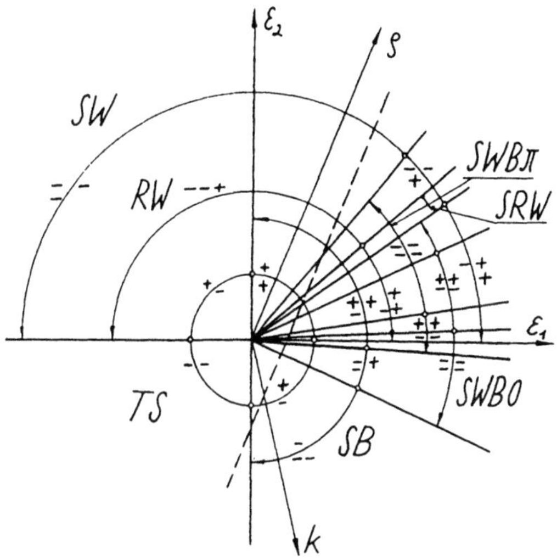

Fig. 6: Bifurcation diagram giving a stratification of the $\varepsilon_1, \varepsilon_2$ respectively ρ, k plane into domains of qualitatively different solutions.

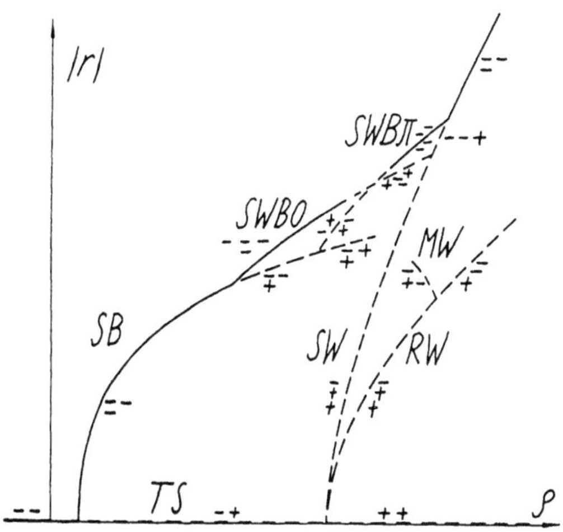

Fig. 7: Bifurcation solutions corresponding to the dotted line in Fig. 6.

Acknowledgement

This work is partly supported by the „Fonds zur Förderung der wissenschaftlichen Forschung" in Austria under project P 5519.

References

1. Paidoussis M. P., *Flow-induced instabilities of cylindrical structures*, Appl. Mech. Rev. 40 (1987) 163–175.

2. Bajaj A. K., Sethna P. R., *Flow Induced Bifurcations to Three-Dimensional Oscillatory Motions in Continuous Tubes*, SIAM J. Appl. Math. 44 (1984) 270–286.

3. Steindl A., Troger H., *Bifurcations of the equilibrium of a spherical double pendulum at a multiple eigenvalue*, Int. Series of Numerical Math. Vol. 79, Birkhäuser Verlag, Basel 1987, 277–287.

4. Carr J., *Applications of Centre Manifold Theory*, Appl. Math. Sciences 35, Springer-Verlag, New York, Heidelberg, Berlin 1981.

5. Lundgren T. S., Sethna P. R., Bajaj A. K., *Stability Boundaries for Flow Induced Motions of Tubes with an Inclined Terminal Nozzle*, J. Sound and Vibration 64 (1979) 553–571.

6. Golubitsky M., Schaeffer D., *Singularities and Groups in Bifurcation Theory*, Appl. Math. Sciences 51, Springer-Verlag, Berlin, Heidelberg, New York 1985.

7. Oberle H. J., Grimm W., Berger E., BNDSCO – *Rechenprogramm zur Lösung beschränkter optimaler Steuerungsprobleme*, Benutzeranleitung, TUM-M8509, Math. Institut, Technische Universität München, 1985.

8. Sugiyama Y., Tanaka Y., Kishi T., Kawagoe H., *Effect of a Spring Support on the Stability of Pipes Conveying Fluid*, J. Sound and Vibration 100 (1985) 257–270.

9. Steindl A., Troger H., *Bifurcations and oscillations in mechanical systems with symmetries*, Proceed. 11th Intern. Conf. on Nonl. Oscilations, (M. Farkas, V. Kertesz, G. Stepan eds.) Janos Bolyai Math. Soc., Budapest 1987, 231–238.

10. Golubitsky M., Stewart I., *Symmetry and Stability in Taylor-Couette Flow*, SIAM J. Math. Anal. 17 (1986) 249–288.

11. Lindtner E., Steindl A., Troger H., *Stabilitätsverlust der gestreckten Lage eines räumlichen Pendels mit elastischer Endlagerung unter einer Folgelast*, ZAMM 67 (1987) T105–T107.

Local and Global Bifurcations of a Three Particle System

S.A. van Gils[1], *E. van Groesen*[1], *and T.P. Valkering*[2]

[1]University of Twente, Dept. of Applied Math.,
P.O. Box 217, NL-7500 AE Enschede, The Netherlands
[2]University of Twente, Center for Theoretical Physics,
P.O. Box 217, NL-7500 AE Enschede, The Netherlands

1. Introduction

Systems of differential equations with symmetry will bifurcate differently compared to systems without symmetry. For instance at a Hopf bifurcation point, where a symmetric steady state looses its stability, more than one branch of periodic states may be produced. Each of these branches can be detected by restricting the state space suitably, and applying standard Hopf bifurcation results within the restricted state space.

We show how the same ideas can be applied to obtain global branches by variational methods. Again, symmetry considerations lead to the choice of suitable functions spaces. The level branches can be continued globally by minimizing suitable functionals. We give the details for a three particle system.

2. Local $D_3 \times S_1$ Equivarant Bifurcations

The general theory of symmetric Hopf bifurcation was developed in Golubitsky and Stewart [4]. A system of N identically coupled particles has the symmetry corresponding to discrete rotation and reflection. Phrased differently, the system of ODE is equivariant with respect to the dihedral group of 2N elements, D_N. The corresponding Hopf bifurcation results have been found independently by Golubitsky and Stewart [5] and Van Gils and Valkering [3].

Here we are interested in a system of 3 particles coupled by a nearest neighbour potential. We review some of the known results on periodic states.

The equations of motion are given by

$$(2.1) \qquad \ddot{y}_n = -\frac{\partial V}{\partial y_n} \qquad (^{\cdot} = \frac{d}{dt}, \quad y_n = y_{n(\mathrm{mod}\,3)}).$$

where V is a nearest neighbour potential determined by a C^∞-function W, $W(0) = W'(0) = 0$, $W''(0) > 0$

$$(2.2) \qquad V = \sum_1^3 W(y_n - y_{n-1}).$$

System (2.1) admits the trivial solution y_n = constant. After transforming (2.1) to a first order ODE, one finds that the linearization around this solution has a double eigenvalue at zero, with geometrical multiplicity one, corresponding to the motion of the mass centre, and a pair of semi-simple purely imaginary complex conjugate eigenvalues which are double, due to the symmetry.

First one needs coordinates in which the symmetry is described naturally. The group D_3 has a two-dimensional irreducible representation. Dividing out the motion of the mass centre (reducing to a four dimensional system) we can thereafter choose coordinates in $\mathbb{R}^4 \simeq \mathbb{C}^2$ such that the action of D_3 on $\mathbb{C}^2$ is generated by the rotation φ and

$$\text{(2.3a)} \qquad \gamma\cdot(z_1,z_2) = \left(e^{i\frac{2\pi}{3}} z_1, e^{-i\frac{2\pi}{3}} z_2\right)$$

$$\text{(2.3b)} \qquad \kappa\cdot(z_1,z_2) = (z_2,z_1)$$

and the vectorfield in the new coordinates commutes with this action. After a near identity transformation, we may also achieve that, up to any finite order in the Taylor expansion in $(z_1,\bar{z}_1,z_2,\bar{z}_2)$ the vector field commutes with the circle action given by the flow generated by the linear part of the vector field. Or, phrased differently, the appropriate bifurcation function obtained by a Liapunov-Schmidt reduction will commute with this circle action. The circle action is given by:

$$\text{(2.3c)} \qquad \theta\cdot(z_1,z_2) = (e^{i\theta}z_1, e^{i\theta}z_2).$$

Given an element $u \in \mathbb{C}^2$, by Σ_u we denote those elements from the group $D_3\times S^1$ which leave u invariant. Fix (Σ_u) is the set of all $z \in \mathbb{C}^2$ which are Σ_u-invariant.

The groep $D_3\times S^1$ acting on $\mathbb{C}^2$ as described in (2.3a-c) has three non trivial isotropy subgroups, each of them has a two dimensional fixed point space. These are listed below.

Isotropy subgroup	Fixed point space	Periodic state
a) $\tilde{\mathbb{Z}}_3 = \{(\gamma, -\frac{2\pi}{3})\}$	$(z_1,0)$	rotating wave.
b) $\mathbb{Z}_2 = \{(id,0),(\kappa,0)\}$	(z_1,z_1)	standing wave.
c) $\mathbb{Z}_2^c = \{(id,0),(\kappa,\pi)\}$	$(z_1,-z_1)$	standing wave.

It is a trivial, but nevertheless important, observation that the vectorfield leaves a fixed point space invariant. As the above spaces are two dimensional, we can apply in each of them Liapunov's Center Theorem to (2.1).

This yields three branches of periodic solutions with the symmetry of the corresponding isotropy subgroup.

a) For the rotating wave the invariance under $\tilde{\mathbb{Z}}_3$ means that

$$y_i(t) = y_{i-1}(t-T/3), \text{ where T is the period.}$$

b) For the standing wave invariant under $\mathbb{Z}_2$ we find that:

$$y_1 = 0; \qquad y_2(t) = -y_3(t).$$

c) For the standing wave invariant under $\mathbb{Z}_2^c$ we find that

$$y_1(t) = -y_1(t+\tfrac{T}{2}), \qquad y_2(t) = -y_3(t+\tfrac{T}{2}).$$

All orbits that bifurcate from the equilibrium state are related to these three by the group action of D_3 and/or a time shift. For instance, applying a reflection to a) yields a wave that rotates the other way around

a') $\quad y_i(t) = y_{i-1}(t+\tfrac{T}{3}).$

Clearly the solutions b) and c) are invariant for reflection about the first particle (apart from time shift in case c). Applying other reflections or rotations one finds solutions that are invariant for reflection with respect to one of the other particles.

Solutions satisfying a)-c) can be given a standard form as follows. a) yields immediately that the wave can be written with one scalar function ξ and a constant a as

(2.4a) $\quad y_i = a + \xi(t+i\tfrac{T}{3}), \quad \overline{\xi} = 0.$

The overbar denotes time averaging. In case b) one obtains similarly

(2.4b) $\quad y_1 = 0, \quad y_2 = a+\xi, \quad y_3 = -a-\xi, \quad \overline{\xi} = 0.$

To find a simple form in case c) introduce the symbol L for the reflection about the first particle

$$(Ly)_1 = -y_1, \quad (Ly)_2 = -y_3, \quad (Ly)_3 = -y_2.$$

Then c) reads $(Ly)(t) = y(t+\tfrac{T}{2})$. Shifting time over $\tfrac{T}{2}$ one finds, using the periodicity of y, $y(t) = (Ly)(t+\tfrac{T}{2})$. Adding these two equalities one obtains

$$(Ly+y)(t) = (Ly+y)(t+\tfrac{T}{2}).$$

This implies that Ly+y is constant, and especially $y_2 - y_3$ is constant. Characterisation c) above yields $\overline{y}_1 = -\overline{y}_1$, $\overline{y}_2 = -\overline{y}_3$ so that finally

(2.4c) $\quad y_1 = \eta, \quad y_2 = \xi+a, \quad y_3 = \xi-a, \quad \overline{\xi} = \overline{\eta} = 0.$

Expressions (2.4a-c) are obtained on the basis of a)-c) only. We now derive further restrictions from the fact that the equations (2) are invariant for time reversal and from the uniqueness of the solutions a)-c).

Consider a solution y(t) of a). Then y(-t) is in the family a'), which is found from a) by applying L. Consequently there is a t_0 so that

$$(2.5) \qquad (Ly)(t+t_0) = y(-t) \quad .$$

Inserting (2.4a) one obtains $a+\xi(t) = -a-\xi(-t+t_0)$ from which it follows that the time origin can be chosen so that ξ is odd

$$(2.6a) \qquad \xi(t) = -\xi(-t), \qquad a = 0.$$

For a remark about the condition a = 0 see below. Time reversed solutions in the cases b) and c) must yield a solution in each family respectively. Consequently in case b) we have $a+\xi(t) = a+\xi(-t+t_0)$ so that we may choose

$$(2.6b) \qquad \xi(t) = \xi(-t).$$

For the last case c) one finds similarly

$$(2.6c^1) \qquad \xi(t) = \xi(-t), \qquad \eta(t) = \eta(-t).$$

Since it also holds that $\eta(t) = -\eta(t+\frac{T}{2})$ and the same for ξ, both functions are odd about $\frac{T}{4}$.

REMARK. Our system has one more symmetry, viz a continuous shift of the coordinates $y_i \to y_i + x$. This implies that to any solution a constant can be added. For the above analysis this has two consequences. The requirement (2.5) is no longer correct: due to the fact that in (2.5) an arbitrary constant (d,d,d) has to be added. Consequently, the condition a = 0 in (2.6a) has to be dropped. In case c) we may require as extra condition that $y_1 + y_2 + y_3 = 0$, so that

$$(2.6c^2) \qquad \xi = -\frac{\eta}{2} \ .$$

3. Global Variational Methods

To extend the local bifurcation results of the foregoing section to global branches of finite amplitude solutions we will use the following variational formulation. On the set of 2π-periodic vector functions x in the first Sobolev space $H = H^1_{per}([0,2\pi],\mathbb{R}^3)$ we consider the functionals

$$(3.1) \qquad K(x) = \frac{1}{2}\int \dot{x}^2 d\tau \quad \text{and} \quad U(x) = \int V(x)d\tau$$

(integration over $[0,2\pi]$) and look for critical points of U on level sets of K. For given $R \in \mathbb{R}_+$ we describe this critical point problem by

$$(3.2) \qquad \mathrm{crit}\{U(x)\,|\,K(x) = R,\ x \in H\}.$$

Such critical points satisfy (under mild conditions) the equations

$$(3.3) \qquad -\ddot{x} = \lambda\frac{\partial V}{\partial x}$$

for some multiplier $\lambda \in \mathbb{R}$, and periodic boundary conditions.

Provided that λ is positive, a simple time-scaling shows that a solution of (3.3) corresponds to a periodic solution of (2.1) with period $\sqrt{\lambda}$. Therefore, periodic solutions of (2.1) will be obtained by proving the existence of critical points of (3.2) in a direct way.

This procedure has been used in this form for the first time by Berger [1,2] to obtain a whole family of periodic solutions, parametrized with the parameter R.

Instead of (3.2) variational principles can be formulated where the period or the total energy are prescribed and treated as parameters. Although closely related to (3.2) the critical points of the functionals in these formulations are generally more difficult to characterize than for (3.2) (see van Groesen [6,7,8,9]).

Returning to (3.2), as is often the case for such variational principles, non-trivial periodic solutions of (2.1) are genuine saddle-points for (3.2). To prove existence, topological methods may occasionally be used. Probably also for the case under consideration (see Rabinowitz e.a. [10] and the references therein for such topological methods which require certain index theories connected to the symmetry of the problem).

Here we prefer to explore a different way (following Berger [1,2], Van Groesen [6,7,8,9] and Valkering [11,12]), which will provide a very explicit characterization of the three different wave motions indicated in the foregoing section.

The general idea is to find a subset X of H such that, first of all, the Euler-Lagrange equation (3.3) is not altered by the restriction to X; for obvious reasons such a set X is called a natural constraint or a naturally constrained set. Moreover, we will be able to find such sets X for which U can in fact be maximized on the level sets of K in X. Hence, a periodic solution of (2.1) is then characterized by a solution of the maximization problem

$$(3.4) \qquad \sup\{U(x) \mid K(x) = R,\ x \in X\},$$

which solution corresponds to a saddle point of (3.2). (Note that the minimization problem for (3.2) always provides a trivial solution x = constant.)

The construction of a naturally constrained set X is by no means straightforward and no general theory is available yet. For the case under consideration we will exploit the fact that we know the symmetries of the system in the construction of the sets X. A consequence of the symmetries is that for each of the three different wave forms described in section 2, the set X of 3-vector functions depend in fact only on a scalar function (cf. (2.6a-c^2)). Explicitly:

$$(3.5)\qquad \begin{aligned} X_a &= \{x = (\xi(t),\xi(t-\tfrac{2\pi}{3}),\xi(t-\tfrac{4\pi}{3}))\,|\,\xi(t) = -\xi(-t),\ \xi \in \hat{H}\},\\ X_b &= \{x = (0,\xi(t)+a,-\xi(t)-a)\,|\,a \in \mathbb{R},\ \xi(t) = \xi(-t),\ \xi \in \hat{H}\},\\ X_c &= \{x = (-\tfrac{2}{3}\xi(t),\tfrac{1}{3}\xi(t)+a,\tfrac{1}{3}\xi(t)-a)\,|\,a \in \mathbb{R},\ \xi(t) = \xi(-t),\ \xi \in \hat{H}\}, \end{aligned}$$

where $\hat{H} = \{\xi \in H^1_{per}([-\pi,\pi],\mathbb{R})\,|\int_{-\pi}^{\pi} \xi = 0\}$.

Simple calculations show that the expresssions for K and U on the sets X_i, $i \in \{a,b,c\}$ reduce to

$$(3.6)\qquad \begin{aligned} K_a &= \tfrac{3}{2}\int \dot{\xi}^2 &&, U_a(\xi) = 3\int_{-\pi}^{\pi} [W(\xi(t)-\xi(t-\tfrac{2\pi}{3}))],\\ K_b &= \int \dot{\xi}^2 &&, U_b(\xi,a) = \int_{-\pi}^{\pi} [2W(\xi(t)+a)+W(-2\xi(t)-2a)], \quad \text{and}\\ K_c &= \tfrac{1}{3}\int \dot{\xi}^2 &&, U_c(\xi,a) = \int_{-\pi}^{\pi} [W(\xi(t)+a)+W(-2a)+ W(-\xi(t)+a)] . \end{aligned}$$

So far we have exploited the symmetries of the problem and reduced the problem (3.2) to a scalar problem.

In a completely standard way, functional analytic methods show that in case a) the maximization problem has a solution for each $R > 0$:

$$\sup\{U_a(\xi)\,|\,K_a(\xi) = R,\ \xi \in \hat{H}\}.$$

The maximization problem for cases b) and c), however, is not well posed (take e.g. ξ fixed, $a \to \infty$). In fact, the behaviour of U with respect to the constant a and with respect to the constrained function ξ, leads one to look for a saddle point of the following explicit form:

$$(3.7)\qquad \sup_{\xi}\ \inf_{a\in\mathbb{R}} \{U_i(\xi,a)\,|\,K_i(\xi) = R,\ \xi \in \hat{H}\}, \qquad i \in \{b,c\}.$$

Under mild conditions on W this minimax problem has indeed a solution. For instance, if W is a convex function, the minimization over $a \in \mathbb{R}$ (for given $\xi \in \hat{H}$) has a unique solution. If we write, for simplicity, $U_i(\xi,a) = \int_{-\pi}^{\pi} \hat{W}_i(\xi,a)$ this optimal value of a is determined by the equation $\int_{-\pi}^{\pi} \frac{\partial \hat{W}_i}{\partial a}(\xi,a) = 0$. This equation, considered as an additional constraint on ξ, can be shown to be a natural constraint (see the papers of Berger and Van Groesen).

Exploiting this fact, the minimax problem (3.7) reduces again to a maximization problem:

$$\sup_{\xi} \left\{ \int_{-\pi}^{\pi} \hat{W}_i(\xi,a) \,\middle|\, K_i(\xi) = R, \int_{-\pi}^{\pi} \frac{\partial \hat{W}_i}{\partial a}(\xi,a) = 0, \ \xi \in \hat{H} \right\}, \qquad i \in \{b,c\}.$$

Elementary methods suffice to prove the existence of a solution of this problem.

In this way we have proved the existence of global branches of 3 different kind of waves by characterizing each element on a branch as a solution of an explicit minimization problem. Besides being able to prove existence, such an extremal characterization gives some additional information. In fact, from the extremal property it follows in a straightforward way that the solutions obtained have in fact 2π as minimal period (in the rescaled time), and not $\frac{2\pi}{k}$ for some $k \in \mathbb{N}$, $k \geq 2$. Moreover, the extremal formulation makes one to study the solution analytically for limiting values of R.

In the case $R \downarrow 0$ one obtains the local bifurcation results of section 2 and if R approaches its maximal value one studies the hard core interaction (Valkering [1982]). Another application is the construction of efficient numerical methods of intermediate values of R.

References

1. M.S. Berger, "On periodic solutions to second order Hamiltonian systems", J. Math. Anal. Appl. 29 (1970) 512-522.
2. M.S. Berger, "Periodic solutions of second order dynamical systems and isoperimetric variational problems", Ame. J. Math. 93 (1971) 1-10.
3. S.A. van Gils and T. Valkering, "Hopf bifurcation and Symmetry: standing and travelling waves in a circular chain", Japan Journal of Appl. Math. 2 (1986) 207-222.
4. M. Golubitsky and I. Stewart, "Hopf bifurcation in the presence of symmetry", Arch. Rational Mech. Anal. 87 (1985) 107-165.
5. M. Golubitsky and I. Stewart, "Hopf bifurcation with dihedral group symmetry: coupled nonlinear oscillators", Contempory Mathematics 56 (1986) 131-173.
6. E. van Groesen, "Duality between period and energy of certain periodic Hamiltonian motions", J. London Math. Soc. 34 (1986) 435-478.
7. E. van Groesen, "Existence of multiple normal mode trajectories on convex energy surfaces of even, classical Hamiltonian systems", J.D.E. 57 (1985) 70-89.
8. E. van Croesen, "Analytical minimax methods for Hamiltonian brake orbits of prescribed energy", J. Math. Anal. Appl. (1988) 131.

9. E. van Groesen, "On small period, large amplitude normal modes of natural Hamiltonian systems", Nonlin. Anal. TMA 10 (1986) 41-53.

10. P.H. Rabinowitz, A. Ambrosetti, I. Ekeland and E.J. Zehnder, "Periodic solutions of Hamiltonian systems and related topics", D. Reidel, Dordrecht (1987).

11. T.P. Valkering, "Periodic permanent waves in an anharmonic chain with nearest-neighbour interaction", J. Phys. A 11 (1978) 1885-1897.

12. T.P. Valkering, "Periodic travelling waves in a non-integrable one-dimensional lattice", Cel. Mech. 28 (1982) 119-131.

Part III

Multibody Dynamics and Control

Multibody System Dynamics
Mathematical Theory and Industrial Applications

W. Schiehlen

Institute B of Mechanics, University of Stuttgart,
Pfaffenwaldring 9, D-7000 Stuttgart 80, Fed. Rep. of Germany

1. Introduction

The dynamics of multibody systems represent a typical interaction of mechanics and mathematics. Multibody systems are introduced as a discrete mathematical model of a real mechanical system resulting in a set of ordinary differential equations to be solved by numerical methods. Therefore, a dynamicist requires profound knowledgde of theoretical mechanics and applied mathematics. In this paper the method of multibody systems and some applications will be treated.

On the basis of the mechanical principles of Newton, Euler, D'Alembert, Lagrange and Jourdain the method of multibody systems has been developed during the last decades. The scientific progress achieved has been represented by proceedings of international symposia edited by Magnus [1], Haug [2] and Bianchi and Schiehlen [3]. Further, the contributions of Wittenburg [4], Schiehlen [5] and Roberson and Schwertassek [6] have to be mentioned.

In addition to the theoretical results computerized formalisms for the generation of equations of motions have been established. Numerical formalisms like ADAMS [7] are to combine the generation and solution of the equations of motion. Symbolical formalisms like NEWEUL [8] supply the equations of motion independent of the integration routine used. Symbolical equations of motion offer a possibility of mechanical interpretation, too.

2. Kinematics of Multibody Systems

Multibody systems are defined by rigid bodies with inertia as well as springs, dampers and actively controlled servomotors without inertia, Fig. 1. The bodies are interconnected by rigid bearings or any other kind of support and subject to additional applied forces and torques. Therefore, the method of multibody systems is well

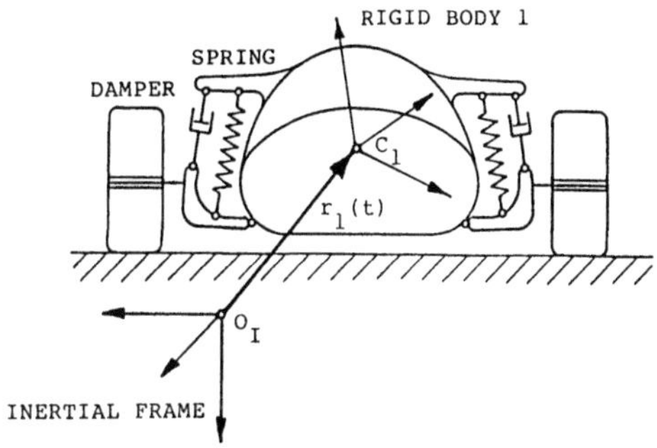

Figure 1. Elements of multibody systems

qualified for the modeling of mechanical systems with complex geometry and great differences in stiffness usually resulting in motions with frequencies less than 50 Hz.
According to the free body principle, each rigid body of the mechanical system is treated separately and all elements without inertia are replaced by forces. The system's position is given relative to an inertial frame by the 3x1-translation vector $r_i(t)$ of the center of mass C_i and the 3x3-rotation tensor $S_i(t)$ written down for each body of the system, $i = 1(1)p$.

A free system of p bodies without any mechanical constraint holds 6p degrees of freedom. Thus, the position of the system can be uniquely described by 6px1-position vector

$$x(t) = [x_1 \; x_2 \; \dots \; x_{6p}]^T . \tag{1}$$

Typical generalized coordinates of a free system are translational coordinates, Euler angles or relative distances.

Then, the system's position can also be represented by

$$r_i = r_i(x) \quad , \quad S_i = S_i(x) \quad , \; i = 1(1)p . \tag{2}$$

Further, the translational und rotational velocity and acceleration, respectively, of the system are found by differentation with respect to the inertial frame as 3x1-vectors:

$$v_i = H_{Ti}(x) \; \dot{x}(t) \quad , \; \omega_i = H_{Ri}(x) \; \dot{x}(t) \quad , \tag{3}$$

$$a_i = H_{Ti}(x) \; x(t) + \frac{\partial v_i}{\partial x^T} \; \dot{x}(t) \quad ,$$

$$\alpha_i = H_{Ri}(x)\ x(t) + \frac{\partial \omega_i}{\partial x^T}\ \dot{x}(t) \quad . \tag{4}$$

The 3x6p-matrices are Jacobians introduced for abbreviation. For more details see Ref. [5].

A <u>holonomic system</u> of p bodies and q holonomic, scleronomic constraints due to rigid bearings results in $f = 6p-q$ positional degrees of freedom. The constraint equation and its derivative

$$x = x(y) \quad , \quad \dot{x} = I(y)\ \dot{y} \tag{5}$$

represent an explicit relation between the 6px1-position vector $x(t)$ and the reduced fx1-position vector

$$y(t) = [\ y_1\ y_2\ \cdots\ y_f]^T \tag{6}$$

summarizing the generalized coordinates of the holonomic system. From (2) and (5) it follows for the system's position

$$r_i = r_i(y) \quad , \quad S_i = S_i(y) \quad , \quad i = 1(1)p \quad , \tag{7}$$

and the accelerations read as

$$a_i = J_{Ti}(y)\ \ddot{y}(t) + \frac{\partial v_i}{\partial y^T}\ \dot{y}(t) \quad ,$$

$$\alpha_i = J_{Ri}(y)\ \ddot{y}(t) + \frac{\partial \omega_i}{\partial x^T}\ \dot{y}(t) \quad . \tag{8}$$

The 3xf-Jacobian matrices can be also obtained from

$$J_{Ti} = H_{Ti}\ I \quad , \quad J_{Ri} = H_{Ri}\ I \tag{9}$$

reducing the computational work in some cases.

Additionally to the holonomic constraints, in <u>nonholonomic systems</u> there exist r nonholonomic constraints. The resulting number of motional degrees of freedom is $g = f-r$. The nonholonomic constraint equation and its derivative

$$\dot{y} = \dot{y}(y,z) \quad , \quad y = K(y,z)\ \dot{z} + \frac{\partial \dot{y}}{\partial y^T}\ \dot{y} \tag{10}$$

show the relation between the fx1-velocity vector $\dot{y}(t)$ and the reduced gx1-velocity vector

$$z(t) = [z_1\ z_2\ \ldots\ z_g]^T \tag{11}$$

characterizing the generalized velocities of the nonholonomic system. From (3), (5) and (10) it follows for the system's velocity

$$v_i = v_i(y,z) \quad , \quad \omega_i = \omega_i(y,z) \tag{12}$$

and for the system's acceleration it remains

$$a_i = L_{Ti}(y,z)\ \dot{z}(t) + \frac{\partial v_i}{\partial y^T}\ \dot{y} \ ,$$

$$\alpha_i = L_{Ri}(y,z)\ \dot{z}(t) + \frac{\partial \omega_i}{\partial y^T}\ \dot{y} \ . \tag{13}$$

The nonholonomic constraints are rarely found in engineering mechanics. Nevertheless, the approach of the generalized velocities can be applied to holonomic systems with great advantage. Generalized velocities are widely known from gyrodynamics and they are also used by Kane and Levinson [9].

The inertial reference frame as introduced above may complicate the choice of the generalized coordinates, the formulation of the forces and the generation of the equations of motion. Then, a moving reference frame is the adequate solution. The related variations in the kinematical equations are presented in detail in Ref. [5].

3. Newton-Euler Equations

For the application of Newton's and Euler's equations to multibody systems, the free body principle has to be used again. For each body of the system these equations read as

$$m_i \, a_i = f_i^a + f_i^r \quad , \; i = 1(1)p \tag{14}$$

$$I_i \, \alpha_i + \tilde{\omega}_i \, I_i \, \omega_i = l_i^a + l_i^r \; . \tag{15}$$

The inertia is represented by the scalar mass m_i and the 3x3-inertia tensor I_i with respect of the center of mass C_i of each body. The forces and torques are 3x1-vectors, all torques have to be related to the center of mass C_i of each body. The applied forces f_i^a and torques l_i^a, respectively, depend on the motion by physical laws. The reaction forces f_i^r and the reaction torques l_i^r, respectively, are due to the constraints given by (5) and/or (10).

The <u>proportional forces</u> are characterized by the system's position and time functions:

$$f_i^a = f_i^a(x,t) \; . \tag{16}$$

Conservative forces due to gravity and springs as well as purely time-varying forces are proportinal forces.

The <u>proportional-differential forces</u> depend on position and velocity:

$$f_i^a = f_i^a(x,\dot{x},t) \tag{17}$$

A parallel spring-damper configuration is a typical example for this class of forces.

<u>The proportional-integral forces</u> are a function of the position and integrals of the position:

$$f_i^a = f_i^a(x,w,t) \; , \quad \dot{w} = \dot{w}(x,w,t) \quad . \tag{18}$$

Here the hx1-Vector $w(t)$ describes the position integrals. E.g. serial spring-damper configurations and the eigendynamics of servomotors result in proportional-integral forces.

The reaction forces and torques originate from bearings and supports. They can be reduced to the generalized constraint forces summarized in a (q+r)x1-vector as

$$g(t) = [g_1\ g_2\ \dots\ g_{q+r}]^T \ . \tag{19}$$

Then, it yields

$$f_i^r = F_i(y,z)\ g(t),\ l_i^r = L_i(y,z)\ g(t) \tag{20}$$

where F_i, L_i are $3x(q+r)$-distribution matrices. The generalized constraint forces are characteristic design parameters of bearings and supports. The distribution matrices can be found by geometrical considerations, too.

The Newton-Euler equations of the global system are summarized in matrix notation as follows. The inertia properties are written in the 6px6p-diagonal matrix

$$\bar{\bar{M}} = \text{diag}\ \{m_1E\ m_2E\ \dots\ I_1\ \dots\ I_p\} \ , \tag{21}$$

the 6px1-force vectors $\bar{q}^c$ and $\bar{q}^a$ represent gyroscopic forces and applied forces, respectively, in the following scheme

$$\bar{q} = [f_1^T\ f_2^T\ \dots\ l_1^T\ \dots\ l_p^T]^T \ . \tag{22}$$

Similar schemes are used for the global 6pxf-matrix $\bar{J}$ and the global 6pxg-matrix $\bar{L}$, respectively as well as for the global 6px(q+r)-distribution matrix $\bar{Q}$.

For holonomic systems from (14), (15), (8), (17), (20) it is obtained

$$\bar{\bar{M}}\ \bar{J}\ \ddot{y}(t) + \bar{q}^c(y,\dot{y}) = \bar{q}^a(y,\dot{y},t) + \bar{Q}\ g(t) \tag{23}$$

and for nonholonomic systems it follows with (10) and (13)

$$\bar{\bar{M}}\ \bar{L}\ \dot{z}(t) + \bar{q}^c(y,z) = \bar{q}^a(y,z,t) + \bar{Q}\ g(t) \tag{24}$$

The Newton-Euler equations represent for all systems 6p scalar algebraic and differential equations. The numerical solution of such equations is not straightforward, further mathematical treatment is recommended.

4. Dynamical Principles

The dynamical principles of D'Alembert and Jourdain result in vanishing virtual work of all constraint forces

$$\delta W = \delta y^T \bar{J}^T \bar{Q} g = 0 \tag{25}$$

and vanishing virtual power

$$\delta P = \delta z^T \bar{L}^T \bar{Q} g = 0 \quad , \tag{26}$$

respectively. Thus, these principles can be used to separate the Newton-Euler equations into purely differential equations and purely algebraical equations for the application of standard solution techniques.

The equations of motion are obtained by premultiplication with the transposed global Jacobian matrix. Then, three advantages are achieved simultaneously: i) symmetrization of the inertia matrix, ii) reduction to minimal order of the differential equation system, iii) elimination of the constraint forces and torques.

Multibody systems are called ordinary if they can be represented by one vector differential equation of the second order with positive definite inertia matrix. Holonomic systems with proportional - differential forces result in ordinary multibody systems. From (23) and (25) the equations of motion are obtained as

$$M(y)\ \ddot{y}(t) + k(y,\dot{y}) = q(y,\dot{y},t) \tag{27}$$

where the fxf-symmetric positive definite inertia matrix M and the fx1-vectors k and q of generalized gyroscopic and applied forces appear.

Multibody systems are called general iff they are not ordinary. Nonholonomic constraints and/or proportional-integral forces produce general multibody systems. From (10), (24) and (26), and (18) the complete set of equations read as

$$\begin{aligned}
&\dot{y} = \dot{y}(y,z), \\
&M(y,z)\ \dot{z}(t) + k(y,z) = q(y,z,w,t) \quad , \\
&\dot{w} = \dot{w}(x,w,t) \quad .
\end{aligned} \tag{28}$$

The number of dynamical equations is further reduced now characterized by the symmetric positive definite gxg-inertia matrix M and the gx1-vectors k and q of the generalized gyroscopic and applied forces.

The <u>equations of reaction</u> are found by premultiplication with the matrix $\bar{Q}^T\bar{\bar{M}}$ according to a proposal by Schramm [10]. E.g., from (23) considering (25), it follows

$$N(y)\, g(t) + \hat{q}(y,\dot{y},t) = \hat{k}(y,\dot{y}) \tag{29}$$

where the symmetric and positive definite qxq-reaction matrix N and the qx1-vectors $\hat{q}$ and $\hat{k}$ of the applied and gyroscopic forces are introduced.

The equations of motion and reaction are strongly simplified if only small motions are considered. A detailed discussion of the linearized equations is given in Ref. [5]. Further, for linear structural vibrations the method of finite element systems and the method of continuous systems can be applied, too. The resulting eigenfrequencies obtained by the method of multibody systems are usually smaller than the real values. Contrarily, the method of finite element systems result in larger eigenfrequencies. For geometrically simple systems with homogeneous stiffness distribution the method of continuous systems can be employed for the identification of the eigenfrequencies.

5. Computerized Derivation of Equations

A main problem in the dynamics of multibody systems is the derivation of the equations of motion and reaction. Computer-aided formalisms represent the adequate solution of the problem.

There are two basic strategies used in computer formalisms. The first strategy applies the Newton-Euler equations according to the free body diagram of the system resulting in

$$\bar{\bar{M}}\,\bar{H}\,x + \bar{q}^c(x,\dot{x}) = \bar{q}^a(x,\dot{x},t) + \bar{Q}\,g(t) \quad . \tag{30}$$

These equations are characterized by a comparatively simple matrix $\bar{H}$ due to the choice of Cartesian coordinates, however, for solution they have to be completed by the constraint equations (5). Thus, a maximal set of 6p+q algebraical and differential equations is obtained which has to be solved numerically in any case.

The second strategy uses immediately the equations of motion (27) and the equations of reaction according to (29). Due to the choice of Lagrangian coordinates a minimal set of 6p-q purely differential equations and q purely algebraical equations are obtained. The partial derivatives connected with Jacobian matrix $\bar{J}$ may be computed symbolically using a formula manipulation routine. for the numerical integration of the symbolical equations of motion any integration routine available can be used.

The first strategy described is implemented in the formalism ADAMS [7] while the second strategy is used by the formalism NEWEUL [8]. The formula manipulation for the equations of motion is in NEWEUL realized by index coding on the basis of FORTRAN 77. This results in an excellent portability of the formalism. Up to date it is not known which of both strategies is more economic. It is conjected that the answer depends on the kind of the application and the computational facilities. However, symbolical equations of motion offer at least the advantage of the mechanical interpretation.

6. Industrial Applications

The method of multibody systems has a broad range of applications in industry and science. Multibody systems are especially qualified for low frequency large displacement motions as found in mechanical and aerospace engineering. Fig. 2 presents typical applications from vehicle dynamics, machine dynamics, robotics and satellite dynamics.

The following steps have to be performed during a dynamical analysis of an industrial problem:

i) Generation of equations of motion (e.g. by NEWEUL),

ii) Integration of equations of motion (e.g. by the routine of SHAMPINE and GORDON [11]),

iii) Animation of motion (computer graphics).

For the animation all necessary information is available from the time-history of the generalized coordinates. For the graphical represenation characteristic points of the surface of each rigid

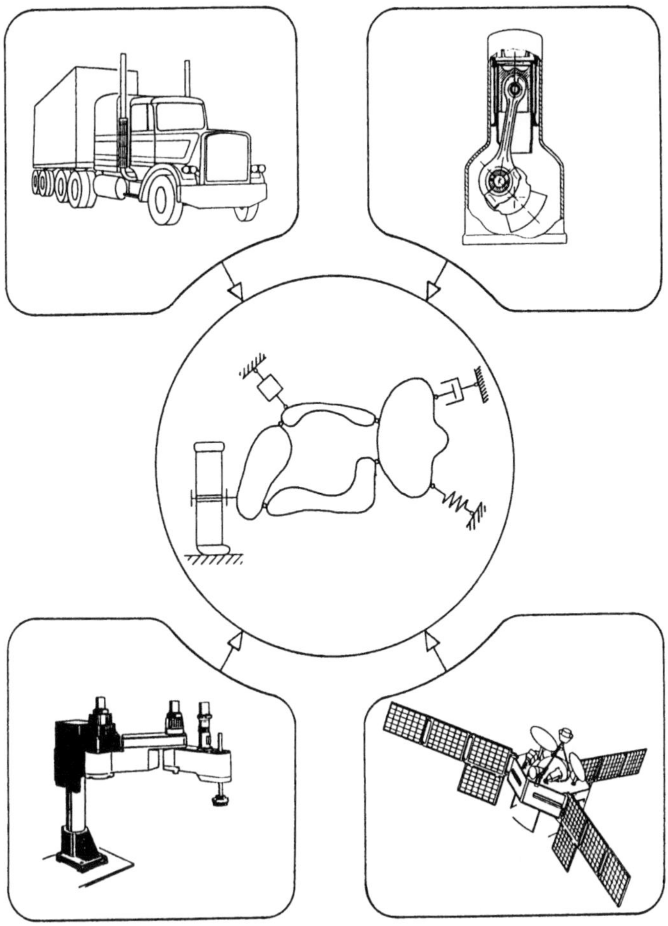

Figure 2. Typical applications of multibody systems

body has to be identified,

$$r_{isurface}(t) = r_i(y) + S_i(y)\rho_{isurface}, \quad i = 1(1)p , \qquad (31)$$

where $r_{isurface}(t_0) = \rho_{isurface}$ follows from the initial position.

The dynamical behavior of a commercial truck has been investigated by Raith [12]. The truck has been modeled as a multibody system with p=8 bodies. Then, all the parameters like masses, moments of inertia, spring and damper coefficients, and the differential equations of the applied tire forces had to be determined. Finally, the motion was presented by computer graphic software, see Fig. 3.

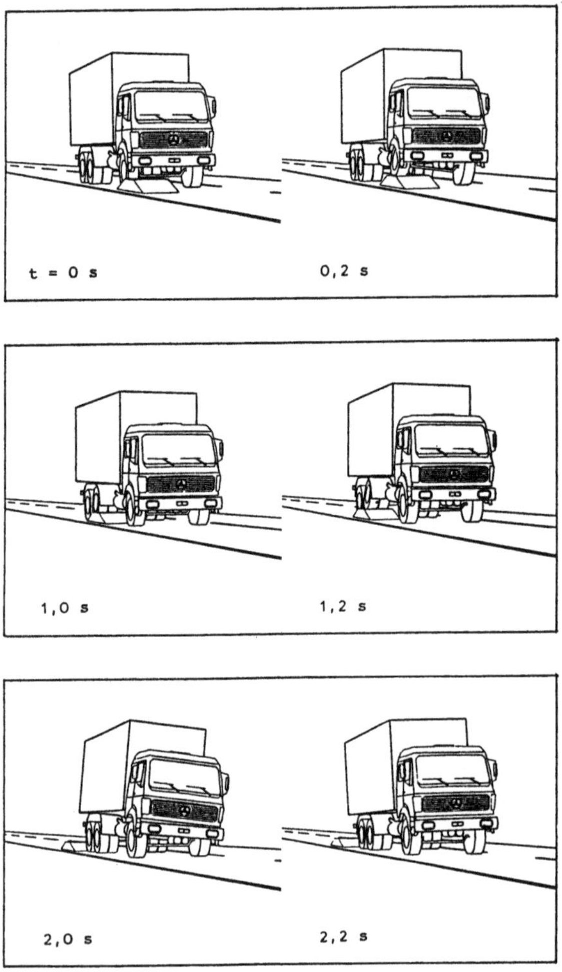

Figure 3. Animation of motion of a commercial truck

7. Conclusion

Mathematical theory and industrial applications are strongly related to each other in the dynamics of multibody systems. Future problems will include computer aided design, hybrid method approach, control by active devices and computer graphics. With respect to the industrial applications the interdisciplinary research has to be extended from mathematics and mechanics to applied informatics. Further, the interaction of the approaches of multibody systems, finite element and boundary element systems, and continuous systems demands more consideration.

Literature

1. K. Magnus, ed.: Dynamics of Multibody Systems (Springer, Berlin/... 1978)

2. E.J. Haug, ed.: Computer Aided Analysis and Opimization of Mechanical System Dynamics (Springer, Berlin/... 1984)

3. G. Bianchi, W. Schiehlen, ed.: Dynamics of Multibody Systems (Springer, Berlin/... 1986)

4. J. Wittenburg: Dynamics of Systems of Rigid Bodies (Teubner, Stuttgart 1977)

5. W. Schiehlen: Technische Dynamik (Teubner, Stuttgart 1985)

6. R.E. Roberson, R. Schwertassek: Dynamics of Multibody Systems (Springer, Berlin/... 1988)

7. ADAMS (Automated Dynamic Analysis of Mechanical Systems). Developed by N. Orlandea, J. Angell, R. Rampalli, Mechanical Dynamics, Ann Arbor, USA.

8. NEWEUL (Newton and Euler based system equations). Developed by E. Kreuzer and W. Schiehlen, University of Stuttgart, Stuttgart, FRG.

9. T.R. Kane, D.A. Levinson: J. Guidance Control Vol. 3, 99 (1980)

10. D. Schramm: Ein Beitrag zur Dynamik reibungsbehafteter Mehrkörpersysteme (VDI-Verlag, Düsseldorf 1986)

11. L.F. Shampine, M.K. Gordon: Computer Solution of Ordinary Differential Equations (Freeman, San Francisco 1975)

12. A. Raith: Modellierung einer Doppelhinterachse und Integration in das Gesamtfahrzeug, DIPL-13 (Institut B für Mechanik, Universität Stuttgart 1986)

A Finite Element Dynamic Analysis of Robotic Manipulators with Flexible Links

J.B. Jonker and C. Keus

Faculty of Mechanical Engineering, Mekelweg 2,
NL-2628 CD Delft, The Netherlands

1. Introduction.

Control of the position and orientation of the end-effector of a manipulator meets with the fundamental problem that the actuators at the joints are separated by the manipulator links, that always are flexible to some extent. When a flexible link manipulator is controlled conventionally by locating every sensor at the actuator (collocated control), then stable control is easy to achieve within a closed-loop bandwidth limited by the first natural frequency of the elastic system, resulting in moderate tracking behaviour of the end-effector. If on the other hand the control forces/torques applied at the joints are based on tip sensor information (non-collocated control) the problem of achieving stability becomes extremely difficult due to nonminimum phase behaviour of the tip [3]. Applying this type of control, however, is an essential step for achieving better control performance. The structural properties of flexible links typically involve the use of a combination of a collocated and a non-collocated control system. The collocated control consists of actuator joint position/rate feedback and is used to stabilize the total closed-loop system. Non-collocated control is applied for feedback of the end-point position error to correct the position and orientation of the end-effector. This control approach which originally has been applied for large space structures [1], has first been applied by CANNON and SCHMITZ [3] for control of the tip of a flexible one-link manipulator. In this paper the above presented control strategy will be applied to a fast moving flexible, three degree of freedom manipulator.

An application is presented of a finite-element based method for the development of (linearized) manipulator models, and for the simulation of flexible manipulator and controller dynamic performance (JONKER [5]). The simulation problem is discussed as an example of tracking a prescribed trajectory of the end-effector.

2. Structure of manipulator control.

The approach that will be followed in this paper is a (standard) two-stage control synthesis [8], consisting of a stage of nominal control input force/torque generation and a stage of perturbed dynamics for system stability and trajectory tracking control. For that purpose the flexible dynamic model is split into a nominal system of low dimension associated with the gross rigid link motion of the manipulator and a linear system linearized about the nominal trajectory. The nominal dynamics stage involves the solution of the inverse problem of kinematics and dynamics, i.e., the determination of the nominal actuator joint coordinates and driving forces for the manipulator with rigid links, to perform the manipulation task. The inverse kinematics deals with the transfer from the nominal trajectory, (path) coordinates ($\mathbf{x}_0{}^p$) and velocities, ($\dot{\mathbf{x}}_0{}^p$) to the nominal actuator joint positions ($\mathbf{e}_0{}^m$) and nominal joint velocities ($\dot{\mathbf{e}}_0{}^m$). The nominal kinematics stage results in a list of nominal coordinates for which a reference input trajectory vector

$$\mathbf{y}_0(t) = [\mathbf{x}_0^{p\,T},\ \mathbf{e}_0^{m\,T},\ \dot{\mathbf{x}}_0^{p\,T},\ \dot{\mathbf{e}}_0^{m\,T}]^T, \tag{1}$$

is specified as function of time t. The vector $\mathbf{y}_0$ contains the setpoint values of the end-effector coordinates and the actuator joint coordinates. Let

$$\mathbf{y}(t) = [\mathbf{x}^{p\,T},\ \mathbf{e}^{m\,T},\ \dot{\mathbf{x}}^{p\,T},\ \dot{\mathbf{e}}^{m\,T}]^T, \tag{2}$$

denote the vector of actual values of the end-point coordinates and the actuator joint coordinates. We will refer to this as the output vector of the manipulator system. The control problem, then is to synthesize a control law for the manipulator such that the output $\mathbf{y}(t)$ tracks the reference input $\mathbf{y}_0(t)$, which implies that the perturbation vector

$$\delta\mathbf{y}(t) = \mathbf{y}_0(t) - \mathbf{y}(t), \tag{3}$$

goes to zero in some specified manner.

Based on the two stage synthesis, the overall control for the manipulator is defined in the form

$$\mathbf{u} = \mathbf{u}_0 + \delta\mathbf{u}, \tag{4}$$

where $\mathbf{u}_0$ is the nominal control input vector and $\delta\mathbf{u}$ is the control input vector synthesized at the stage of perturbed dynamics. The nominal control input is used as feedforward compensation for the gross rigid link motion of the manipulator. The main part of the nonlinear dynamic effects due to changes of the manipulator configuration as well as the dynamic coupling between the rigid link modes is predicted and cancelled by the nominal feedforward compensation.

3. Linearized manipulator models.

Linearized manipulator models are obtained through linearization of the nonlinear manipulator equations of motion around the nominal trajectory. The time varying linearized equations may be converted in piecewise linearized time-invariant equations by dividing the total trajectory time T, in m convenient intervals, where $t=t_i(i=1,2,\ldots,m)$ denote the time at the end of interval i. We approximate the linear time varying system in the interval $(t_i+t_{i-1})/2<t<(t_i+t_{i+1})/2$ by the simplified linear time invariant system at $t=t_i$,[5]

$$\mathbf{M}_0\begin{bmatrix}\delta\ddot{\mathbf{e}}^m\\ \delta\ddot{\boldsymbol{\varepsilon}}^m\end{bmatrix}+\mathbf{K}_0\begin{bmatrix}\delta\mathbf{e}^m\\ \delta\boldsymbol{\varepsilon}^m\end{bmatrix}=-\delta\boldsymbol{\sigma}^m, \tag{5}$$

where the vectors $\mathbf{e}^m$ and $\boldsymbol{\varepsilon}^m$ represent the manipulator joint coordinates and the flexible deformation mode coordinates, respectively. The symmetric matrices $\mathbf{M}_0$ and $\mathbf{K}_0$, evaluated along the nominal trajectory, have the partioned structure

$$\mathbf{M}_0=\begin{bmatrix}\mathbf{M}_0^{ee} & \mathbf{M}_0^{e\epsilon}\\ \mathbf{M}_0^{\epsilon e} & \mathbf{M}_0^{\epsilon\epsilon}\end{bmatrix},\qquad \mathbf{K}_0=\begin{bmatrix}0 & 0\\ 0 & \mathbf{K}^{\epsilon\epsilon}\end{bmatrix}. \tag{6}$$

The matrices $\mathbf{M}_0^{e\epsilon}$ and $\mathbf{M}_0^{\epsilon e}$ represent the dynamic coupling between the gross rigid link motion and the flexible deformation of the links. The corrective driving forces and torques, represented by the vector $\delta\boldsymbol{\sigma}^m$, are applied only at the actuator joints and if the actuator dynamics is not considered, then there is a simple linear relation between the vector of control inputs $\delta\mathbf{u}$ and the vector $\delta\boldsymbol{\sigma}^m$

$$\delta\boldsymbol{\sigma}^m=-\mathbf{B}\delta\mathbf{u},\quad \text{where } \mathbf{B}=\begin{bmatrix}\mathbf{I}\\ 0\end{bmatrix}. \tag{7}$$

Equation (7) represents the input equation and the matrix B is called the input matrix. The formulation must be completed by specifying the output equation, also called the sensor equation. If we assume that both position and velocity sensors are available, then the output equation is of the form

$$\delta \mathbf{y} = \begin{bmatrix} \mathbf{C}_p \\ \mathbf{0} \end{bmatrix} \begin{bmatrix} \delta \mathbf{e}^m \\ \delta \boldsymbol{\varepsilon}^m \end{bmatrix} + \begin{bmatrix} \mathbf{0} \\ \mathbf{C}_v \end{bmatrix} \begin{bmatrix} \delta \dot{\mathbf{e}}^m \\ \delta \dot{\boldsymbol{\varepsilon}}^m \end{bmatrix} . \tag{8}$$

If the components of the output vector **y** represent the positions at the sensor locations collocated with the actuator joint positions, then the output matrix is of the form

$$\mathbf{C}_p = \mathbf{C}_v = \mathbf{B}^T . \tag{9}$$

If on the otherhand the positions of the end-effector are available, then for the case of rigid links the output matrix becomes

$$\mathbf{C}_p = \mathbf{C}_v = \mathrm{DF}_0^{xp} , \tag{10}$$

where DF_0^{xp} is referred to as the Jacobian matrix of the manipulator relating the nominal end-effector velocities $\dot{\mathbf{x}}_0{}^p$ to the nominal actuator velocities $\dot{\mathbf{e}}_0{}^m$, i.e.,

$$\dot{\mathbf{x}}_0^p = \mathrm{DF}_0^{xp} \dot{\mathbf{e}}_0^m . \tag{11}$$

Equations (5), (7) and (8) form the linearized manipulator model which is available to the control designer.

4. Controllability and observability.

For testing controllability and observability of flexible link manipulators the criteria of HAUTUS [4] are applied. TRUCKENBRODT [7], showed that the Hautus criterion for controllability leads to the requirement that constrained modes may not coincide with unconstrained modes. Constrained modes are determined from a modal analysis of the homogeneous system

$$\mathbf{M}_0^{\varepsilon\varepsilon} \delta \ddot{\boldsymbol{\varepsilon}}^m + \mathbf{K}_0^{\varepsilon\varepsilon} \delta \boldsymbol{\varepsilon}^m = \mathbf{0}, \tag{12}$$

which is obtained when the rigid link motion of the manipulator is fixed and the links vibrate about a certain prescribed configuration. Unconstrained or system modes are obtained from a modal analysis of the complete homogeneous system (5). If some of the constrained eigenvalues coincide with the unconstrainted or system eigenvalues, then the system (5,7) is **not** completely controllable. Physically this implies that these modes cannot be influenced by any of the actuators. The treatment of observability is similar in several respects to the above discussion for controllability. If $\mathbf{C}_p = \mathbf{C}_v = \mathbf{B}^T$ (collocated control), then the conditions for observability are identical with those obtained for controllability. If tip position feedback is applied then the rigid link system is observable provided that $DF_0{}^{xp}$ is not singular. For a flexible link manipulator we have the additional condition that a sensor position may not coincide with the nodes of vibration that will be controlled.

5. Actuator joint feedback control.

This control utilizes proportional plus derivative feedback (PD), where the position sensors are collocated with the actuators. The control law is given by

$$\delta \mathbf{u} = -\mathbf{K}_p \delta \mathbf{e}^m - \mathbf{K}_v \delta \dot{\mathbf{e}}^m, \tag{13}$$

where $\mathbf{K}_p$ and $\mathbf{K}_v$ are the position and rate feedback gain matrices. A straightforward way to achieve decoupling for the special case of rigid links is to specify the gain matrices $\mathbf{K}_p$ and $\mathbf{K}_v$ as

$$\mathbf{K}_p = \mathbf{M}_0^{ee} \boldsymbol{\Omega}^2 \tag{14}$$

and

$$\mathbf{K}_v = 2\mathbf{M}_0^{ee} \boldsymbol{\beta}\, \boldsymbol{\Omega}, \tag{15}$$

where

$$\boldsymbol{\Omega}^2 = \begin{bmatrix} \omega_1^2 & & 0 \\ & \ddots & \\ 0 & & \omega_n^2 \end{bmatrix}, \quad \text{and} \quad \boldsymbol{\beta} = \begin{bmatrix} \beta_1 & & 0 \\ & \ddots & \\ 0 & & \beta_n \end{bmatrix}. \tag{16}$$

The diagonal matrices $\boldsymbol{\Omega}$ and $\boldsymbol{\beta}$ contain the desired servo loop frequencies and the corresponding active damping ratios. These control laws were first applied by BOOK [2] to investigate the limitations of rigid link based control laws in designing control schemes for flexible link manipulators. In order to obtain adequate damping of the lowest flexible mode a bandwidth of approximately $\omega_i = 0.5(\omega_1)_c$ and a relative damping $\beta_i \simeq 0.7-0.9$ suffices, where $(\omega_1)_c$ is the lowest frequency of the corresponding constrained mode.

6. End-point feedback control.

In the type of end-point feedback control used, the deviation of the tip positions/rates and the joint deviation rates are fed back according to the control law

$$\delta \mathbf{u} = -\mathbf{K}_{pt} \delta \mathbf{x}^p - \mathbf{K}_{vt} \delta \dot{\mathbf{x}}^p - \mathbf{K}_v \delta \dot{\mathbf{e}}^m. \tag{17}$$

The tip feedback gain matrices $\mathbf{K}_{pt}$ and $\mathbf{K}_{vt}$ are given by

$$\mathbf{K}_{pt} = \mathbf{M}_0^{ee}\boldsymbol{\Omega}^2[DF_0^{xp}]^{-1}, \tag{18}$$

and

$$\mathbf{K}_{vt} = 2\mathbf{M}_0^{ee}\beta_t\boldsymbol{\Omega}\ [DF_0^{xp}]^{-1}, \tag{19}$$

where, DF_0^{xp} is defined in Eq. (11). The achievable closed-loop bandwidth of the tip position feedback is limited by the nonminimum phase behaviour of the system. KEUS [6], shows that for increasing $\mathbf{K}_v$ stability and adequate damping is obtained with a closed-loop bandwidth that is approximately $\omega_i=0.25(\omega_1)_u$, where $(\omega_1)_u$ is the lowest (non-zero) frequency of the corresponding unconstrained mode. The combination of end-point feedback with actuator joint rate feedback should demonstrate some improvement to justify the increased complexity.

7. The simulation model.

Fig. 1 illustrates a finite element representation of the example manipulator. Note that both links and joints are modelled as specific finite elements. The revolute joints of this manipulator are modelled by the hinge elements (3), (5) and (7), where the hinge axis of the elements (5) and (7) are parallel providing for the in-plane motion of the end-effector. The in-plane motion is driven by the linear actuators (1) and (2). The three major links of the manipulator are referred to as the shoulder (link 4), upperarm (link 6) and forearm (link 8). The shoulder can rotate relatively to the inertial reference frame (x,y,z) about the vertical axis of hinge element (3). This element is driven by a torque source. The links (10) and (11) are rigidly connected with the upperarm. In the

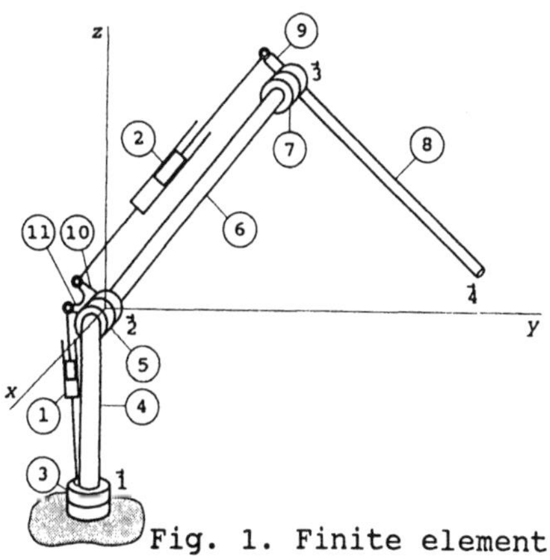

Fig. 1. Finite element representation of example manipulator.

Table 1. Kinematic and dynamic link parameters.

link number	4	6	8	9	10	11
length ℓ [m]	0.4	0.7	0.7	0.1	0.1	0.1
bending stiffness EI/ℓ^3 [N/m]	∞	82500	13000	∞	∞	∞
torsion stiffness GI_p/ℓ^3 [N/m]	∞	62750	∞	∞	∞	∞
mass per unit length m [kg/m]	0	4	2	0	0	0

presented configuration of Fig. 1 the position of the end-effector is determined by the vector $\mathbf{x}^4=[0,\ 1\text{m},\ 0.1\text{m}]^T$. The upperarm and the forearm have a uniform circular cross section and are assumed to be flexible. Referring to Fig. 1, table 1 lists the kinematic and dynamic properties of the links. The lumped mass of the bearing assembly and the end-effector mass are respectively m^3=10kg and m^4=30kg. The gravity loads are evaluated as an applied force in the negative z-direction at each mass, equal in magnitude to the weight of the masses. Each individual link is modelled by a single beam element. This corresponds to 9 flexible degrees of freedom (ε^m) in addition to the actuator joint coordinates e^1, e^2 and e^3 representing the elongation of the linear actuators (1), (2) and the relative rotation of actuator (3). The actuators are modelled as pure force/torque sources without dynamics. For a detailed describtion of the finite elements the reader referred to [5]. The mode shapes with the corresponding natural frequencies (ω_u,ω_c) for the unconstrained and constrained manipulator conditions are determined for three different manipulator configurations as shown in table 2. One can observe from this table that the lowest natural frequencies (ω_c) for both the in-plane and the out-of-plane motion are more or less invariant with respect to the manipulator configuration and vary slightly about 30 rad/sec. Fig. 2 shows a contour plot of the effective working space for the manipulator. When controller designs are applied with a bandwidth beyond the lowest constrained natural frequencies, special attention must be given to the unconstrained and constrained mode shapes of the out-of-plane motion which correspond to the same natural frequency. Fig. 2 indicates two shaded regions within the working space where the second unconstrained mode has a frequency $(\omega_2)_u$ which differs less than ten percent from the corresponding frequency of the second constrained mode $(\omega_2)_c$ (region I) and the first constrained mode $(\omega_1)_c$ (region II). Inside of region I there is a subregion,

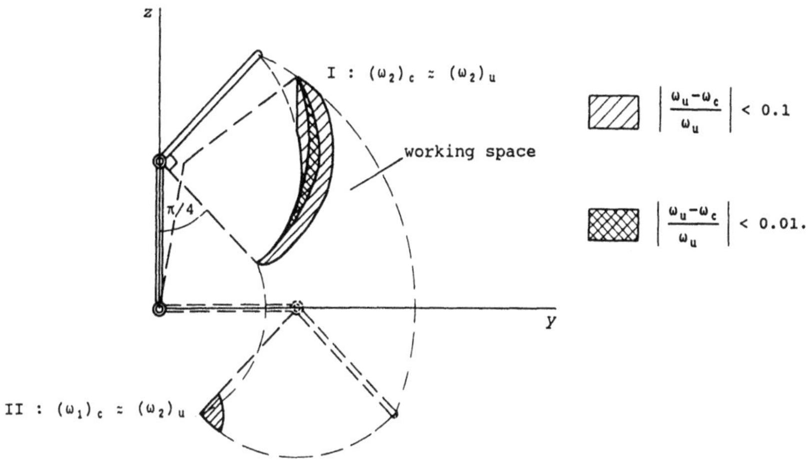

Fig. 2. Working space and ill-controllable regions.

Table 2. Mode shapes and natural frequencies ω_u, ω_c [rad/sec]

	I		II		III	
IN-PLANE MODES (CONSTRAINED)	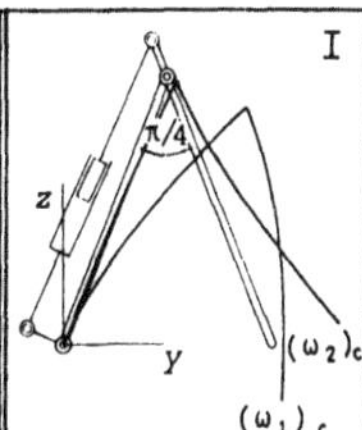		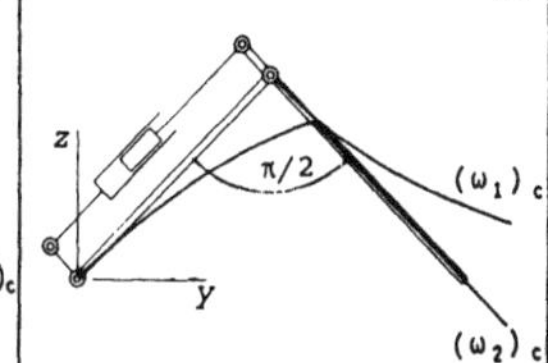		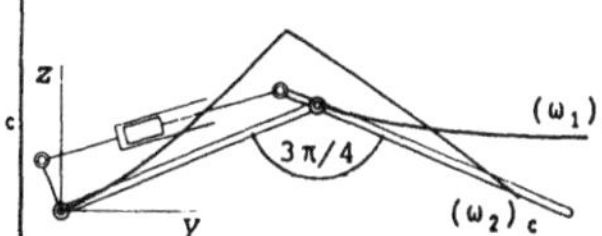	
mode nr.	1	2	1	2	1	2
ω_u	0	0	0	0	0	0
ω_c	33	106	36	77	36	100
OUT-OF-PLANE MODES (CONSTRAINED)	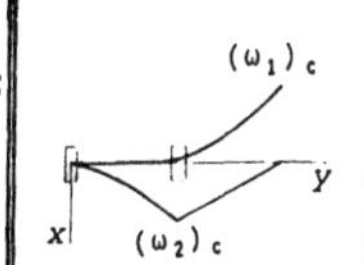		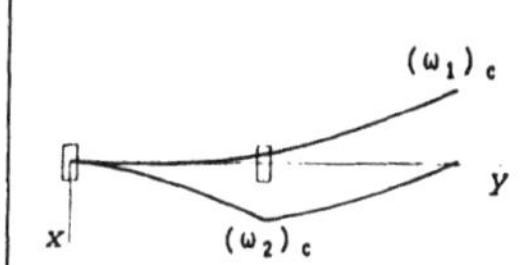		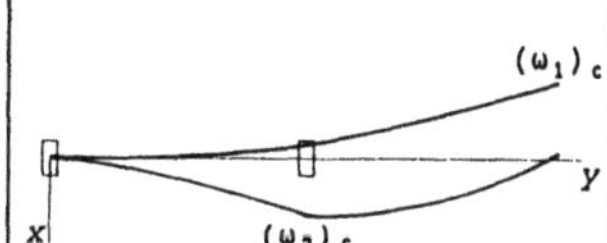	
mode nr.	1	2	1	2	1	2
ω_u	0	84	0	90	0	103
ω_c	31	149	27	156	25	182

indicated by a double shading, where the difference is even less than one percent. For the regions indicated, it is difficult to design stabilizing controllers for the out-of-plane motion that incorporate some of the uncontrollable modes.

8. Simulation results.

The manipulation task implies to transfer the manipulator tip along a straight line with a trapeziodal velocity profile (Fig. 3). The initial and final manipulator configuration correspond with the con-

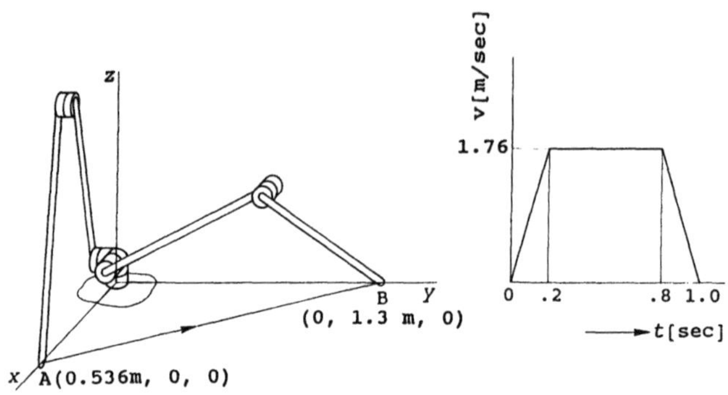

Fig. 3. Motion trajectory and velocity profile.

figurations I and III in table 2. Note that the trajectory path is chosen outside of the ill-controllable regions indicated in Fig. 2. Because of the configuration dependence of the inertia properties, both the mass matrix $\mathbf{M}_0^{ee}$ and the diagonal matrix $\boldsymbol{\Omega}$ in the equations for the gain matrices $\mathbf{K}_p$, $\mathbf{K}_v$ in Eqs. (14,15) and the matrices $\mathbf{K}_{pt}$, $\mathbf{K}_{vt}$ in Eqs. (18,19) must be updated during the simulation. However, it was already indicated from table 2 that the lowest natural frequency $(\omega_1)_c$ is almost invariant with respect to the manipulator configuration. Therefore it is sufficient to update only the mass matrix $\mathbf{M}_0^{ee}$ and to take a constant magnitude for the servo loop frequencies, being the diagonal elements of the matrix $\boldsymbol{\Omega}$. Tables 3 and 4 list the numerical values of the elements of the matrices $\boldsymbol{\Omega}$ and $\boldsymbol{\beta}$ used for the collocated joint feedback controller and for the non-collocated tip controller. The matrices $\boldsymbol{\Omega}$, $\boldsymbol{\beta}$ and $\boldsymbol{\beta}_t$ are selected by examining root loci for various feedback gain values [6].

Table 3. Feedback gain matrices $\boldsymbol{\Omega}$ and $\boldsymbol{\beta}$ (collocated joint control).

actuator	collocated joint control	
	$\boldsymbol{\Omega}$ =	$\boldsymbol{\beta}$ =
1	15	.85
2	15	.85
3	15	.85

Table 4. Feedback gain matrices $\boldsymbol{\Omega}$, $\boldsymbol{\beta}$ and $\boldsymbol{\beta}_t$ (non-collocated tip control).

actuator	non-collocated tip control		
	$\boldsymbol{\Omega}$ =	$\boldsymbol{\beta}$ =	$\boldsymbol{\beta}_t$ =
1	50	.04	.85
2	50	.04	.85
3	20	.3	.55

Note that for the in-plane motion the closed-loop frequencies (ω_1,ω_2) of the non-collocated control are far beyond the lowest constrained natural frequency $((\omega_c)_1 \approx 30$rad/sec) of the system. They are limited by the maximum allowable deformation of the links. However, the increase in the position bandwidth for the out-of-plane motion is much smaller since it is limited by the non-minimum phase behaviour. The trajectory has been performed by taking 13 linearizations. The initial condition disturbances at the joint positions are determined by $\delta e^1=0$, $\delta e^2=-0.01$m and $\delta e^3=-0.005$ rad, corresponding with initial tip position deviations $\delta x^4=0.098$m, $\delta y^4=-0.0022$m and $\delta z^4=0.033$m. In Fig. 4 the trajectory tracking errors of the tip/joint positions and the actuator force/torque tracking are shown for the collocated joint feedback control and for the non-collocated tip control. It is obvious from Fig. 4 that the collocated joint control is not able to bring the system sufficiently close to the desired path. Also deviations of the tip position caused by bending of the links due gravity forces can be observed at the time interval between t=0.2 sec. and t=0.8 sec. The non-collocated tip controller yields a rapid transient response. The tracking forces of actuators (1) and (2) in the case of end-point control are 10 times as large as the tracking forces in the case of joint feedback control for these actuators. The tip position deviations at t=0.2 sec and t=0.8 sec of the out-of-plane direction are caused by the transitions of the velocity profile. The transient response of the joint position deviations of the actuators (1) and (2) show the whip effect which appears to be a condition for fast tip positioning. The gravity effect on the position deviation of actuator (2) is clearly illustrated at the time interval between t=0.2 sec. and t=0.8 sec.

9. Some concluding remarks.

For the flexible link manipulator considered fast and precise closed-loop tip-position control has been achieved by applying simple control laws for the in-plane motion. The achievable bandwidth is limited to $\omega_i \approx 0.25(\omega_1)_u$ where $(\omega_1)_u$ represents the lowest (non-zero) frequency of the corresponding mode. The ultimate improvement for the in-plane motion is limited by the maximum allowable deformation of the links. However, with the application of these control laws no significant improvement could be obtained for the out- of plane direction. The improvement for the out-of-plane motion is limited by the non-minimum phase behaviour.

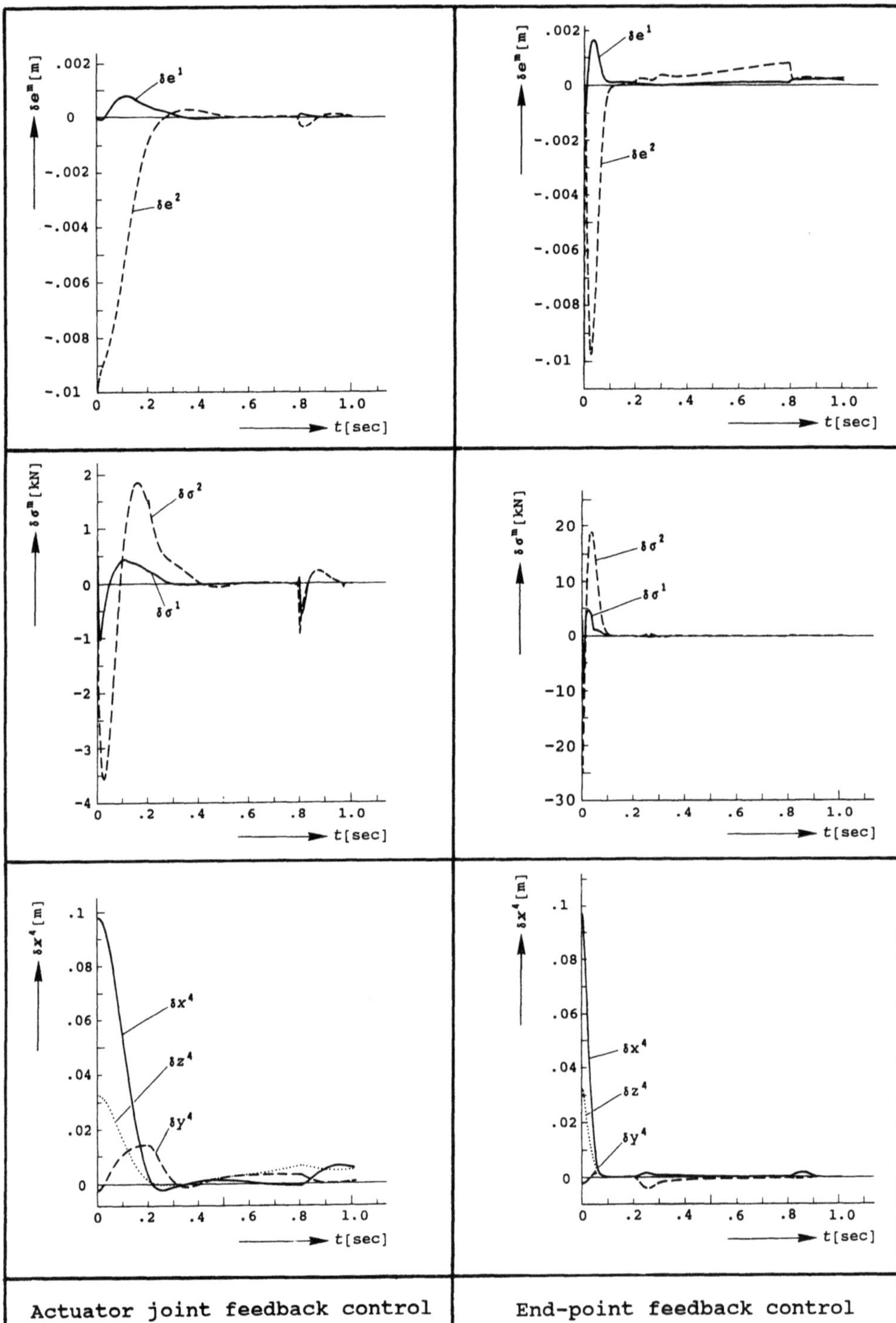

Fig. 4. Trajectory tracking errors of tip/joint positions and actuator force tracking errors of actuators 1 and 2

<u>References.</u>

[1] Aubrun, J.N., "Theory of the Control of Structures by Low-Authority Controllers", Journal of Guidance and Control, Vol. 3, nr. 5, 1980.

[2] Book, W.J., Maizza-Neto, O. and Whitney, D.E., "Feedback Control of Two Beam, Two Joint Systems with Distributed Flexibility", ASME Journal of Dynamic Systems, Measurement and Control, Vol. 97, nr. 4, 1975.

[3] Cannon, R.H. and Schmitz, E., "Initial Experiments on the End-point Control of a Flexible One-Link Robot", Int. J. Robotics Research, 3, 3 (1984).

[4] Hautus, M.L.J., "Controllability and Observability Conditions of linear Autonomous Systems", Indagationes Mathematicae 31, 1969, pp. 443-448.

[5] Jonker, J.B., "A Finite Element Dynamic Analysis of Flexible Spatial Mechanisms and Manipulators", Doctors Thesis, Delft University of Technology, Faculty of Mechanical Engineering, 1988.

[6] Keus, C., "End-point Position Control of Flexible Link Manipulators", Engineers Thesis, Lab. for Measurement and Control, Delft Univ. of Techn., The Netherlands, report nr. A- 434, 1988.

[7] Truckenbrodt, A., "Bewegungsverhalten und Regelung Hybrider Mehrkörpersysteme mit Anwendung auf Industrieroboter", Fortschritt-Berichte der VDI zeitschriften, Reihe 8, Nr. 33, 1980. Verlag des Vereins Deutsher Ingenieure.

[8] Vukobratovic, M. and Stokic, D., "Control of Manipulation Robots", Springer Verlag, Berlin, 1982.

Part IV

Mechanics and Mathematics of Non-Classical Materials

Mathematical Modelling for Non-Classical Materials

V. Tvergaard

Department of Solid Mechanics, The Technical University of Denmark, DK-2800 Lyngby, Denmark

Current developments in material modelling are discussed, with special emphasis on damage and final failure. For ductile metals a model that accounts for the nucleation and growth of microscopic voids is presented. Here, a finite strain formulation is necessary to represent fracture by void coalescence. For a single phase polycrystalline ceramic an analysis of stress singularities due to elastic anisotropies of the grains, and their effect on micro-cracking, is discussed.

1. Introduction

The application of mathematical models to describe the behaviour of materials is classical in mechanics. Most models have focussed on the deformations of a solid under given loads, e.g. elasticity, plasticity, visco-elasticity or creep. However, in the last ten years there has been an increasing interest in also modelling damage, including the effect of damage on deformations, and final material failure. Thus, the general understanding of material behaviour has been much improved.

The application of materials such as metals, ceramics, polymers or fiber reinforced composites in engineering is no novelty; but the properties of these materials have been strongly improved, and detailed mathematical models have helped, both in the material development and in the effective use of the materials. The types of phenomena that have been modelled include cavitation or micro-cracking in metals, crazing in polymers, phase transformations or micro-cracking in ceramics, and delamination, fiber breakage or matrix cracking in composites.

In the present paper some of these new features of materials mechanics will be illustrated by two examples. First, an elastic-plastic material model is discussed that accounts for the nucleation and growth of micro-voids in ductile metals (GURSON [1,2]). This model has been developed on the basis of a number of micro-mechanical studies, and has been further extended to account for void coalescence (TVERGAARD and NEEDLEMAN [3]). The plastic dilatancy that is the apparent macroscopic effect of porosity results in early loss of ellipticity of the governing equations, which has a significant effect on final failure.

The second example concerns micro-cracking in ceramics. These materials have superior properties in a number of respects, such as resistance to wear or to creep at high temperatures; but brittle fracture gives a severe limitation to the applicability. The effect of crystal anisotropies on microcracking in the grain boundaries has been analysed (EVANS [4]; TVERGAARD and HUTCHINSON [5]), to get a better understanding of the grain size dependence of cracking.

2. Porous Ductile Materials

Classical elastic-plastic material models rely on the assumptions of normality of the plastic flow rule, incompressibility of plastic flow and insensitivity to a superposed hydrostatic pressure, which are in good agreement with experimental observations for metals (HILL [6]), and also agree with the micro-mechanics based on slip in single crystals. If the material contains small holes, these holes can grow during plastic deformation, resulting in compressibility and pressure sensitivity of the macroscopic plastic flow. Models for such materials will be discussed here.

Porosities in ductile materials play a significant role at ductile fracture, where microscopic voids nucleate from second phase particles and subsequently grow to coalescence; but also the effect of porosities in powder compacted metals has attracted much interest. GURSON [1] has focussed on the ductile fracture aspect and has used micro-mechanical studies to derive an approximate yield condition of the form $\Phi(\sigma^{ij},\sigma_M,f) = 0$, where f is the void volume fraction, σ^{ij} is the average macroscopic Cauchy stress tensor, and σ_M is an equivalent tensile flow stress representing the actual microscopic stress-state in the matrix material. Approximate yield conditions of this form have also been proposed for powder metallurgical materials on a purely empirical basis (e.g. see SHIMA and OYANE [7]).

The physical model used by GURSON [1] is a unit cube similar to that of BISHOP and HILL [8], but considering an aggregate of voids and ductile matrix rather than a polycrystalline aggregate. The macroscopic strains are defined in terms of the displacements on the surface of the unit cube, and the macroscopic stresses are defined as the area average of the microscopic stresses over the appropriate face of the unit cube. Then, BISHOP and HILL [8] have proved that these macroscopic stresses and strains are work conjugate, and that they satisfy a maximum plastic work principle on the macroscale. BERG [9] has pointed out that the analysis can be extended to apply to the void-matrix aggregate, and GURSON [1] has used approximate upper-bound solutions on the microscopic level to draw similar conclusions for macroscopic aggregate behaviour.

The approximate yield condition derived by GURSON [1] on the basis of a rigid-plastic upper bound analysis for a spherical volume element with a concentric spherical void is of the form

$$\Phi = \frac{\sigma_e^2}{\sigma_M^2} + 2q_1 f^* \cosh\left\{\frac{\sigma_k^k}{2\sigma_M}\right\} - \left[1 + (q_1 f^*)^2\right] = 0 \tag{2.1}$$

for $q_1 = 1$ and $f^* = f$. Here, $\sigma_e = (3s_{ij}s^{ij}/2)^{½}$ is the macroscopic effective Mises stress, and $s^{ij} = \sigma^{ij} - G^{ij}\,\sigma_k^k/3$ is the stress deviator. The effect of f on the yield condition (2.1) is illustrated in Fig. 1.

A Lagrangian formulation of the field equations is used in this section, with the metric tensors in the current configuration and the reference configuration denoted by G_{ij} and g_{ij} , respectively, the Lagrangian strain tensor $\eta_{ij} = \frac{1}{2}(G_{ij} - g_{ij})$, and the contravariant stress components σ^{ij} on current base vectors. Indices range from 1 to 3, and the summation convention is adopted for repeated indices.

The plastic part of the macroscopic strain-rate for the porous ductile material is taken to be given by

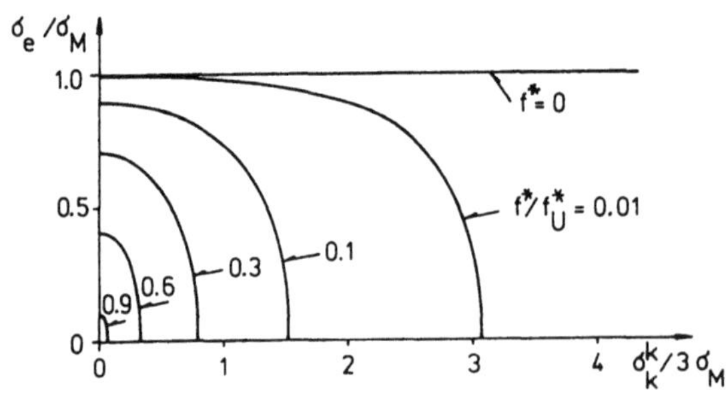

Fig. 1. Yield surface dependence on the hydrostatic tension for various values of the function f^* in (2.1).

$$\dot{\eta}^{P}_{ij} = \Lambda \frac{\partial \Phi}{\partial \sigma^{ij}} \quad (2.2)$$

since normality of the plastic flow rule for the matrix material implies macroscopic normality (BERG [9]; BISHOP and HILL [8]).

The rate of growth of the void volume fraction is written on the form

$$\dot{f} = (\dot{f})_{growth} + (\dot{f})_{nucleation} \quad (2.3)$$

As the matrix material is plastically incompressible, the increment due to growth is given by

$$(\dot{f})_{growth} = (1-f)\ G^{ij}\dot{\eta}^{P}_{ij} \quad (2.4)$$

NEEDLEMAN and RICE [10] have proposed the following expression for the increment due to nucleation

$$(\dot{f})_{nucleation} = \mathcal{A}\ \dot{\sigma}_M + \mathcal{B}\ (\sigma^k_k)^{\bullet}/3 \quad (2.5)$$

which can describe experimentally observed nucleation mechanisms.

It is assumed that the rate of equivalent plastic work in the matrix material equals the macroscopic rate of plastic work

$$\sigma^{ij}\dot{\eta}^{P}_{ij} = (1-f)\sigma_M \dot{\epsilon}^{P}_{M} \quad (2.6)$$

Then, using the uniaxial stress-strain relationship $\dot{\epsilon}^P_M = (1/E_t - 1/E)\dot{\sigma}_M$, the increment of the equivalent tensile flow stress in the matrix material is found as

$$\dot{\sigma}_M = \frac{EE_t}{E - E_t} \frac{\sigma^{ij}\dot{\eta}^{P}_{ij}}{(1-f)\sigma_M} \quad (2.7)$$

Now, the value of the parameter Λ in (2.2) is determined from the consistency condition $\dot{\Phi} = 0$, required during plastic loading. Substituting (2.3)-(2.5) and (2.7) into this condition gives

$$\dot{\eta}^{P}_{ij} = \frac{1}{H} m^G_{ij} m^F_{k\ell} \overset{\nabla}{\sigma}{}^{k\ell} \quad (2.8)$$

where $\overset{\nabla}{\sigma}{}^{k\ell}$ is the Jaumann co-rotational rate of the Cauchy stress tensor, H is a scalar hardening function, and

$$m_{ij}^{G} = \frac{3}{2}\frac{s_{ij}}{\sigma_M} + \alpha\, G_{ij} \quad , \quad m_{ij}^{F} = \frac{3}{2}\frac{s_{ij}}{\sigma_M} + \beta\, G_{ij} \tag{2.9}$$

$$\alpha = \frac{f^*}{2} q_1 \sinh\left[\frac{\sigma_k^k}{2\sigma_M}\right] \quad , \quad \beta = \alpha + \frac{\mathcal{B}\sigma_M}{6}\frac{\partial\Phi}{\partial f} \tag{2.10}$$

Continued plastic loading requires $\Phi = 0$ and $m_{k\ell}^{F}\, \overset{\nabla}{\sigma}{}^{k\ell}/H \geq 0$ (for more details, see TVERGAARD [11]).

Finally, assuming that the total strain rate is the sum of an elastic part and the plastic part (2.8), the incremental stress-strain relationship is found of the general form

$$\dot{\tau}^{ij} = \left[\mathcal{L}^{ijk\ell} - \mu\, M_G^{ij} M_F^{k\ell}\right]\dot{\eta}_{k\ell} \tag{2.11}$$

where τ^{ij} is the Kirchhoff stress tensor, $\mathcal{L}^{ijk\ell}$ are elastic instantaneous moduli, and μ is non-zero during plastic yielding.

According to (2.9) the normal, m_{ij}^{F}, of the yield surface differs from the direction, m_{ij}^{G}, of the plastic strain increment in cases where $\beta \neq \alpha$ (i.e. when $\mathcal{B} \neq 0$) . Thus, while voids are nucleating with $\mathcal{B} \neq 0$ the current yield surface differs from the surface $\Phi = 0$, which is characterized as non-normality of the plastic flow rule. This type of behaviour has the significant consequence that HILL's [12] theory for the uniqueness and bifurcation in elastic-plastic solids does not apply. Upper and lower bounds to the point, at which uniqueness of the incremental solution is lost, have been suggested by RANIECKI and BRUHNS [13], but it appears that particularly the lower bound is a rather crude approximation. It is noted that such non-normality of the plastic flow rule is also found for geological materials (see RUDNICKI and RICE [14]).

Loss of ellipticity of the governing equations, leading to the formation of localized shear bands and subsequent shear fracture, is very sensitive to deviations from the classical elastic-plastic material model [14,15]. Thus, both the macroscopic plastic dilatancy and the non-normality of plastic flow exhibited by the porous ductile material model tend to significantly reduce the critical strain for the onset of shear localization. Consequently, the model helps explain the formation of shear bands often observed in experiments for various ductile materials.

The prediction of ductile fracture in structural elements or test specimens is the main purpose of developing a material model that accounts for the nucleation and growth of voids to coalescence. This ability of the model is illustrated by a numerical analysis of necking and failure in a round tensile test specimen carried out by TVERGAARD and NEEDLEMAN [3]. Initially the bar has essentially uniform thickness and no voids, but after some straining bifurcation into a necking mode takes place, and subsequently nucleation and growth of voids starts to occur in the localized neck region.

Figure 2 shows contours of constant void volume fraction f in the final stages of the test. It is seen that coalescence of voids leads to the formation of a penny-shaped crack that grows from the centre of the neck. At some stage the direction of growth starts to zig-zag, leading to a final fracture mode that is in very good agreement with the well known cup-cone fracture observed experimentally. Thus, it appears that the material model gives a realistic description of actual material behaviour. This conclusion has recently been confirmed by more detailed comparisons of theoretical and experimental results.

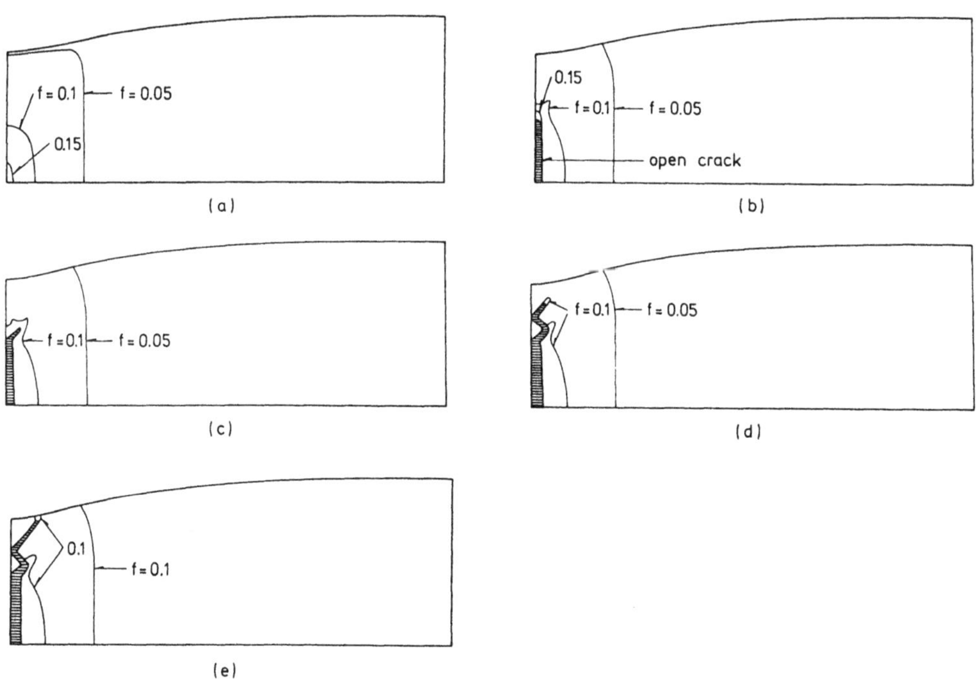

Fig. 2. Curves of contant void volume fraction in round tensile bar during final failure [3].

3. Micro-Cracking in Ceramics

Advanced ceramic materials are mostly produced by powder compaction at high temperatures. When the sintered material is cooled from the fabrication temperature, appreciable thermal residual stresses develop in the material matrix, and these residual stresses contribute significantly to the micro-cracking that characterizes the failure process in many brittle polycrystalline ceramics. The cracks occur at the particle-matrix interface in two phase ceramics or along the grain boundaries for single phase ceramics.

Analyses of the grain boundary stresses that develop in non-cubic materials due to thermal expansion anisotropy have been carried out by EVANS [4] and FU and EVANS [16]. They considered a group of isotropic elastic, planar, hexagonal grains embedded in an infinite isotropic elastic matrix, which allowed for a closed form solution. It was found that the stresses near a triple grain junction exhibit a logarithmic singularity, and this is important in considering micro-crack nucleation.

Recently, TVERGAARD and HUTCHINSON [5] have studied the effect of elastic anisotropy as well as thermal expansion anisotropy of the grains. Here, a planar array of hexagonal grains is used as a 2D model of the polycrystalline solid (Fig. 3a). Each grain is assumed to be elastic orthotropic, but the orientation of the principal axes of the orthotropy (x_i^*) differs from grain to grain. On the principal axes the nonzero in-plane components of the tensor $C_{ijk\ell}$ of orthotropic elastic moduli are denoted by $C_{11} = C_{1111}$, $C_{22} = C_{2222}$, $C_{12} = C_{1122}$ and $C_{44} = C_{1212}$, and the corresponding principal values of the thermal expansion coefficients are denoted by $\alpha_1 = \alpha_{11}$ and $\alpha_2 = \alpha_{22}$. The stress-strain relations are of the form

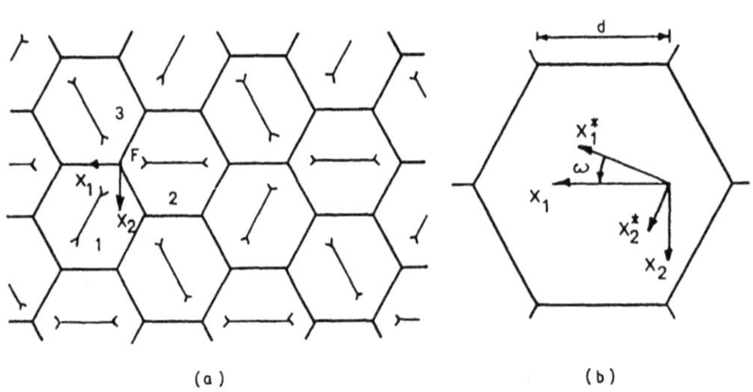

Fig. 3. Planar array of hexagonal grains, with principal axes x_α^* of the anisotropies. The direction of the x_1^*-axis is indicated in each grain.

$$\sigma_{ij} = C_{ijk\ell}(\epsilon_{k\ell} - \epsilon^T_{k\ell}) \quad , \quad \epsilon^T_{k\ell} = -\Delta T \, \alpha_{k\ell} \tag{3.1}$$

where the thermal contraction $\epsilon^T_{k\ell}$ results from cooling below the fabrication temperature, and ΔT denotes the cooling range.

A particular crystal anisotropy considered is defined by

$$C_{11} = f \, C^I_{11} \; , \; C_{22} = \frac{1}{f} \, C^I_{22} \; , \; C_{12} = C^I_{12} \; , \; C_{44} = C^I_{44} \tag{3.2}$$

where C^I_{11} , C^I_{22} etc. denote isotropic elastic constants. The amount of thermal expansion anisotropy is specified by

$$\Delta\alpha = \frac{1}{2}(\alpha_1 - \alpha_2) \tag{3.3}$$

Thus, the grain anisotropies are specified by the values of two parameters f and $\Delta\alpha$. Furthermore, in the special model aggregate considered there are three different crystal orientations, specified by the angles $\omega_2 = 0^\circ$ and $\omega_3 = -\omega_1 = 60^\circ$, corresponding to the grains numbered 1, 2 and 3, respectively, in Fig. 3a. For the purpose of analysis it is convenient to rotate the stress-strain relations (3.1) to the axes of an appropriate global coordinate system.

The stress singularity at a grain boundary triple-point junction is analysed for the particular triple-point junction denoted by F in Fig. 3a. Using a Williams-type [17] singularity analysis, the stress variation in the near vicinity of point F is assumed to be of the form

$$\sigma_{ij} = k \, r^\lambda \, \tilde{\sigma}_{ij}(\theta) \tag{3.4}$$

with the associated variation of the displacement components

$$(u,v) = k \, r^{\lambda+1} \left[\tilde{u}(\theta) \; , \; \tilde{v}(\theta)\right] \tag{3.5}$$

Here, r and θ are polar coordinates centered at the junction, and k is an amplitude factor. Substituting (3.5) into the equilibrium equations, and the appropriate jump conditions at the grain boundaries, gives a set of homogeneous ordinary differential equations for $\tilde{u}(\theta)$ and $\tilde{v}(\theta)$, and λ is found as an eigenvalue that allows for non-trivial solutions of these homogeneous equations.

The variation of the eigenvalues with elastic anisotropy of the type specified by (3.2) is shown in Fig. 4, and Fig. 5 shows an

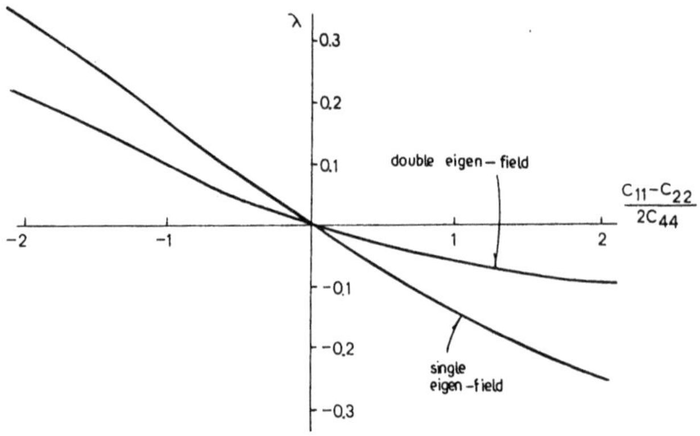

Fig. 4. Exponent of the r-variation of the two most singular, physically relevant eigenfields around point F [5].

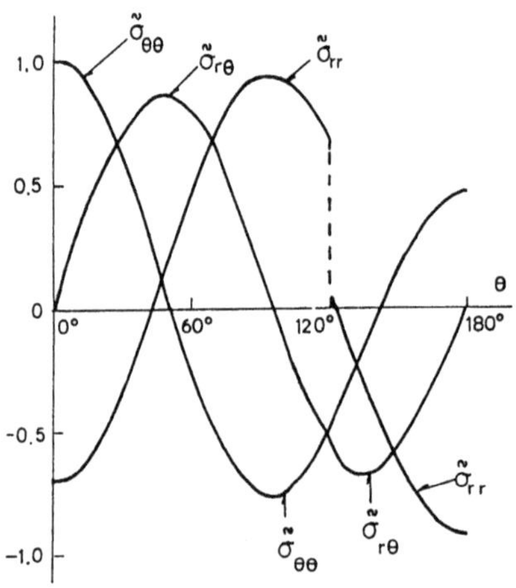

Fig. 5. Variation of $\tilde{\sigma}_{ij}(\theta)$ for the symmetric field associated with the double eigenvalue at $(C_{11} - C_{22})/2C_{44} = 2$, where $\lambda = -0.091$.

example of the corresponding variation of the stress components $\tilde{\sigma}_{ij}(\theta)$. It is seen in Fig. 4 that the stresses associated with the eigenfields are unbounded $(\lambda < 0)$ at the junction when $C_{11} > C_{22}$, whereas for $C_{11} < C_{22}$ the stresses go to zero at the junction $(\lambda > 0)$. These examples of grain configurations near a triple-point junction are far from being exhaustive, but they do suggest that a triple-point junction can be a vicinity of significant stress concentration, which will enhance the possibility of micro-crack nucleation.

The value of the amplitude factor k is not determined by the eigenvalue problem. This value depends on the stress field away from the triple-grain junction and can be determined by matching (3.4) and (3.5) with this stress field. In [5] the type of stress singularity encountered, represented by the value of λ , is the main interest.

The condition for the onset of grain boundary micro-fracture depends on the size of small defects in the boundary. The most critical location of such defects is at a triple-grain junction where a high stress peak occurs. The critical size of such defects is analysed numerically in [5] by considering a small crack of length $2a$ (see Fig. 6), extending from a triple-grain junction as that denoted by F in Fig. 3a.

The stress intensity at the crack tip is characterized by the path-independent J-integral [18], defined by

$$J = \int_{\Gamma} (Wn_1 - \sigma_{ij}n_j u_{i,1})ds \quad , \quad W(\epsilon_{k\ell}) = \int_0^{\epsilon_{k\ell}} \sigma_{ij} d\epsilon_{ij} \tag{3.6}$$

Here, u_i are the displacement components, Γ is any counterclockwise contour surrounding the crack tip, n_j is the outward normal of Γ, and ds is the arc length along Γ. The small crack will grow along the grain boundary, when the value of J exceeds the grain boundary fracture energy for the particular ceramic material considered.

For the aggregate subject to uniaxial plane strain tension, $\Sigma_{22} > 0$, $\Sigma_{11} = 0$, $\Delta T = 0$ (where Σ_{ij} are the average stresses), elastic anisotropy of the crystals is found to affect the J-value, so that a defect of a given length $2a$ is much more critical at a triple-grain junction corresponding to a negative λ-value according to Fig. 4, where the stresses would be unbounded even in the absence of a small grain boundary defect.

The influence of thermal expansion anisotropy is illustrated in Fig. 7, for the aggregate subject to cooling $\Delta T > 0$, $\Sigma_{ij} = 0$. It is seen that the value of J/a increases for a decreasing value of the ratio $2a/d$ ($\bar{E}$ is the uniaxial plane strain tension modulus of the aggregate). Thus, a defect of a given length $2a$ is more critical the larger the grains are. This agrees perfectly with experimental observations [19,20], which show that cooling from the fabrica-

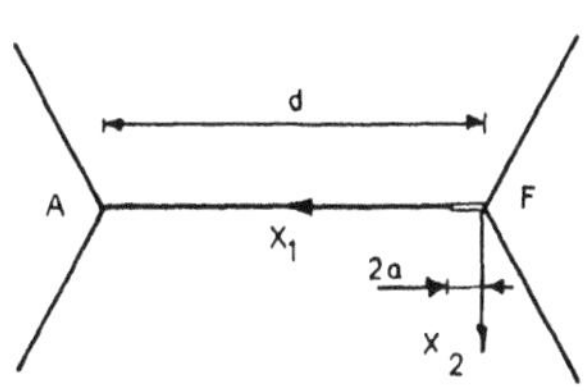

Fig. 6. Grain boundary crack of length $2a$ at the triple point F.

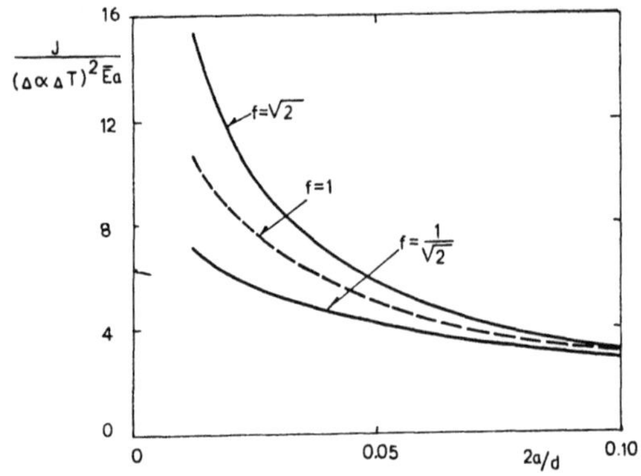

Fig. 7. J-integral as a function of crack length for ceramic material with anisotropic grains, subject to thermal contraction [5].

tion temperature can result in spontaneous cracking along grain boundaries, if the grains are relatively large, while fine grained materials do not crack. It is also noted that if the defects embedded in the residual stress fields alone are close to the critical size, a superposed external stressing of the polycrystalline ceramic will easily lead to micro-cracking.

The type of behaviour found in Fig. 7 for isotropically elastic crystals $(f = 1)$ agrees with the results of EVANS [4]. As is seen in the figure, elastic anisotropies have a significant influence on the J-values found. Thus, for $f = \sqrt{2}$ the critical size of a defect that will result in a fully cracked grain boundary facet is significantly smaller than that for $f = 1$. As would be expected this agrees with the result of the singularity analysis, where $C_{11} > C_{22}$ (i.e. $f > 1$) results in unbounded stresses at the triple-grain junction.

The type of micro-cracking discussed above would correspond to closely spaced cracks all over the aggregate and would make the material useless. However, in a very small region of high stress concentration such micro-cracking can have the beneficial effect of lowering the macroscopic stress peak and thus increasing the overall strength of the material. The toughening of ceramics associated with a micro-crack process zone around the tip of a macroscopic crack has attracted a great deal of interest recently [21,22,23].

References

1. A.L. Gurson: J. Engng. Mater. Tech. 99, 2 (1977).
2. A.L. Gurson: In Proc. Int. Conf. Fracture, ed. by D.M.R. Taplin, Vol. 2A (Pergamon Press, Oxford 1977) 357.

3. V. Tvergaard, A. Needleman: Acta Metall. 32, 157 (1984).
4. A.G. Evans: Acta Metall. 26, 1845 (1978).
5. V. Tvergaard, J.W. Hutchinson: Micro-cracking in ceramics induced by thermal expansion or elastic anisotropy. Danish Center for Appl. Math. and Mech., Report No. 351 (1987).
6. R. Hill: The mathematical theory of plasticity (Clarendon Press, Oxford 1950).
7. S. Shima, M. Oyane: Int. J. Mech. Sci. 18, 285 (1976).
8. J.F.W. Bishop, R. Hill: Phil. Mag. 42, 414 (1951).
9. C.A. Berg: In Inelastic Behaviour of Solids, ed. by M.F. Kanninen *et al.*, (McGraw-Hill, New York 1970) 171.
10. A. Needleman, J.R. Rice: In Mechanics of Sheet Metal Forming, ed. by D.P. Koistinen *et al.*, (Plenum Press, Oxford 1978) 237.
11. V. Tvergaard: J. Mech. Phys. Solids 30, 399 (1982).
12. R. Hill: J. Mech. Phys. Solids 6, 236 (1958).
13. B. Raniecki, O.T. Bruhns: J. Mech. Phys. Solids 29, 153 (1981).
14. J.W. Rudnicki, J.R. Rice: J. Mech. Phys. Solids 23, 371 (1975).
15. J.R. Rice: In Theoretical and Appl. Mech., ed. by W.T. Koiter (North Holland, Amsterdam 1977) 207.
16. Y. Fu, A.G. Evans: Acta Metall. 33, 1515 (1985).
17. M.L. Williams: J. Appl. Mech. 19, 526 (1952).
18. J.R. Rice: J. Appl. Mech. 35, 379 (1968).
19. J.A. Kuszyk, R.C. Bradt: J. Am. Ceram. Soc. 56, 420 (1973).
20. R.W. Rice, R.C. Pohanka: J. Am. Ceram. Soc. 62, 559 (1979).
21. M. Ortiz: J. Appl. Mech. 54, 54 (1987).
22. J.W. Hutchinson: Crack tip shielding by micro-cracking in brittle solids. To appear in Acta Metall. (1987).
23. M. Rühle, A.G. Evans, R.M. McMeeking, P.G. Charalambides, J.W. Hutchinson: Microcrack toughening in alumina/zirconia. To be published (1987).

Nonstandard Modelling of Certain Quasi-Stationary Debonding Processes in Laminates

C. Woźniak

Department of Mathematics, Comp. Sci. and Mechanics,
University of Warsaw, Warsaw, Poland

In this contribution we show how certain homogenized models of quasi-stationary debonding processes in linear-elastic laminates can be obtained. We use the method of microlocal modelling, [1], which is based on the concepts of nonstandard analysis, [2]. The "discrete" model of debonding, proposed in [3], is taken as a starting point of the analysis.

Preliminaries

We shall deal with the multilayered composite (laminate) in the form of a thick shell, which in its undeformed state occupies the region Ω in R^3, cf. Fig. 1. Let Λ_K, $K = 1, \ldots, S\text{-}1$, stand for the sheets of the composite and $\Pi_K = (\partial\Lambda_K \cap \partial\Lambda_{K-1})$, $K = 2, \ldots, S\text{-}1$, be the interlaminae surfaces. We assume that every sheet is homogeneous and linear-elastic; the adjacent layers are bonded together on interfaces by very thin layers of the bonding material. Hence every such bonding layer will be modelled (in the undeformed state) as a surface Π_K, $K = 2, \ldots, S\text{-}1$. We also assume that before the possible delamination, the bonding material behaves as linear elastic but can sustain only restricted values of the interlaminae tractions. Thus, during the deformation, the displacement fields $\boldsymbol{u}(\cdot, \tau)$ ($\tau \in [0, \tau_f]$ is the time coordinate) in the laminae can suffer discontinuities $[\![\boldsymbol{u}(\cdot, \tau)]\!]$ across the interlaminae surfaces Π_K, $K = 2, \ldots, S\text{-}1$. Let $\boldsymbol{T}(x, \tau)$, $\boldsymbol{b}(x, \tau)$, $\varrho(x, \tau)$ stand for the Cauchy stress, body force and mass density, respectively, defined (almost everywhere) on Ω. Moreover,

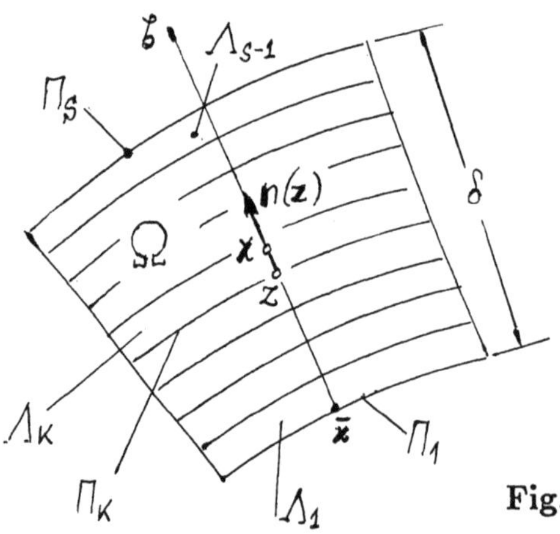

Fig. 1. Multilayered composite

let $t(z,\tau)$ be the interlaminae traction vector field, defined (almost everywhere) on $\cup\, \Pi_K$. For the sake of simplicity assume taht the trace of the displacement field on the boundary $\partial\Omega$ is known. Then for the quasi-stationary processes we obtain the principle of virtual work in the known form

$$\int_{\Omega} \mathrm{tr}\{\boldsymbol{T}\nabla \boldsymbol{v}\}\, dV + \int_{\Pi} \cdot [\![\boldsymbol{v}]\!]\, dA = \int_{\Omega} \varrho \boldsymbol{b}\cdot \boldsymbol{v}\, dV \quad , \quad \Pi \equiv \bigcup_{K=2}^{S-1} \Pi_K \quad , \tag{1}$$

which has to hold for every test function defined (almost everywhere) on Ω, which can suffer discontinuities $[\![\boldsymbol{v}]\!]$ across Π and such that $\boldsymbol{v} = 0$ on $\partial\Omega$. The stress-strain relation in every layer is given by

$$\begin{aligned} &\boldsymbol{T}(x,\tau) = \boldsymbol{C}_K[L\boldsymbol{u}](\boldsymbol{x},\tau) \quad ; \quad L\boldsymbol{u} \equiv \tfrac{1}{2}(\nabla \boldsymbol{u} + \nabla \boldsymbol{u}^T) \quad , \quad \boldsymbol{x} \in \Lambda_K \quad , \\ &K = 1, \ldots, S-1 \quad , \end{aligned} \tag{2}$$

where $\boldsymbol{C}_K$ is the (constant) tensor of elastic modulae in K-th layer. Let

$$\begin{aligned} &\pi_0([\![\boldsymbol{u}]\!]) = \tfrac{1}{2}\gamma_N [\![\boldsymbol{u}]\!]_N^2 + \tfrac{1}{2}\gamma_\tau [\![\boldsymbol{u}]\!]_T^2 - \alpha \quad , \\ &[\![\boldsymbol{u}]\!]_N = [\![\boldsymbol{u}]\!]\cdot \boldsymbol{n} \quad , \quad [\![\boldsymbol{u}]\!]_T \equiv [\![\boldsymbol{u}]\!] - \boldsymbol{n}[\![\boldsymbol{u}]\!]_N \quad , \end{aligned}$$

be the strain energy function of the bonding material before the possible delamination, where γ_N, γ_T and α are known positive reals. Following [3] we assume that the delamination may take place if $\pi_0([\![\boldsymbol{u}]\!](z,\tau)) = 0$. Setting

$$\pi([\![\boldsymbol{u}]\!] \equiv \begin{cases} \pi_0([\![\boldsymbol{u}]\!]) & \text{if } \pi_0([\![\boldsymbol{u}]\!] < 0 \quad , \\ 0 & \text{if } \pi_0([\![\boldsymbol{u}]\!] \geq 0 \quad , \end{cases}$$

and

$$\theta(\boldsymbol{z},\tau) \equiv \begin{cases} 1 \ \text{ if } \pi([\![\boldsymbol{u}]\!](\boldsymbol{z},\sigma)) < 0 & \text{for every } \sigma \in [0,\tau) \quad , \\ 0 \ \text{ if } \pi([\![\boldsymbol{u}]\!](\boldsymbol{z},\sigma)) = 0 & \text{for some } \sigma \in [0,\tau) \quad , \end{cases}$$

we conclude that the interlaminae strain energy function can be assumed as a product $\theta(\boldsymbol{z},\tau)\pi([\![\boldsymbol{u}]\!](\boldsymbol{z},\tau))$. Function $\pi(\,\cdot\,)$ is non smooth and non convex; taking also into account the inpenetrability condition $[\![\boldsymbol{u}]\!]_N(\boldsymbol{z},\tau) \geq 0$ where $[\![\boldsymbol{u}]\!]_N = [\![\boldsymbol{u}]\!]\cdot n$, we obtain the following form of the local interlaminae traction condition

$$\boldsymbol{t}(\boldsymbol{z},\tau) \in \theta(\boldsymbol{z},\tau)\overline{\partial}\pi([\![\boldsymbol{u}]\!](\boldsymbol{z},\tau) + \partial\, \mathrm{ind}_{\overline{R}_+}([\![\boldsymbol{u}]\!]_N(\boldsymbol{z},\tau)) \quad , \quad \boldsymbol{z} \in \Pi \quad , \tag{3}$$

where $\overline{\partial}\pi$ is so called Clarke's generalized gradient, [4]. For the particulars the reader is referred to [3].

It is easy to see that the quasi-stationary model of debonding processes, given by (1–3), is not plausible from the point of view of the possible engineering applications due to the large number S-2 of interlaminae surfaces Π_K on which the displacement fields can suffer discontinuities. The aim of this contribution is to obtain homogenized models of the debonding processes under consideration. To this end we shall apply the method of microlocal (nonstandard) modelling, [1], under the assumption that the composite has a periodic structure in the direction normal to interlaminae surfaces.

Nonstandard Modelling

Now assume that the composite is made of a large number of repeating basic shell units of a thickness ε. Every such unit comprises s thin homogeneous sheets with $\mathbb{C}_a$, ϱ_a, $a = 1, \ldots, s$ as (constant) tensors of elastic modulae and (constant) mass densities, respectively, cf. Fig. 2. Let $\zeta \in [0, \delta]$ be the coordinate shown in Fig. 1.

We have $\zeta = (\boldsymbol{x} - \overline{\boldsymbol{x}}) \cdot \boldsymbol{n}(\boldsymbol{x})$ where $\overline{\boldsymbol{x}}$ is a projection of $\boldsymbol{x}$ onto Π_1, cf. Fig. 1. It is easy to see that the composite under consideration is $(0, \varepsilon)$-periodic in the direction normal to the layers.

Let $\mathcal{P} = \mathcal{P}^{(1)}$ denote a problem of finding displacements $\boldsymbol{u} = \boldsymbol{u}^{(1)}$, strains $\boldsymbol{E} = \boldsymbol{E}^{(1)}$ and stresses $\boldsymbol{T} = \boldsymbol{T}^{(1)}$ on the basis of (1–3) for the periodic composite we deal with. Let $\mathcal{P}^{(n)}$, $n = 2, 3, 4, \ldots$, stand for the problems obtained from $\mathcal{P}$ by substituting $(0, \varepsilon/n)$ periodic structure in the place of $(0, \varepsilon)$-periodic structure; all other entities describing the problem $\mathcal{P}$ remain unchanged. Thus we pass from the periodic composite under consideration to a certain composite with n-times thinner basic layers. Let $\boldsymbol{u}^{(n)}$, $\boldsymbol{E}^{(n)}$, $\boldsymbol{T}^{(n)}$ be the solution of the problem $\mathcal{P}^{(n)}$; we tacitly assume that such a solution exists (cf. [3]). The microlocal modelling (nonstandard homogenization) we are going to apply is based on two fundamental assumptions, namely on the Homogenization Hypothesis and on the Microlocal Approximation Assumption. The Homogenization Hypothesis states that the solution of $\mathcal{P}$ can be approximated by the solution of $\mathcal{P}^{(n)}$ for every positive integer n:

$$\boldsymbol{u} \sim \boldsymbol{u}^{(n)} \quad , \quad \boldsymbol{E} \sim \boldsymbol{E}^{(n)} \quad , \quad \boldsymbol{T} \sim \boldsymbol{T}^{(n)} \quad , \quad n \in N_+ \quad , \tag{4}$$

where if $\boldsymbol{u}$, $\boldsymbol{E}$ and $\boldsymbol{T}$ are taken for some $\zeta = m\varepsilon + \Delta\zeta$ [m is a positive integer and $\Delta\zeta \in (0, \varepsilon)$] then $\boldsymbol{u}^{(n)}$, $\boldsymbol{E}^{(n)}$, $\boldsymbol{T}^{(n)}$ have to be taken for $\zeta = m\varepsilon + k\varepsilon/n + \Delta\zeta/n$, $k = 0, 1, \ldots, n$. Roughly speaking, this hypothesis states that the composite under consideration has sufficiently thin layers and hence its macro-properties can be approximately treated as independent of the parameter ε. From the so-called transfer principle of the nonstandard analysis (cf. [2] for the foundations of the nonstandard analysis and [5] for its applications to mechanics) it follows that the problem $\mathcal{P}$ can

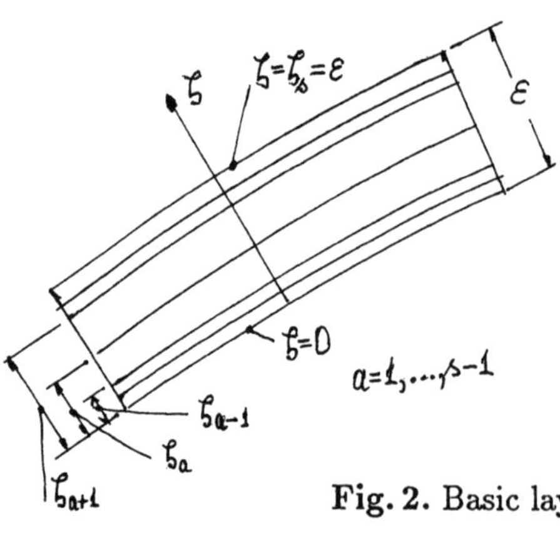

Fig. 2. Basic layer

be approximated by the (nonstandard) problem $\mathcal{P}^{(\breve{\omega})}$, where $\breve{\omega}$ is an arbitrary but fixed unfinite positive integer. Using the notation introduced above, we conclude that

$$\boldsymbol{u} \sim \boldsymbol{u}^{(\breve{\omega})} \quad , \quad \boldsymbol{E} \sim \boldsymbol{E}^{(\breve{\omega})} \quad , \quad \boldsymbol{T} \sim \boldsymbol{T}^{(\breve{\omega})} \quad . \tag{5}$$

In the problem $\mathcal{P}^{(\breve{\omega})}$ we deal with a certain "composite" made of infinite number of infinitely thin basic layers (having the thickness $\varepsilon/\breve{\omega}$).

In order to formulate the second basic assumption we shall introduce the functions h_a: $\boldsymbol{R} \to \boldsymbol{R}$, $a = 1, \ldots, s-1$, and i_A: $\mathbb{R} \to \boldsymbol{R}$, $A = 1, \ldots, s$, which are $(0, \varepsilon)$-periodic and given by

$$h_a(b) \equiv \begin{cases} 0 & \text{if } b \in [0, b_{a-1}] \cup [b_{a+1}, \varepsilon] \quad , \\ (b - b_{a-1})/\eta_a & \text{if } b \in [b_{a-1}, b_a] \quad , \\ (b_{a+1} - b)/\eta_a & \text{if } b \in [b_a, b_{a+1}], \end{cases}$$

$$i_A(b) \equiv \begin{cases} -b + b_a - \dfrac{\delta}{2} & \text{if} \quad b \in [0, b_a) \quad , \\ -b + b_a + \dfrac{\delta}{2} & \text{if} \quad b_a \in (b_a, \varepsilon] \quad , \end{cases}$$

where $\eta_a \equiv (b_{a+1} - b_a)/\varepsilon$. For every $b \in \boldsymbol{R}$ we have $h_a(b) = h_a(b + \varepsilon)$, $i_A(b) = i_A(b + \varepsilon)$. It has to be emphasized that the functions $i_A(\cdot)$ are non-continuous.

The Microlocal Approximation states that the displacements $u^{(\breve{\omega})}$ in the problem $\mathcal{P}^{(\omega)}$ can be expected in the class of functions given by (summation convention with respect to a and A holds):

$$\boldsymbol{u}^{(\breve{\omega})}(\boldsymbol{x}, \tau) =^* \boldsymbol{\omega}(\boldsymbol{x}, \tau) + \frac{1}{\breve{\omega}} {}^*h_a(\breve{\omega} b)^* \boldsymbol{q}^a(\boldsymbol{x}, \tau) + \frac{1}{\breve{\omega}} {}^*i_A(\breve{\omega} b)^* \boldsymbol{d}^A(\boldsymbol{x}, \tau) \quad , \tag{6}$$

where ${}^*\boldsymbol{\omega}$, ${}^*\boldsymbol{q}^a$, ${}^*\boldsymbol{d}^A$ are the standard unknown functions, satisfying the appropriate regularity conditions. Setting

$$\lambda_{ab} \equiv \begin{cases} -1/\eta_e & \text{if } a = b \quad , \\ 1/\eta_a & \text{if } a = b + 1 \quad , \\ 0 & \text{if otherwise} \quad , \end{cases} \qquad \begin{aligned} &\boldsymbol{d} \equiv \boldsymbol{d}^1 + \ldots + \boldsymbol{d}^s \quad , \\ &N\boldsymbol{v} \equiv \tfrac{1}{2}(\boldsymbol{n} \otimes \boldsymbol{v} + \boldsymbol{v} \otimes \boldsymbol{n}) \quad , \quad \boldsymbol{v} \in \mathbb{R}^3 \quad , \end{aligned}$$

we can show that the strains in ath material component are determined by the formula

$$\boldsymbol{E}_a(\boldsymbol{x}, \tau) = L\boldsymbol{\omega}(\boldsymbol{x}, \tau) + \lambda_{ab} N \boldsymbol{q}^b(\boldsymbol{x}, \tau) - N\boldsymbol{d}(\boldsymbol{x}, \tau) \quad . \tag{7}$$

The functions $\boldsymbol{\omega}$ are called the macrodisplacements; the functions $\boldsymbol{q}^a$ describe the effect of the jump of material properties on the strain distribution and are referred to as the microlocal parameters. The functions $\mathbb{D} = N\boldsymbol{d}$ describe the possible jumps of the displacements on the interfaces of sheets and will be called the strain incompatibilities. The left-hand side of Eq. (7) is called the constitutive strain and is a sum of the macrostrain $L\boldsymbol{\omega}$, which is called the microlocal strain and the incompatibility strain (taken with the minus sign).

The main feature of the microlocal modelling is that the nonstandard problem $\mathcal{P}^{(\breve{\omega})}$ under the (nonstandard) constraints (6) leads to a certain problem $\tilde{\mathcal{P}}$ for the macrodisplacements $\boldsymbol{\omega}$, microlocal parameters $\boldsymbol{q}^a$ and displacement incompatibilities $\boldsymbol{d}$; problem $\tilde{\mathcal{P}}$ involves only "standard" mathematical entities and represents a certain homogenized model of the $(0,\varepsilon)$-periodic composite under consideration. For the particulars the reader is referred to [1] and to the related papers. Here we put aside all calculations leading from the nonstandard problem $\mathcal{P}^{(\breve{\omega})}$ [under constraints (6)] to the "standard" problem $\tilde{\mathcal{P}}$, and we formulate below only the resulting relations.

Results

The system of governing relations for the problem $\tilde{\mathcal{P}}$ will be given by the (quasi-stationary) equations of motion

$$\operatorname{div}\tilde{\boldsymbol{T}} + \tilde{\varrho}\boldsymbol{b} = 0 \quad , \quad \tilde{\boldsymbol{S}}_b = 0 \quad ; b = 1, \ldots, s-1 \quad , \tag{8}$$

where $\tilde{\boldsymbol{T}}$ and $\tilde{\boldsymbol{S}}_b$ are called the mean stress and the stress excess, respectively, by the constitutive equations

$$\tilde{\boldsymbol{T}} = \sum_{a=1}^{s-1} \eta_a \mathbb{C}_a[\boldsymbol{E}_a] \quad , \quad \tilde{\boldsymbol{S}}_b = \sum_{a=1}^{s-1} \eta_a \lambda_{ab} \mathbb{C}_a[\boldsymbol{E}_a]\boldsymbol{n} \quad , \tag{9}$$

where

$$\boldsymbol{E}_a = \boldsymbol{L}\boldsymbol{\omega} + \lambda_{ab}\boldsymbol{N}\boldsymbol{q}^b - \boldsymbol{N}\boldsymbol{d} \quad , \quad a = 1, \ldots, s-1 \quad , \tag{10}$$

is the constitutive strain, and by the conditions which describe the interaction between the adjacent layers

$$\tilde{\boldsymbol{T}}\boldsymbol{n} \in \partial \operatorname{ind}_{\overline{\boldsymbol{R}}_+}(d_N) + \theta\overline{\partial}\pi(\boldsymbol{d}) \quad , \quad d_N \equiv \boldsymbol{d}\cdot\boldsymbol{n} \quad . \tag{11}$$

It has to be emphasized that all fields in formulae (8–11) are defined (almost everywhere) on the region Ω occupied by the composite under consideration in its undeformed state. These formulae represent the quasi-stationary homogenized model of the periodic multilayered composite. The obtained model takes into account the effect of bonding of adjacent layers and the possibility of delamination.

At the end of the note we shall give the functional form of the homogenized model obtained above. To this aid we have to observe that the microlocal parameters $\boldsymbol{q}^b$ can be eliminated from the governing relations. Using equations $\tilde{\boldsymbol{S}}_b = 0$, $b = 1, \ldots, s-1$, combined with (9) and (10), we obtain

$$\boldsymbol{E}_a = \mathbb{G}_a[\boldsymbol{L}\boldsymbol{\omega} - \boldsymbol{N}\boldsymbol{d}] \quad , \quad a = 1, \ldots, s-1 \quad ,$$

where $\mathbb{G}_a$ are the known linear mappings. Now defining

$$\tilde{\mathbb{C}}[\cdot] \equiv \sum_{a=1}^{s-1} \eta_a \mathbb{C}_a[\mathbb{G}_a[\cdot]] \tag{12}$$

we obtain the constitutive equations in the form

$$\tilde{T} = \tilde{\mathbb{C}}[\tilde{E}] \quad , \quad \tilde{E} = L\omega - Nd \quad . \tag{13}$$

Substituting the right-hand sides of Eqs. (13) into (8) and (11), we arrive at the sytem of governing relations for $\omega(\cdot)$ and $\boldsymbol{d}(\cdot)$.

Let $\mathcal{W}$, $\mathcal{E}$, $\mathcal{D}$ stand for the linear topological normed spaces of functions $\omega(\cdot,\tau)$, $\boldsymbol{E}(\cdot,\tau)$, $\boldsymbol{d}(\cdot,\tau)$, respectively. Moreover, let $(\mathcal{W}, \langle\cdot,\cdot\rangle_{\mathcal{W}}, \mathcal{W}^*)$, $(\mathcal{E}, \langle\cdot,\cdot\rangle_{\mathcal{E}}, \mathcal{E}^*)$, $(\mathcal{D}, \langle\cdot,\cdot\rangle_{\mathcal{D}}, \mathcal{D}^*)$ be the pertinent dual pairings. Now we see that $L : \mathcal{W} \to \mathcal{E}$ and $N : \mathcal{D} \to \mathcal{E}$ are linear continuous operators; their duals will be denoted by L^*: $\mathcal{E}^* \to \mathcal{W}^*$ and N^*: $\mathcal{E}^* \to \mathcal{D}^*$, respectively. Let us also introduce the linear continuous operator $\mathbb{C} : \mathcal{E} \to \mathcal{E}^*$, defined by means of Eq. (12), as well as the entities listed below

$$\begin{aligned}
&\mathbb{K} \equiv \{d \in \mathcal{D} : d_N(x) \geq 0 \text{ for almost every } x \in \Omega\} \quad , \\
&\mathcal{J}(\boldsymbol{d};\theta) \equiv \int_\Omega \theta\pi(\boldsymbol{d})dV, \ \boldsymbol{d} \in \mathcal{D} \quad , \\
&\langle \boldsymbol{v}, f^*(\tau)\rangle_{\mathcal{W}} \equiv \int_\Omega \tilde{\varrho}\boldsymbol{b}\cdot\boldsymbol{v}\, dV \quad , \ \boldsymbol{v} \in \mathcal{W} \quad ; \\
&\langle \boldsymbol{E}, T^*(\tau)\rangle_{\mathcal{E}} \equiv \int_\Omega \mathrm{tr}\{\boldsymbol{E}\tilde{\boldsymbol{T}}\}\, dV \quad , \ \boldsymbol{E} \in \mathcal{E} \quad .
\end{aligned}$$

Then the global form of the governing relations will be given by the (quasi-stationary) equations of motion

$$L^*T^*(\tau) - f^*(\tau) = 0 \quad , \tag{14}$$

by the constitutive equations

$$T^*(\tau) = \mathbb{C}\boldsymbol{E}(\tau) \quad , \quad \boldsymbol{E}(\tau) = L\omega(\tau) - N\boldsymbol{d}(\tau) \quad , \tag{15}$$

and by the interlaminae bonding conditions

$$N^*T^*(\tau) \in \partial\,\mathrm{ind}_K(d(\tau)) + \overline{\partial}\mathcal{J}(d(\tau) \quad ; \quad \theta(\tau)) \quad , \tag{16}$$

where the meaning of function $\theta(\cdot,\tau)$ was explained in Preliminaries. Hence we arrive finally at the system of relations

$$\begin{aligned}
&L^*\mathbb{C}L\omega(\tau) - L^*\mathbb{C}N\boldsymbol{d}(\tau) = f^*(\tau) \quad , \\
&N^*\mathbb{C}N\boldsymbol{d}(\tau) - N^*\mathbb{C}L\omega(\tau) \in -\partial\,\mathrm{ind}_{\mathbb{K}}(\boldsymbol{d}(\tau)) - \overline{\partial}\mathcal{J}(\boldsymbol{d}(\tau));\theta(\tau) \quad ; \\
&\tau \in [0,\tau_f] \quad ,
\end{aligned} \tag{17}$$

for the macrodisplacements $\omega(\cdot,\tau)$ and the displacement incompatibilities $d(\cdot,\tau)$. The problem of finding ω and d from (17) is non-local in time due to the presence of the functional θ which depends on the history of the process. However, using the line of approach proposed in [3], after the time discretization of the interval $[0,\tau_f]$ we arrive at the sequence of the time local problems

$$L^*\mathbb{C}L\omega(\tau_m) - L^*\mathbb{C}Nd(\tau_m) = f^*(\tau_m) \quad ,$$

$$N^*\mathbb{C}Nd(\tau_m) - N^*\mathbb{C}L\omega(\tau_m) \in -\partial \operatorname{ind}_{\mathbb{K}}(d(\tau_m)) - \overline{\partial}\mathcal{J}(d(\tau_m)) \quad ; \quad \theta(\tau_{m-1})) \quad , \tag{18}$$

where $m = 1, \ldots, M$ and $\tau_0 = 0$, $\tau_M = \tau_f$. Since $\theta(\tau_0) = 0$, then all these problems can be successively solved for $m = 1, \ldots, M$.

The analysis and the possible applications of the homogenized model of the quasi-stationary processes for the laminates under consideration will be given in the separate paper.

References

1. Cz. Woźniak: A nonstandard method of modelling of thermoelastic periodic composites, Int. J. Eng. Sci. **25**, 483–498 (1987)
2. A. Robinson: *Non-Standard Analysis* (North Holland Elsevier, Amsterdam, 1974) (revised edition)
3. Z. Naniewicz, Cz. Woźniak: On the quasi-stationary models of debonding processes in layered composites, Ing. Arch., in press
4. F.H. Clarke: Generalized gradients of Lipschitz functionals, Adv. in Math. **40**, 52–67 (1981)
5. Cz. Woźniak: Nonstandard analysis in mechanics, Adv. in Mechanics **9**, 3–35 (1986)

Multilevel Model of Micro-Macro Interactions and Residual Stresses in Physical Systems

B. Orłowska

Construction Industry Computing Services, ETOB Warsaw Computer Center, Bema 65, PL-01-244 Warszawa, Poland

Many physical phenomena possess simultaneous characteristics on various levels or scales. In some detail this fact is explained on the example of plastic deformation, where four levels, named macro, meso, micro and atomic, play a role. An own space is assigned to each level, and all these spaces are combined to a multilevel space in the form of Carthesian product space. In this space a Lagrangian description, which also contains interactions between the levels is introduced, and it is shown how physical observations can be projected from lower levels to the macrolevel. The so obtained macrodescription can include irreversible processes, although Lagrange formalism is used. It is also shown that macroscopic residual stresses arise as the result of microscopic motions which are incompatible with the macrodeformation.

1. Motivation and Introduction

Thorough investigations of physical processes often require observations on various levels or scales, consider for instance a solid body under inelastic deformation. It is very common to make phenomenological tests in machines which allow us to measure the development of strains under the application of prescribed and controlled forces. If marks are attached to the surface of the specimen, then the successively progressing deformation can be seen with our eyes or felt with our hands - from there the term "phenomenological". In this way information about the behaviour of the investigated material can be obtained. This phenomenological information might suffice for certain applications of such materials. It is certainly not enough for an understanding of materials which are for instance needed to develop new materials with properties that are better for some intended application.

Helpful for a better understanding are observations using a light microscope. Such techniques allow us to detect certain structures of the body, for instance, a grain structure or a slip line pattern which tells us something about the mode of deformation.

If we look with higher magnification, for instance with an electron microscope, then we detect even finer structures. These often form smaller grains (subgrains) within the grain structure, but also other figures are frequently seen.

We know nowadays that the mentioned structures are consequences of the fact that the bodies in which we seen them are crystalline or polycrystalline. To verify this, observations on even smaller scales are necessary. In fact, the crystallinity of matter is proved by x-ray experiments where wave lengths of the order of atomic distances are used. These are lenghts which are small compared to the wave lenghts used in electron and light microscopes.

In the above treatment of inelastic behaviour of matter we have introduced four levels of description to which we might refer as phenomenological, or "macro", as "meso", as "micro" and as "atomic". If four observers on these levels describe what they see, then they report completely different phenomena. Nevertheless they all speak of inelastic deformation. This means that, notwithstanding the different appearance of the results, there must be close connections between them. The laying open of these connections is a task of fundamental importance. The attempts towards this aim are innumerous, and still are we remote from this aim.

In our present contribution we present a frame within which the mentioned connections have their place. We propose a multilevel description which comprises the physical situation on all levels. In other words, we introduce one multilevel observer which can discribe the situation on every level simultaneously. The four-level problem sketched above will serve us for illustration only, at least as far as the physics is concerned.

The very concept of the multilevel theory can be sketched by the following considerations. Let us take the inelastic deformation as an example. It is clear that a theory which permits to calculate the motion of all elementary particles, in particular of the atoms which build up the body, is complete in the sense that in principle all observations can be derived from it. Here the term "in principle" implies that in reality this cannot be done. Therefore, a description of inelasticity which is restricted to the atomic level, is rather useless. Since it, nevertheless, contains somehow all phenomena observed on the other levels, too, we may speak of a description of inelasticity which is a projection onto the atomic level from the ensemble of all levels.

On the other hand, we may think of a purely phenomenological description, which we understand as a projection onto the macroscale from the ensemble of all levels. Unlike in the case of the projection onto the atomic level the projection onto the macrolevel goes with loss of information and, therefore, introduces thermodynamic variables like entropy, temperature and dissipation. In other words, it is the projection from lower to higher levels which brings in the irreversibility into our description of inelasticity.

The application of the sketched concepts requires firstly a description of the physical situation on the various levels, secondly a specification of the interaction between these levels and thirdly a prescription of how the projections onto a selected level from the ensemble of all levels is to be done. In the following sections we give a scheme, which admittedly is somewhat formalistic and does not replace the proper physical considerations needed to specify the physical situation.

Nevertheless, this scheme can well provide some directives for the proceeding in the treatment of problems with multilevel characteristics.

2. The Structural Model of Multilevel System

For simplicity we shall restrict ourselves to problems of mechanics and thermodynamics. The main feature of our situation is then that "something moves". This "something" will often be an assembly of particles which should be identified appropriately on the various levels. The observers on these levels make their measurements in their own physical spaces in analogy to Kröner's internal observer, see ref. [1]. Let us specify the levels by figures i = 1,2 ..., n, where n is the number of levels considered. For simplicity assume that the observer spaces are Euclidean, not necessarily all of the same dimension. It is convenient to include the time as an additional dimension. Note that the time scales on the various levels may differ from each other. For instance, some observers may find distinguished relaxation times which they would like to use for the description of the observed phenomena. Denoting now the space-time of the ith level as E^i we propose to construct a N-dimensional space time E as the cartesian product of the level space-times:

$$E = E^1 \times E^2 \times \ldots \times E^n \ , \qquad (1)$$

where $N = \sum_i^n d_i$ and d_i is the dimension of the ith space-time.

For more general case of differentiable manifold we obtain

$$M = M^1 \times M^2 \ \ldots \ \times M^n \ .$$

The introduction of the Cartesian product (1) implies that also the coordinate systems X^i (one system for level i) used to specify points in E^i and assumed here to include the time are combined to an N-dimensional coordinate system (X_A) by a Cartesian product:

$$X = (X_A) = X^1 \times X^2 \ \ldots \ \times X^n \ , \quad A = 1,2, \ldots, N \ . \qquad (2)$$

To introduce the cartesian products does not mean a restriction in generality. It may be, however, that in particular circumstances other types of combinations of the level spaces can be more conventient. So far we have no experience in this point. In analogy to a fibre bundle model the multilevel product space is rather a base space structure than a trivial bundle.

In the space E^i of the ith level the physical system occupies a certain material space Ω^i, say. All Ω^i's are then combined, again as cartesian products:

$$\Omega = \Omega^1 \times \Omega^2 \times \ldots \times \Omega^n \ . \qquad (3)$$

3. The Motion of a Multilevel Dynamical System

The first step in the theory of multilevel systems is that the various observers build up a model which to sufficient accuracy describes the physical situation in their level. Of course, the goodness of the results will depend on that of the models introduced. In our example of inelastic deformation such models can be as follows:

On the macrolevel, i.e. level no. 1, we see a deformable solid body which on this scale obeys the equations of continuum mechanics. We define particles at points X^1 (which might include time t^1) and describe their motion by (one or more) dynamical variables

$$x^1 = \psi^1 (X^1) . \tag{4}$$

On the mesolevel, i.e. level no. 2, we see grains with slip lines in them. When the deformation proceeds, then the motion can be read off from the change in the slip line pattern. Hence we can describe this motion by (one or more) dynamical variables

$$x^2 = \psi^2 (X^2) , \tag{5}$$

where X^2 now represents position and time in level no. 2.

On the microlevel we see the development of dislocation structures which we describe by dynamical variables

$$x^3 = \psi^3 (X^3) . \tag{6}$$

The X^3 must now represents functions which characterize the dislocation structure. They might be numerous when a detailed description is intended.

As to the atomic level (no. 4), it might often suffice to regard only the consequence that the body is crystalline. In many situations of inelastic deformation the approximation that "point defects, which can be seen with highest resolution only, do not contribute to the inelastic behaviour" may be good enough. This is particularly true in the so-called time-dependent plasticity (always an approximation), whereas in other deformation modes, for instance in high temperature creep, the role of the point defects can be decisive. In such cases dynamical variables

$$x^4 = \psi^4 (X^4) \tag{7}$$

would be introduced where X^4 characterizes position and time on the atomic scale and the time scale is e.g. determined by the jump frequencies of the point defects.

As another example consider the physical system "liquid crystal". This system consist usually of elongated molecules which possess two classes of the degrees of freedom (translations and rotations). Imagine two observers, one who by his measuring apparatus can perceive only translations, whereas the other sees only rotations. In such a situation the system is described on two levels which are both macroscopic. Motion occurs on both levels and is described by the dynamical variables, say

$$x^1 = \psi^1 (X^1), \quad x^2 = \psi^2 (X^2), \tag{8}$$

where ψ^1 and ψ^2 are now interpreted as displacement and rotation field, respectively. X^1 as well as X^2 denote simply coordinate systems, which coincide for the two levels. This implies that the two observers also see the same volume filled with their physical system. The variables x^2 may, for instance, be the three Eulerian angles.

Combining (4-7) or (8) we can comprise the whole motion, i.e. the motion as it occurs simultaneously in all levels, into the Cartesian product

$$x = x^1 \times x^2 \times \dots x^n, \tag{9}$$

where n = 4 in the first example and n = 2 in the second example. The Cartesian product (9) implies that the motion on all levels can be observed by one multilevel observer and comprised into one complex equation

$$x = \psi^1 (X^1) \times \psi^2 (X^2) \times \dots \times \psi^n (X^n) \equiv \psi (X^1 \dots X^n) . \tag{10}$$

Note that $\psi(X^1 \dots X^n)$ can be written as $\psi(X)$ in the case of motions which are described by homeomorphic functions ψ only.

4. Action Functional and Noether's Theorem, Conservation laws

The very first and most fundamental assumption of the structural theory is that every isolated dynamical system can be described by the characteristic action functional W. Note that the multilevel dynamical system is isolated as whole but each particular level is open. If the observation occurs on n levels, then W has the form

$$W = \int_\Omega L [X, \psi (X^1 \dots X^n) , \psi_{,A} (X^1 \dots X^n)] \, d\Omega , \tag{11}$$

where, A denotes differentiation with respect to X_A (A=1 .... N).

The indicated form of the Lagrangian permits us in particular to include also interactions between the levels. Such interactions are, in general, to be expected. The Euler-Lagrange equations of the pertaining variational problem are

$$\frac{\partial L}{\partial \psi_{\mathcal{L}}} - \frac{\partial}{\partial X_A} \frac{\partial L}{\partial \psi_{\mathcal{L},A}} = 0 \quad , \tag{12}$$

where $\mathcal{L} = 1,2,3, \ldots, N$. For details of the derivation of eq. (12) see ref.[2].

The conservation laws express the fact that certain physical quantities, e.g. momentum, energy, angular momentum, are constant in time. These laws follow as a consequence from (12) or directly from the action functional (11). This has to do with the structure of this functional and with its symmetry properties in the multilevel product space-time E.

Introducing in the usual manner the energy-momentum tensor with components

$$T \equiv T_{AB} \equiv \frac{\partial L}{\partial \psi^{\mathcal{L}}_{,A}} \psi^{\mathcal{L}}_{,B} + L\, \delta_{AB} \quad (A,B, \ \mathcal{L} = 1 \ldots N) \ , \tag{13}$$

we obtain the following form of the conservation laws:

$$\int_{\Omega} \left(\frac{\partial L}{\partial \psi^{\mathcal{L}}_{A}} \psi^{\mathcal{L}}_{B} - L\, \delta_{AB} \right)_{,A} \equiv \int_{\Omega} \nabla_{\mu} T \, d\Omega \equiv \int_{\Omega} \mathrm{Div}_N T \, d\Omega = 0 \ . \tag{14}$$

Here ∇_N is the N-dimensional Nabla vector with components $\frac{\partial}{\partial X_A}$.

With $\mathrm{Div}_N = \mathrm{div}^1 + \mathrm{div}^2$, $d\Omega = d\Omega^1 \, d\Omega^2$ and by Stokes' and Fubini theorem application (14) becomes

$$\int_{\partial\Omega} T\, d(\partial\Omega) = \int_{\partial\Omega^1} \int_{\Omega^2} T\, d(\partial\Omega^1)\, d\Omega^2 + \int_{\Omega^1} \int_{\partial\Omega^2} T\, d\Omega^1 \, d(\partial\Omega^2) = 0 \tag{15}$$

The projection from two-level $\Omega^1 \times \Omega^2$ system onto one macrolevel is modeled by

$$\lim_{\Omega^2 \to 0} \frac{\int_{\Omega^2} T\, d\Omega^2}{\int_{\Omega^2} d\Omega^2} = \overline{T}^2$$

and (16)

$$\lim_{\Omega^2 \to 0} \frac{\int_{\partial\Omega^2} T d(\partial\Omega^2)}{\int_{\Omega^2} d\Omega^2} = \overline{\mathrm{div}\, T^2}$$

Introducing mean values $\overline{T}^2$ and $\overline{\mathrm{div}\, T^2}$ taken over Ω^2 we obtain

$$\int_{\Omega^1} \mathrm{div}^1 \, \overline{T}^2 \, d\,\Omega^1 = - \int_{\Omega^1} \overline{\mathrm{div}^2 \, T^2} \, d\,\Omega^1 \tag{17}$$

If it is now permitted to interpret $\overline{T}^2$ as the energy-momentum tensor of the macrolevel obtained after projection to this level from the multilevel, then (17) expresses an interesting result: In the now phenomenological description energy and momentum are not conserved because in general the left hand side of (17) does not vanish. We rather see that energy and momentum flows to the macrolevel from the microlevel according to the right hand side of (17). A similar statement can be derived for the angular momentum.

The question arises: are we entitled to interpret $\overline{T}^2$ in the above sketched manner? Assume that the level no. 2 is a real microlevel so that $\Omega^2 \ll \Omega^1$. In this case each point X^1 in the macrolevel is included a small region Ω^2, the microlevel. If Ω^2 is small enough then it makes sense to speak of energy and momentum which include the microenergy of Ω^2 at each point X^1 in Ω^1. If, however, Ω^2/Ω^1 is not small, then the above interpretation is not meaningful. This applies to the liquid crystal if its description is chosen as sketched above.

The energy transfer from level 2 to level 1 can proceed in the form of heat but might also just increase the internal energy. Thus the projection from micro- to macrolevel leads us to thermodynamical considerations; this may be understood that the first level and its region Ω^1 is open toward the second level and its region Ω^2 and vice versa. (The exchange of matter and energy is possible between the two levels). Of course, to establish equations which solve our problem completely, we must specify the physical situation on the various levels.

5. The Residual Stress Problem

Residual, or internal, stresses are those which persist in a body without the presence of any external action (of forces, temperature etc.). They occur, in particular, after plastic deformation. A simple illustration concerns the deformation of a polycrystal in which the grains suffer distorsions that are different from the mean distorsions of the whole body. This occurs frequently, because the resistance against flow depends on the grain orientation, which varies from one grain to the other. In this section we consider only the macrolevel, 1, of the whole body and the microlevel, 2, of the grains. If these were isolated from each other, then the same stress in all grains would lead to different strains such that the grains would no longer fit together. These strains are usually called plastic. Because of the nonfitting they are incompatible strains.

In real experiments the grains are not first isolated from each other, but cohere during the whole deformation process. The coherence is then enforced by the cohesive forces which give rise to a superimposed elastic strain. This strain is also incompatible such that the total strain, which in the linearized theory is simply the sum of plastic and elastic strain, is compatible. In fact, the body is supposed to remain compact under the deformation.

We see from this discussion that the residual stress problem is related to the incompatibility problem. A convenient description of the kinematics of deformable media is given in the language of differential geometry. There the strain is represented by the metric tensor, and the compatibility equations are equivalent to the vanishing of the (Riemann) curvature tensor formed with the metric tensor.

Consider now a two-level system as the one just described. For simplicity we omit the time dimension, so that our system is embedded in the sixdimensional Euclidean product space $E = E^1 \times E^2$, 1 and 2 referring to the two levels. In cartesian coordinates the sixdimensional metric tensor has the form δ_{AB} (A,B = 1....6). Now deform the multilevel system by means of the 6 x 6 matrix

$$F^{\mathcal{L}}_{A}(X) = \partial x^{\mathcal{L}} / \partial X^A \tag{17}$$

($\mathcal{L}$ = 1....6). Here it is essential, that the matrix F depends on all coordinates X^A. The deformation (17) carries over the sixdimensional line elements dX^A into elements $dx^{\mathcal{L}}$. The sixdimensional distance between the two points separated by dX^A, namely

$$ds_0^2 = \delta_{AB} \, d X^A \, d X^B \tag{18}$$

becomes, with the usual summation convention,

$$ds^2 = \delta_{AB} \, F^A_{\mathcal{L}} \, F^B_{\beta} \, dx^{\mathcal{L}} dx^{\beta} = g_{\mathcal{L}\beta} \, dx^{\mathcal{L}} dx^{\beta} , \tag{19}$$

where the metric tensor of the deformed state is

$$g_{\mathcal{L}\beta} = \delta_{AB} \, F^A_{\mathcal{L}} \, F^B_{\beta} \tag{20}$$

In another representation, $g_{\mathcal{L}\beta} = g_{\mathcal{L}} \cdot g_{\beta}$, with $g_{\mathcal{L}}$ ($\mathcal{L}$ = 1....6) as base vectors deformed from the original cartesian base vectors of the initial state. Written in dyadic form the metric tensor of the deformed system is

$$\begin{aligned} g &= \sum_{\mathcal{L}\beta}^{6} g_{\mathcal{L}\beta} \, g^{\mathcal{L}} g^{\beta} \qquad\qquad (g^{\mathcal{L}} \cdot g_{\beta} = \delta^{\mathcal{L}}_{\beta}) \\ &= \left(\sum_{\mathcal{L}=1}^{3} \sum_{\beta=1}^{3} + \sum_{\mathcal{L}=1}^{3} \sum_{\beta=4}^{6} + \sum_{\mathcal{L}=4}^{6} \sum_{\beta=1}^{3} + \sum_{\mathcal{L}=4}^{6} \sum_{\beta=4}^{6} \right) g_{\mathcal{L}\beta} \, g^{\mathcal{L}} \, g^{\beta} \end{aligned} \tag{21}$$

In matrix representation this can be written as

$$(g) = \begin{vmatrix} g^{11} & , & g^{12} \\ g^{21} & , & g^{22} \end{vmatrix} \tag{22}$$

Note that in general all components of (g) depend on all space coordinates. As is well-known, the curvature tensor constructed with a metric of the form (20), where F is a gradient, vanishes. Thus we conclude that the total curvature of our six-dimensional two-level system vanishes. the basic conclusion can be drawn already from the linearized curvature tensor:

$$R_{\mathcal{L}\beta\gamma\delta} = \frac{1}{2}\,(g_{\mathcal{L}\delta,\beta\gamma} + g_{\beta\gamma,\mathcal{L}\delta} - g_{\beta\delta,\mathcal{L}\gamma} - g_{\mathcal{L}\gamma,\beta\delta}) \tag{23}$$

($\mathcal{L}, \beta, \gamma, \delta = 1....6$). Here the comma denotes differentiation. Using the decomposition (21) or (22) of the metric tensor, we find that also the two-level curvature tensor splits into four contributions so that

$$R = R^{11} + R^{12} + R^{21} + R^{22} = 0 \tag{24}$$

Similarly as in section 4 we make the projection onto the macrolevel by taking the average over Ω^2. Then

$$\overline{R^{11}}^{\,2} = -\,(\overline{R^{12}}^{\,2} + \overline{R^{21}}^{\,2} + \overline{R^{22}}^{\,2}). \tag{25}$$

$\overline{R^{11}}^{\,2}$ can be considered as the pure macropart of the curvature tensor. As in section 4 we argue, that this interpretation is meaningful only, if $\Omega^2 \ll \Omega^1$. In fact, in our example, where X^1 gives the position of a macropoint, i.e. a grain, it is sensible to specify the strain, or curvature, situation in this grain, but not that in other grains, which belong to other macropoints.

Eq. (25) implies that, although the two-level curvature tensor vanishes, the projected macrotensor $\overline{R^{11}}^{\,2}$ does not in general disappear. It only vanishes, when the microdeformation is compatible with the macrodeformation. This situation is an exception in the field of plastic deformation.

Generalizing this result to multilevel systems, we arrive at the following conclusion: If the motions on the small scales are incompatible with the motion on the macrolevel, then they produce a macrostrain incompatibility leading to the generation of macroscopic residual stresses. Thus the theory has given us results that are confirmed by the physics of the problem.

We emphasize that these results have the character of a frame within which the residual stress problem can be treated. This frame alone does not suffice to solve actual problems of this kind. This is clear, because we have not included more detailed physical quantities such as distributions of grain sizes, shapes and orientations.

6. Comparison with Homogenization of Microperiodic Media

To illustrate the procedure of the projection between levels further, consider a system which is observed on a threedimensional macrolevel 1 and microlevel 2. To have something definite in mind, let us think of an elastic body which has elastic moduli that vary periodically over the body with a wavelenght small compared to the whole body. Introduce a macroscale and a microscale together with the pertinent observers. The elastic problem of such a body is the object of the so-called homogenization theory. We introduce macro- and microcoordinates X^1 and X^2 respectively. The (three dimensional) macrovolume Ω^1 and the microvolume Ω^2 are the volumes occupied respectively by the whole body and the elementary cell which is repeated periodically through the body..

It is shown in homogenization theory that the problem can be split into a purely macroscopic problem of a body with homogeneous elastic properties and a purely periodic problem, see ref.ref.[3, 4, 5]. More strictly, an alternating repeated solution of homogeneous and periodic problems is needed. The phenomenological description requires averaging over the elementary cell (Ω^2) like in our previous example. Since the body is assumed to be purely elastic, the energy flux from micro to macro serves to increase the internal energy rather than to heat the body up. It is planned to give a detailed description of the incorporation of homogenization theory into the general scheme of the multilevel theory in a later article. This example is helpful for understanding and illustrating of the basic ideas of the multilevel theory, because here an explicit formulation of the basic equations is possible.

7. Conclusion

Physical phenomena with multilevel characteristics are usually very complex. To solve problems of this kind requires a detailed knowledge of the physical processes going on the various levels. These processes are not studied in the present work. Instead we have tried to drow conclusions from the very fact that the physics of some problem has multilevel character. The theory built up in this way can only form a frame comprising conditions which are satisfied by all, or almost all, multilevel systems in physical situations (we do not yet exclude the possibility of certain exceptions). Note that many theories whithin physics form frames rather than have detailed physical contents. This is true for thermodynamics where the complex physical information is contained in the equation of state, which stand outside the actual thermodynamics, or for Maxwell's equations which need to be complemented,by, in general complex constitutive laws. The cooperations of micro- and macroprocesses is of great concern to those interested in a deeper understanding of the physics governing the behaviour of materials, for instans in plastic deformation. In fact, having a frame can give us ideas how to proceed toward the uniting of the microscopic and macroscopic processes.

It was shown that it is the interaction of the micro- and macroprocesses, which leads to the irreversible behaviour found so often in the phenomenological description; and similarly, the occurrence of macroscopic residual stresses could be explained. It remains to examine already existend attempts to unite micro- and macro-descriptions of phenomena with multilevel characteristics, and to see, how they fit into the scheme developed here.

Acknowledgement

The author is greatly indebted to Prof. E. Kröner and Prof. Cz. Woźniak for valuable discussions on the subject considered in this note.

References

1. E. Kröner: Continuum Theory of Defects, in Physics of Defects, Les Houches Session XXXV, 1980, R. Balian et al. eds., (North Holland, Amsterdam 1981)
2. B. Orłowska: Nonlocal Theory of Continuous Media in Structural Interpretation, 24 Polish Solid Mechanics Conference 1983, avail. from the author on request.
3. B. Orłowska: Macroscopic Properties of Continuous Media with Periodic Structure, Bulletin de l'Acad. Pol. Sci., Vol. XIX, 1 (1971)
4. B. Orłowska: Macroscopic Model of Hyperelastic Bodies System with Periodic structure, Bulletin de l'Acad. Pol. Sci., Vol. XIX, 2 (1971)
5. E. Sanchez-Palencia: Non-Homogeneous Media and Vibration Theory, (Springer, Berlin, Heidelberg 1980)

Hemivariational Inequalities and Their Applications to Composite Material Structures

P.D. Panagiotopoulos

Dept. of Civil Eng., Aristotle University, GR-54006 Thessaloniki, Greece and Institute for Technical Mechanics, RWTH, D-5100 Aachen, Fed. Rep. of Germany

The behaviour of composite material structures is described by nonmonotone possibly multivalued phenomenological material laws, which have a "sawtooth" form because of the arising limit phenomena, as e.g. the cracking or crushing of one phase of the composite, the interface bonding failure, the delamination effect etc. These problems cannot be treated by the classical methods of structural analysis, which give rise to variational equality formulations. Moreover, they cannot be formulated in terms of variational inequalities due to the lack of monotonicity, but they give rise to a new type of variational expressions the hemivariational inequalities.

1. Stress-Strain Laws in Composite Material Structures

The mechanical behaviour of the composite materials is very complicated because of the possibility of limit behaviour in one ore more components of the material, as, for instance, cracking or crushing of one phase of the composite, microbuckling and/or debonding of the fibers, slip and delamination effects between the laminae etc. Therefore the classical stress-strain diagrams for composite materials have a "sawtooth" form, i.e. they are nonmonotone and possibly multivalued. In the peaks of the diagrams fiber failure takes place, in the downward parts of the diagram the load is transferred to the rest of the fibers and in the upward parts stress redistribution and strength recovery takes place (see e.g. /1/ ÷ /5/.

The arising nonmonotone stress-strain relations are derived by nonconvex superpotentials introduced first in 1981 in /6/,/7/,/8/ via the notion of generalized gradient /9/ and give rise to hemivariational inequalities and to substationarity principles /7/,/10/÷/14/. Convexification of the superpotentials is not permitted because the most important positions of equilibrium are lost as we shall see further. The hemivariational inequalities express the principle of virtual and/or complementary virtual work in inequality form for the whole structure obeying this nonconvex superpotential law.

Note that the superpotential laws describe, besides the composite material behaviour, several other phenomena appearing in composite material structures, such as the unilateral contact and friction between interfaces, the delamination and slip between adjacent laminae of a laminate etc. Let us assume that in a body Ω

there is a surface A along which debonding and slip of the material is possible. The surface A may be e.g. an interface between the adjacent laminae in laminated structures, or between a fiber and the matrix in a fiber reinforced structure. Until now much research work has been done on the description of the behaviour of an interface and on its influence on the equilibrium of the whole composite structure. Here a model based on nonconvex superpotentials is proposed. We assume that the interface can be simulated by joint elements having two strain components: the normal strain ε_N and the shearing strain ε_T with respect to A. Let σ_N and σ_T be the corresponding stresses. Then we may assume between σ_N and ε_N, and between σ_T and ε_T the laws of Fig. 1. This approach assumes A as a part of a given body Ω with different material behaviour. There is also another method more suitable for laminated plates, shells etc. We consider that each interface side belongs to a different body say Ω_1 and Ω_2 and we assume laws as the ones of Figs. 1b÷1d between $-S_N$ and $[u_N]$ and $-S_T$ and $[u_T]$. Here S_N and S_T is the normal and the tangential boundary traction of each one of the two bodies separated by the interface A and $[u_N]$ and $[u_T]$ the relative normal and tangential displacement. Thus superpotential laws of the form

$$-S_N \in \bar{\partial} j_N([u_N]) \quad (1a) \qquad \text{and} \qquad -S_T \in \bar{\partial} j_T([u_T]) \quad (1b)$$

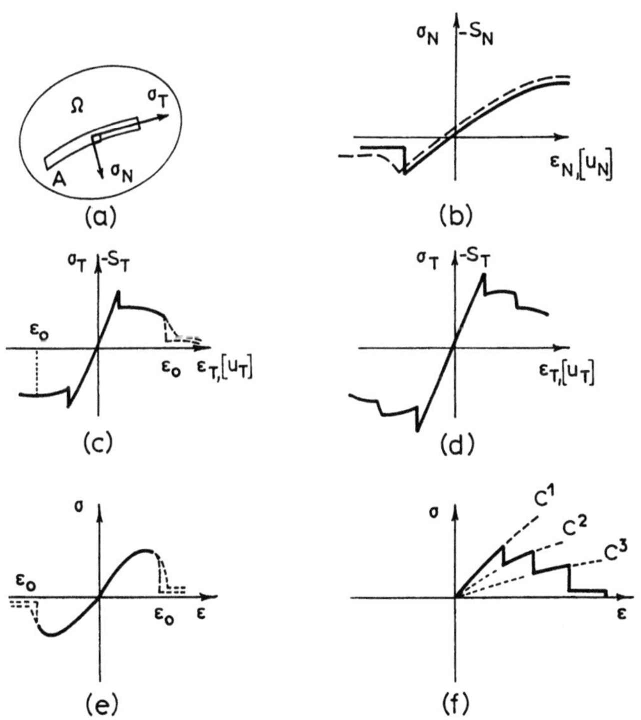

Fig. 1 Interface laws in composite structures. Stress-strain diagrams of composite materials

are considered, or, if the uncoupling of the normal and the tangential action is not permitted, of the form

$$\{-S_N,-S_T\}\in\bar{\partial}j(\{u_N,u_T\}) \tag{2}$$

where $\bar{\partial}$ denotes the generalized gradient. Thus (1a) for instance is equivalent by definition to

$$j_N^o(u_N,u_N^*-u_N) \geq -S_N(u_N^*-u_N) \quad \forall u_N^*\in \mathbb{R} , \tag{3}$$

where $j_N^o(.,.)$ is the directional differential of Clarke /9/. Analogously we can put all the aforementioned sawtooth onedimensional laws in the form

$$\sigma\in\bar{\partial}w(\varepsilon) , \tag{4}$$

where w is a nonconvex superpotential. The law (4) is defined for each element of the structure and describes a stress-strain relation on the whole length of the strain axis, i.e. (4) is a "complete"law. If we want to study in a global manner the behaviour of a structure under a given loading we may consider stress-strain diagrams of each element defined on the whole of the strain axis. If this is not the case and if only the final values ε_0 and ε_0' are known (cf. Fig. 1e), we may render the respective law complete by assuming a brittle (vertical lines) or a semibrittle behaviour (inclined lines) beyond ε_0 and ε_0'. A complete law /15/ has as a main advantage the consideration of several nonlinearities like cracking, crushing, delamination, slip etc. using a static or a quasi-static method. This approach may be applied to composite material structures, because it provides an estimation of the behaviour of the structure through purely static methods, as happens in several engineering theories, e.g., the calculation of reinforced concrete structures etc. But this is possible only using nonconvex superpotentials, since the sawtooth form is predominant in the stress-strain curves obtained experimentally in composite materials. In (4) w denotes the area between the graph and the strain axis.

Generalizations of the above onedimentional examples for threedimensional continua can be constructed in analogy to the ideally plastic and locking materials. Let us e.g. construct a threedimensional analogon to the onedimensional stress-strain diagram of Fig. 1f. One method would be to consider four elastic energy structures which intersect each other and form reentrant corners causing the stress jumps. Moreover the last energy surface should guarantee that $\sigma = 0$ after the last jump (cf. Fig. 2a). However the superpotential of a composite material will be defined using a more general method based on the consideration of certain surfaces in the strain space which lead in a natural way to the threedimensional analogon of the vertical jumps of Fig. 1f. We refer at this point to the parallel research in the theory of plasticity based on "yield" surfaces in the strain space /16/,/17/. It is worth mentioning that the concept of consecutive "yield" surfaces in the strain space seems to be the only natural way for the threedimensional description of the law of Fig. 1f: Let us define in the space $\mathbb{R}^6$ of the symmetric strain tensors $\varepsilon = \{\varepsilon_{ij}\}$ certain surfaces $F_i(\varepsilon) = 0$ $i=1,2,\dots,m$, which are continuously differentiable, and $K_1 = \{\varepsilon | F_1(\varepsilon) \le 0\}$ are closed sets. If $\varepsilon \in \mathrm{int} K_1 = \{\varepsilon | F_1(\varepsilon) < 0\}$ then we have a linear elastic behaviour governed by the elasticity tensor $C^1 = \{C^1_{ijhk}\}$. If ε satisfies $F_1(\varepsilon) = 0$ then brittle fracture appears (vertical segment in Fig. 1f) and we assume that (Fig. 2d)

$$\sigma = \{\sigma_{ij}\} = \{C^1_{ijhk}\varepsilon_{hk}\} + \lambda_1 \mathrm{grad} F_1(\varepsilon), \quad \lambda_1 \in [-a_1, 0] \ . \tag{5}$$

Here $a_1 \geq 0$ satisfies the relation

$$\{C^1_{ijhk}\varepsilon_{hk}\} - a_1 \mathrm{grad} F_1(\varepsilon) = \{C^2_{ijhk}\varepsilon_{hk}\} \ , \tag{6}$$

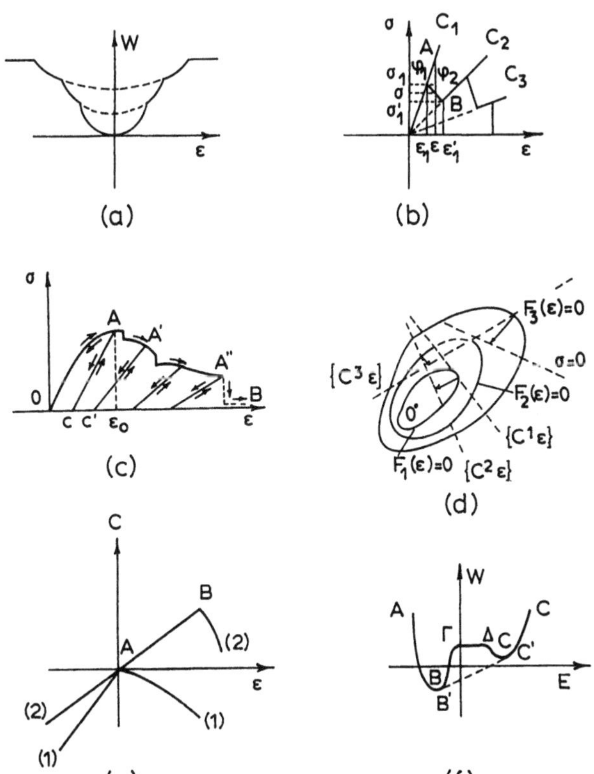

Fig. 2 3D-generalization of the sawtooth laws. Unloading and convexification

where $C^2 = \{C^2_{ijhk}\}$ is the elasticity tensor of the next linear elastic region. This region lies within the admissible strain set $K_2 = \{\varepsilon | F_2(\varepsilon) \leq 0\}$. For $\varepsilon \in \{\varepsilon | F_1(\varepsilon) > 0,\ F_2(\varepsilon) < 0\}$ we have that $\sigma_{ij} = C^2_{ijhk}\varepsilon_{hk}$ and for ε lying on the boundary of K_2 we can again assume a law of the form (5). We continue in this way until the last brittle fracture surface, after which the stress is zero, i.e. the right-hand-side in (6) is zero. The aforementioned procedure can be modified to include the case of semibrittle fracture (Fig. 2b): First we note that AB in Fig. 2b is equal to $(\varepsilon - P_{K_1}\varepsilon)(tg\phi_1 + tg\phi_2)$, where $P_{K_1}(\varepsilon)$ denotes the projection of ε onto $K_1 = \{\varepsilon | \varepsilon \leq \varepsilon_1\}$; in our case $P_{K_1}(\varepsilon) = \varepsilon_1$. Therefore we may write the law of Fig. 2b in three dimensions by assuming that for $F_1(\varepsilon) \geq 0$ we have

$$\sigma = \{\sigma_{ij}\} = \{C^1_{ijhk}\varepsilon_{hk}\} - \mu_1\{\varepsilon_{ij} - (P_{K_1}\varepsilon)_{ij}\} \quad \mu_1 > 0 \ . \tag{7}$$

Moreover for $F_1(\varepsilon) < 0$ we have a linear elastic behaviour governed by Hooke's tensor C^1. Relation (7) holds for all $\{\varepsilon_{ij}\}$ which satisfy the inequality

$$\{C^1_{ijhk}\varepsilon_{hk}\} - \mu\{\varepsilon_{ij} - (P_{K_1}\varepsilon)_{ij}\} \geq \{C^2_{ijhk}\varepsilon_{hk}\} \ . \tag{8}$$

Let us denote their set by S^1. Let $\{\varepsilon'_{1ij}\}$ be the value for which (8) is satisfied as equality. This procedure continues until the final surface $\sigma = 0$ is reached. Here we may assume convexity and closedness of K_1, K_2 etc. in order to have uniqueness of the projection. For further information on the semibrittle fracture see /12/. All the aforementioned threedimensional generalizations of the sawtooth one-dimensional laws for composite materials, due to their formulation, can be put in the general form $\sigma \in \bar{\partial} w(\varepsilon)$, where w is a nonconvex superpotential and $\bar{\partial}$ denotes the generalized gradient. The form of w is given e.g. in the case of (7) by the expression

$$w(\varepsilon) = \begin{cases} \frac{1}{2}\varepsilon_{ij}C^1_{ijhk}\varepsilon_{hk} - \mu_1 \|\varepsilon - P_{K_1}\varepsilon\|^2 (= A_1) \text{ if } \varepsilon \in S^1 \\ A_1 - \frac{1}{2}\varepsilon'_{1ij}C^2_{ijhk}\varepsilon'_{1hk} + \frac{1}{2}\varepsilon_{ij}C^2_{ijhk}\varepsilon_{hk} - \mu_2 \|\varepsilon - P_{K_2}\varepsilon\|^2 \text{ if } \varepsilon \notin S^1, \varepsilon \in S^2 \\ \text{etc.} \end{cases} \tag{9}$$

A deeper discussion would impose further restrictions on the forms of F_i, C^1 etc.; e.g. if one wants to assure the feasibility of the previous geometric constructions (i.e. nonvoid intersections of hypersurfaces, finite a_i in (6) etc.) that $\bar{\partial} w(\varepsilon)$ is nonempty and bounded for each ε and that w is locally Lipschitzian.

Until this point we have considered only the case of loading. A more general approach which takes into account all the possibilities of loading and/or unloading (not a given unloading path) leads to appropriately defined multifunctions involving ε, σ and the time increments $\mathring{\varepsilon}$ and $\mathring{\sigma}$ /12/. We refer also to /12/ for the discussion of certain numerical aspects and the comparison of the method of hemivariational inequalities with the classical incremental iterative methods. It is worth noting that even the application of an incremental iterative method of numerical analysis leads in a natural way to a hemivariational inequality. Indeed let us consider a truss obeying the loading and unloading diagram of Fig. 2c. Suppose that at the end of a load increments sequence some truss elements have stress and strains on the loading branch, say at A, and some on unloading branches, say at B. Then for the next load increment Δp we have to solve a truss problem where the truss elements at A obey the law (1) of Fig. 2e and those at B the law (2) of Fig. 2e respectively. Due to the lack of monotonicity the overall behaviour of the structure is described by a hemivariational inequality /7/,/10/. The use of the diagrams of Fig. 2e "solves" the "decision" problem concerning the branch to be followed. Appropriate approximation of the curved parts of the diagrams by lines leads to variational inequalities equivalent to linear complementarity problems (L.C.P); a part only of the "information" included into the inequalities of a L.C.P is actually used in a classical incremental iterative procedure for the loading-unloading problem. An extension of the previous thoughts to three dimensions is obvious. In Fig. 2f we show by means of the simple example of a rod-element having the strain energy w that the convexification of the nonconvex superpotential eliminates several positions of equilibrium.

2. Delamination of Laminated Plates. The Coercive Problem

The delamination is caused mainly by the interlayer normal stresses and is one of the main causes for the strength-degradation of composite plates. Here we shall describe the behaviour of the interlayer binding material together with the limit phenomena of cracking and crushing by a nonmonotone possibly multivalued law between the interlaminar bonding forces and the corresponding relative deflections. Each plate is considered as orthotropic subjected to classical boundary conditions which guarantee the coerciveness of the bending energy $a_j(\zeta_j,\zeta_j)$ of the plate Ω_j, $j=1,\dots,r$. The interlaminar binding material occupies a subset Ω'_m of Ω_j and Ω_{j+1} such that $\bar{\Omega}'_m \cap \Gamma_j = \emptyset$ and $\bar{\Omega}'_m \cap \Gamma_{j+1} = \emptyset$ where Γ_j is the boundary of the plate Ω_j assumed to be of the $C^{0,1}$-type. Index m numbers the interfaces and takes values from 1 to $m'=r-1$. We denote by $f^{(j)}$ the distributed vertical load of the j-plate. In order to model the action of the interlaminar normal stresses σ_{33}, $f^{(j)}$ is splitted into $\bar{f}^{(j)}$, which describes the interaction of the two plates, and $\bar{\bar{f}}^{(j)} \in L^2(\Omega_j)$, which represents the external loading of the j-th plate:

$$f^{(j)} = \bar{f}^{(j)} + \bar{\bar{f}}^{(j)} \quad \text{in } \Omega_j, \quad j=1,2 . \tag{10}$$

If f denotes the stress in the interlaminar binding layer, we have that $f = \bar{f}^{(1)} = -\bar{f}^{(2)}$ in Ω'_1. Then a phenomenological law conncting f with the relative deflection of the plates $[\zeta]=\zeta^{(1)}-\zeta^{(2)}$ is introduced. Let f be a multivalued non-monotone function $\hat{\beta}$ of $[\zeta]$ i.e.

$$-f\in\hat{\beta}([\zeta]) \quad \text{in } \Omega_1' . \tag{11}$$

Note that cracking as well as crushing effects of brittle or semi-brittle nature can be described by this law. The impenetrability restriction would imply a downward vertical branch for $[\zeta]=0$. However, a lightly inclined branch is considered here in order to take into account the slight compression of each plate in the vertical direction. We assume further for $\overline{f}^{(j)}$ that

$$\overline{f}^{(j)}=0 \quad \text{in } \Omega_j-\Omega_1', \quad j=1,2 . \tag{12}$$

The foregoing relations are assumed to hold on every interface between the consecutive plates. Now we shall express relation (11) in terms of a nonconvex superpotential. Thus we assume that $\beta:\mathbb{R}\to\mathcal{P}(\mathbb{R})$ is defined in the following way. Let $\beta\in L^\infty_{loc}(\mathbb{R})$ and let

$$\overline{\beta}_\varepsilon(\xi)=\operatorname*{ess\,sup}_{|\xi_1-\xi|\leq\varepsilon}\beta(\xi_1) \text{ and } \underline{\beta}_\varepsilon(\xi)=\operatorname*{ess\,inf}_{|\xi_1-\xi|\leq\varepsilon}\beta(\xi_1) . \tag{13}$$

These are increasing and decreasing functions of ε, respectively, for any $\varepsilon>0$, $\xi\in\mathbb{R}$; thus, their limits for $\varepsilon\to0$ are well-defined and let these limits be denoted by $\overline{\beta}(\xi)$ and $\underline{\beta}(\xi)$ respectively. The multivalued function $\hat{\beta}:\mathbb{R}\to\mathcal{P}(\mathbb{R})$ is now defined by the interval-expression

$$\hat{\beta}(\xi)=[\underline{\beta}(\xi), \overline{\beta}(\xi)] . \tag{14}$$

If $\beta(\xi_{\pm0})$ exists for every $\xi\in\mathbb{R}$, then a locally Lipschitz function $j:\mathbb{R}\to\mathbb{R}$ can be determined up to an additive constant, such that $\hat{\beta}(\xi)=\overline{\partial}j(\xi)$. Thus (11) is now written in the form

$$-f\in\hat{\beta}([\zeta])=\overline{\partial}j([\zeta]) . \tag{15}$$

Here the directional differential of Clarke, $j^o(.,.)$ is

$$j^o(\xi,\xi_1)=\limsup_{\lambda\to0_+,\, h\to0}\frac{1}{\lambda}\int_{\xi+h}^{\xi+h+\lambda\xi_1}\beta(t)dt . \tag{16}$$

We denote further by Z_j a closed linear subspace of $H^2(\Omega_j)$ defined by the boundary conditions of the j-plate by $||\cdot||$ the Z_j-norm and by $(.,.)$ the L^2-product. Then the problem takes the following form (cf. e.g. /11/,/12/).

Problem P: Find $\zeta^{(j)} \in Z_j$ $j=1,2,\dots,r$ such that

$$\sum_{j=1}^{r} a_j(\zeta^{(j)}, z^{(j)}-\zeta^{(j)}) + \sum_{m=1}^{m'} \int_{\Omega'_m} j_m^o([\zeta]^{(m)}, [z]^{(m)} - [\zeta]^{(m)}) d\Omega \geq$$

$$\geq \sum_{j=1}^{r} (\bar{\bar{f}}^{(j)}, z^{(j)} - \zeta^{(j)}) \qquad \forall z^{(j)} \in Z_j . \tag{17}$$

The basic assumption for the existence and the approximation of the solution of our problem, concerns the interface law: For some $\xi \in \mathbb{R}$

$$\operatorname*{ess\,sup}_{(-\infty,-\xi)} \beta^{(m)}(\xi) \leq \operatorname*{ess\,inf}_{(+\xi,+\infty)} \beta^{(m)}(\xi) , \quad m=1,2,\dots,m' , \tag{18}$$

i.e. that the nonmonotone $\beta^{(m)}$ increases "ultimately" to the right. Obviously by performing an appropriate translation of the coordinate axis, one can assume that for some $\xi \in \mathbb{R}$

$$\operatorname*{ess\,sup}_{(-\infty,-\xi)} \beta^{(m)}(\xi) \leq 0 \leq \operatorname*{ess\,inf}_{(+\xi,+\infty)} \beta^{(m)}(\xi), \quad m=1,2,\dots,m' . \tag{18a}$$

Let p be a mollifier ($p \in C_c^{\infty}(-1,+1)$, $p \geq 0$ with $\int_{-\infty}^{+\infty} p(\xi) d\xi = 1$), and let $\beta_\varepsilon = p_\varepsilon * \beta, \varepsilon > 0$, where $p_\varepsilon(\xi) = (1/\varepsilon) p(\xi/\varepsilon)$ and * denotes the convolution product. Also a Galerkin basis of Z_j, $j=1,\dots,r$ is introduced. Let Z_{jn} be the corresponding n-dimensional subspace of Z_j. The finite dimensional Problem $P_{\varepsilon n}$ reads:

Problem $P_{\varepsilon n}$: Find $\zeta_{\varepsilon n}^{(j)} \in Z_{jn}$, $j=1,2,\dots,r$ such that

$$\sum_{j=1}^{r} a_j(\zeta_{\varepsilon n}^{(j)}, z^{(j)}) + \sum_{m=1}^{m'} \int_{\Omega'_m} \beta_\varepsilon^{(m)}([\zeta_{\varepsilon n}]^{(m)})[z]^{(m)} d\Omega = \sum_{j=1}^{r} (\bar{\bar{f}}^{(j)}, z^{(j)}) ,$$

$$\forall z^{(j)} \in Z_{jn}, \quad j=1,\dots,r . \tag{19}$$

Proposition 1: Let assumptions (18a) hold. Then for $\bar{\bar{f}}^{(j)} \in L^2(\Omega_j)$, $j=1,\dots,r$ problem P has at least one solution.

Proof: We sketch only the proof here (cf. also /11/,/12/). From the assumptions (18a) we show the estimates

$$\int_{\Omega'_m} \beta_\varepsilon^{(m)}([\zeta_{\varepsilon n}]^{(m)})[\zeta_{\varepsilon n}]^{(m)} d\Omega \geq -\rho_1^{(m)} \rho_2^{(m)} \operatorname{mes} \Omega'_m , \tag{20}$$

where $\rho_1^{(m)} > 0$, $\rho_2^{(m)} > 0$. Then from (20) and the coerciveness of $a_j(.,.)$ we obtain by Brouwer's fixed point theorem that $P_{\varepsilon n}$ has a solution $\zeta_{\varepsilon n}^{(j)}$ and that $\|\zeta_{\varepsilon n}^{(j)}\| < c$ $j=1,\dots,r$. Thus $\zeta_{\varepsilon n}^{(j)} \to \zeta^{(j)}$ weakly in Z_j and strongly in $C^0(\bar{\Omega}_j)$ for $j=1,\dots,r$ due to the compact imbedding $H^2(\Omega_j) \subset C^0(\bar{\Omega}_j)$. Since $\beta_\varepsilon \in C^{\infty}(\mathbb{R})$ we have that

$\sup_{\Omega_m'}|\beta_\varepsilon^{(m)}([\zeta_{\varepsilon n}]^{(m)})| < c$ where c is independant of ε and n. Thus $\beta_\varepsilon^{(m)}([\zeta_{\varepsilon n}]^{(m)}) \to \chi^{(m)}$ weakly in $L^1(\Omega_m')$. Now we pass to the limit $n\to\infty$, $\varepsilon\to 0$ in (19) and we obtain

$$\sum_{j=1}^{r} a_j(\zeta^{(j)},z^{(j)}) + \sum_{m=1}^{m'} \int_{\Omega_m'} \chi^{(m)}[z]^{(m)}d\Omega = \sum_{j=1}^{r} (\bar{\bar{f}}^{(j)},z^{(j)}) \quad \forall z^{(j)}\in Z_j \quad j=1,\dots,r\,. \tag{21}$$

Now we prove as in /10/ p.275 (the only difference in the proof here is that $\omega=\emptyset$) that

$$\chi^{(m)}\in\hat{\beta}^{(m)}([\zeta]^{(m)}) \quad \text{a.e. on } \Omega_m' \quad m=1,\dots,m' \tag{22}$$

and the proof is completed, q.e.d.

3. The Semicoercive Problem

Suppose here that the boundary conditions of the plates Ω_j permit a rigid-plate deflection (if no interlaminar stresses are present) i.e. that $\ker a_j = \{q^{(j)}|a_j(q^{(j)},q^{(j)}) = 0\} \neq \{\emptyset\}$ and let $\ker a_j$ be finite dimensional for $j=1,\dots,r$. Then $a_j(.,.)$ is coercive on the quotient space $H^2(\Omega_j)/\ker a_j$. Moreover $||z^{(j)}||$ is equivalent to $a_j(\tilde{z}^{(j)},\tilde{z}^{(j)})^{1/2} + |q^{(j)}|_2$, where $z^{(j)} = \tilde{z}^{(j)} + q^{(j)}$, $q^{(j)}\in\ker a_j$, $\tilde{z}^{(j)}\in\ker a_j^{\perp}$ (i.e. $(\tilde{z}^{(j)},q^{(j)}) = 0 \ \ \forall q^{(j)}\in\ker a_j$). $|\cdot|_2$ denotes the L^2-norm and $j=1,\dots,r$. We denote further by $[q]$ the relative rigid deflection and by $[q]_+$ and $[q]_-$ the expressions $([q]+|[q]|)/2$ and $(-[q]+|[q]|)/2$ respectively. Note that in the plate theory $q^{(j)}$ is a polynomial of first degree at most.

The following proposition gives a necessary condition for the existence of the solution. The notation

$$\beta^{(m)}(-\infty) = \limsup_{\xi\to-\infty} \beta^{(m)}(\xi) \text{ and } \beta^{(m)}(\infty) = \liminf_{\xi\to\infty} \beta^{(m)}(\xi) \tag{23}$$

is used.

Proposition 2: Let

$$\beta^{(m)}(-\infty) \le \beta^{(m)}(\xi) \le \beta^{(m)}(\infty) \quad \forall\xi\in\mathbb{R}\,, \quad m=1,\dots,m'\,. \tag{24}$$

Then a necessary condition for the existence of a solution of Problem P is the inequality

$$\sum_{m=1}^{m'} \int_{\Omega_m'} [\beta^{(m)}(-\infty)[q]_+^{(m)} - \beta^{(m)}(\infty)[q]_-^{(m)}]d\Omega \le \sum_{j=1}^{r} (f^{(j)},q^{(j)}) \le$$

$$\leq \sum_{m=1}^{m'} \int_{\Omega'_m} [\beta^{(m)}(\infty)[q]_+^{(m)} - \beta^{(m)}(-\infty)[q]_-^{(m)}] d\Omega \quad \forall q^{(j)} \in \ker a_j \ . \tag{25}$$

If (24) holds as a strict inequality (with < instead of ≦) the same is the case for the necessary condition (25).

Proof: It results easily by setting in (17) $z^{(j)} - \zeta^{(j)} = \pm q^{(j)}$ $j=1,\ldots,r$, using the definition of $j_m^o(.,.)$, the fact that $z \to j_m^o(.,z)$ is positively homogeneous and the assumption (23). Thus we get that for q

$$\int_{\Omega'_m} j_m^o([\zeta]^{(m)},[q]^{(m)}) d\Omega \leq \int_{\Omega'_m} [\beta^{(m)}(\infty)[q]_+^{(m)} - \beta^{(m)}(-\infty)[q]_-^{(m)}] d\Omega \tag{26}$$

and analogously for $\int j_m^o([\zeta]^{(m)}, -[q]^{(m)}) d\Omega$. q.e.d.

Proposition 3: Let

$$\beta^{(m)}(-\infty) < \beta^{(m)}(\infty), \quad m=1,\ldots,m' \ . \tag{27}$$

Then if

$$\sum_{m=1}^{m'} \int_{\Omega'_m} [\beta^{(m)}(-\infty)[q]_+^{(m)} - \beta^{(m)}(\infty)[q]_-^{(m)}] d\Omega < \sum_{j=1}^{r} (f^{(j)}, q^{(j)}) <$$

$$< \sum_{m=1}^{m'} \int_{\Omega'_m} [\beta^{(m)}(\infty)[q]_+^{(m)} - \beta^{(m)}(-\infty)[q]_-^{(m)}] d\Omega \quad \forall q^{(j)} \in \ker a_j,\ q^{(j)} \neq 0 \ , \tag{28}$$

problem P has a solution.

Proof: The proof is analogous to the one of /11/ p.94 (see also /12/) and differs from the proof of prop. 1 because, instead of the estimate (20), the better estimate

$$\int_{\Omega'_m} \beta_\varepsilon^{(m)}([\zeta_{\varepsilon n}]^{(m)})[\zeta_{\varepsilon n}]^{(m)} d\Omega \geq (1-\tfrac{1}{N}) \int_{\substack{|[\tilde{\zeta}_{\varepsilon n}(x)]^{(m)}| < \frac{\delta\alpha}{N} \\ |[q_{\varepsilon n}(x)]^{(m)}| > \delta\alpha}} \beta_\varepsilon^{(m)}([\zeta_{\varepsilon n}]^{(m)})[q_{\varepsilon n}]^{(m)} d\Omega \ -$$

$$-\rho_1^{(m)}\rho_2^{(m)} \operatorname{mes} \Omega'_m \quad \text{for } m=1,\ldots,\ m'=r-1 \tag{29}$$

is used, where $\delta \in (0,1]$, $N>1$, $\alpha>0$ are appropriately chosen numbers. For the proof of (29) see /11/ p.95-97. q.e.d.

Acknowledgement

This research project was supported by the Scientific Division of NATO (grant no. 86/0694).

References

1. G.S. Holister, C. Thomas: Fibre Reinforced Materials (Elsevier, London 1966)
2. N.S. Shidharan: Elastic and Strength Properties of Continuous/Chopped Glass Fiber Hybrid Sheet Molding Compounds. Short Fiber Reinforced Composite Materials, ASTM STP 772, ed. by B.A. Sanders (ASTM, Philadelphia 1982)
3. J.G. Williams, M.D. Rhodes: Effect of Resin on Impact Damage Tolerance of Graphite/Epoxy Laminates, Proc. 6th Int.Conf. on Composite Materials, Testing and Design, ASTM STP 787, ed. by I.M. Daniel (ASTM, Philadelphia 1982)
4. M.M. Schwartz: Composite Materials Handbook (McGraw-Hill, N.York 1984)
5. C.C. Baniotopoulos, P.D. Panagiotopoulos: In Engineering Analysis of New Composites (Omega Press, London 1988)
6. P.D. Panagiotopoulos: Mech.Res.Comm. 8, 335 (1981)
7. P.D. Panagiotopoulos: Acta Mech. 48, 111 (1983)
8. P.D. Panagiotopoulos: C.R.Acad.Sc. Paris 296II, 1105 (1983)
9. F.H. Clarke: Trans.A.M.S. 205, 247 (1975)
10. P.D. Panagiotopoulos: Inequality Problems in Mechanics and Applications. Convex and Nonconvex Energy Functions (Birkhäuser Verlag, Basel, Boston 1985) Russian Transl. (MIR Publ., Moscow 1988)
11. P.D. Panagiotopoulos: In Topics in Nonsmooth Mechanics, ed. by J.J. Moreau, P.D. Panagiotopoulos, G. Strang (Birkhäuser Verlag, Basel, Boston 1988) p.75
12. P.D. Panagiotopoulos: In Nonsmooth Mechanics and Applications, ed. by J.J. Moreau, P.D. Panagiotopoulos, Springer CISM lect.notes Vol.302 (Springer, Wien, N.York)
13. Z. Naniewicz, Cz. Woźniak: Ing. Archiv 1988 (to appear)
14. Z. Naniewicz: J.Nonl.Anal.Appl. 12 (1988) (to appear)
15. P.D. Panagiotopoulos, C. Baniotopoulos: Eng. Analysis 1, 20 (1984)
16. J. Casey, P.M. Naghdi: In Constitutive Equations: Macro and Computational Aspects, ed. by K.J. Willam (A.S.M.E., N.York 1984) p.53
17. J. Casey, P.M. Naghdi: In Mechanics of Engineering Materials, ed. by C.S. Desai and R.H. Gallagher (J. Wiley, 1984) p.61

Poiseuille Flows of Incompressible Simple Fluids

G. Mayne

Département de Mathématique, Université Libre de Bruxelles,
Campus Plaine C.P. 218/1 – Boulevard du Triomphe,
B-1050 Bruxelles, Belgium

1. Introduction

In this paper, we shall use a geometrical method to show that the Poiseuille flow of any simple fluid in Noll's meaning, is generally impossible. This geometrical method will be used to solve the Poiseuille problem when it is possible.

We shall consider rectilinear, steady flows of incompressible simple fluids in a tube of any cross-section. Taking cartesian coordinates with axis x_3 parallel to the flow direction, the velocity field of the flow is defined by one function ϕ of two variables :

$$v_1 = v_2 = 0 \qquad v_3 = 2\phi(x_1,x_2) \tag{1}$$

It can be shown [1] that the three equations of motion reduce to two relations

$$\begin{cases} -c = \operatorname{div}(k_1 \operatorname{grad} \phi) \\ R'(\phi) = \operatorname{div}(k_2 \operatorname{grad} \phi) \end{cases} \tag{2}$$

where k_1, k_2 are constitutive functions depending on the modulus K of the gradient of ϕ :

$$k_i = k_i(K) \qquad K = |\operatorname{grad} \phi|$$

$$-c = \frac{\partial p}{\partial x_3}$$

The constant $-c$ is the longitudinal pressure gradient and $R(\phi)$ is an unknown function of ϕ which appears in the expression of the hydrostatic pressure p. R' means first derivative.

2. A Geometrical Way to Find the Equation of the Poiseuille Problem

In the cross-section of the tube, we shall consider orthogonal curvilinear coordinates (y_1,y_2) whose curves y_1 = constant are the curves of constant speed of the flow . Of course, these coordinates are unknown and in the change from cartesian coordinates (x_1,x_2) to coordinates (y_1,y_2) which is

$$y_1 = \phi(x_1,x_2)$$
$$y_2 = \psi(x_1,x_2) \tag{3}$$

y_1 is the same function $\phi(x_1,x_2)$ as in (1).
In these coordinates, equations (2) become

$$g_{1,1} = -\lambda - \mu R'$$
$$g_{2,1} = g_2(\eta + \nu R') \tag{4}$$

where

$$g_1 = \sqrt{g_{11}} \qquad g_2 = \sqrt{g_{22}}$$

$$\lambda = \frac{cg_1^2 k_2}{k_1'k_2 - k_1k_2'} \qquad \mu = g_1^2 \frac{k_1}{k_1'k_2 - k_1k_2'}$$

$$\eta = +\, cg_1 \frac{-k_2 + g_1k_2'}{k_1'k_2 - k_1k_2'} \qquad \nu = g_1 \frac{g_1k_1' - k_1}{k_1'k_2 - k_1k_2'} \tag{5}$$

$_{,1}$ and $_{,2}$ denote partial derivatives with respect to y_1 and y_2.
g_{11} and g_{22} are the non zero components of the metric tensor in the coordinates (y_1,y_2). The constitutive functions k_1 and k_2 are given functions of K. As

$$K = |\text{grad}\ \phi| = g_1^{-1}$$

λ, μ, η, ν are known functions of g_1.
If $k_1'k_2 - k_1k_2' = 0$, k_1 and k_2 are proportional and system (2) is trivially compatible.
Since (y_1,y_2) are plane curvilinear coordinates, g_1 and g_2 have to verify a zero total curvature condition which is :

$$\left(\frac{g_{2,1}}{g_1}\right)_{,1} + \left(\frac{g_{1,2}}{g_2}\right)_{,2} = 0 \tag{6}$$

The mechanical problem which is : find the longitudinal velocity 2ϕ of the Poiseuille flow is replaced by a geometrical one which is : find plane curvilinear coordinates whose metric tensor satisfies equations (4) and (6). This system contains three equations for two unknowns; fortunately, it is possible to obtain [2] an explicit compatibility condition which is

$$(\alpha + \beta R')(S - \gamma - \delta R' - \varepsilon R'^2 - \varphi R'') + (\lambda + \mu R')(\gamma' + \delta'R' + \varepsilon'R'^2 + \varphi'R'') + S' - \delta R'' - 2\varepsilon R'R'' - \varphi R''' = 0 \tag{7}$$

where

$$\alpha = 2(\eta + \lambda') \qquad \beta = 2(\mu' + \nu) \tag{8}$$

$$\begin{aligned}
\gamma' &= c^2 \frac{g_1 k_2^2}{k_1' k_2 - k_1 k_2'} \left(\frac{g_1}{k_2} \frac{k_2 - g_1 k_2'}{k_1' k_2 - k_1 k_2'} \right)' \\
\delta' &= 2c \frac{g_1 k_1^2}{k_1' k_2 - k_1 k_2'} \left(\frac{g_1}{k_1} \frac{k_2 - g_1 k_2'}{k_1' k_2 - k_1 k_2'} \right)' \\
\varepsilon' &= \frac{g_1 k_1^3}{2k_2(k_1' k_2 - k_1 k_2')} \left(\frac{2g_1 k_1 k_2 - g_1^2 k_1' k_2 - g_1^2 k_1 k_2'}{k_1^2 (k_1' k_2 - k_1 k_2')} \right)' \\
\varphi' &= \frac{g_1 k_1' - k_1}{k_1' k_2 - k_1 k_2'}
\end{aligned} \tag{9}$$

In (7), S is an arbitrary function of y_1, S' is its first derivative and the greek small letters are known functions of g_1; R is a function of y_1 defined previously.

3. Theorem

If the compatibility condition (7) depends explicitely on g_1, the curves of constant speed of the flow are either concentric circles or parallel straight lines.

Proof : if (7) depends explicitely on g_1, g_1 only depends on y_1 and equations (4) and (6) reduce to the following :

$$g_1' \overset{\text{def}}{=} \frac{dg_1}{dy_1} = -\lambda - \mu R' \tag{10}$$

$$g_{2,1} = g_2(\eta + \nu R') \tag{11}$$

$$g_{2,1} = g_1 F_2(y_2) \tag{12}$$

Setting $g_1 = H_1'(y_1)$, (9) gives :

$$g_2 = F_2 H_1 + G_2 \tag{13}$$

where G_2 is an arbitrary function of y_2.
From (11), it follows that

$$(\ell n\, g_2)_{,12} = 0 \tag{14}$$

Putting (13) in (14), we get

$$H_1'(F_2' G_2 - F_2 G_2') = 0 \tag{15}$$

If $F_2 \neq 0$, $G_2 = hF_2$

where h is an arbitrary constant.
The metric of the plane in the coordinates (y_1, y_2) is given by

$$ds^2 = H_1'^2 dy_1^2 + F_2^2 (H_1 + h)^2 dy_2^2 \tag{16}$$

which can be put in the form

$$ds^2 = d\bar{y}_1^2 + \bar{y}_1^2 d\bar{y}_2^2 \tag{17}$$

through the coordinate transformation

$$\bar{y}_1 = H_1 + h \qquad \bar{y}_2 = \int F_2 dy_2 \tag{18}$$

$(\bar{y}_1, \bar{y}_2)$ are plane polar coordinates which are related to cartesians' ones (x_1, x_2) through

$$\begin{aligned} x_1 &= \bar{y}_1 \cos \bar{y}_2 \\ x_2 &= \bar{y}_1 \sin \bar{y}_2 \end{aligned} \tag{19}$$

The lines $\bar{y}_1$ = constant or y_1 = constant are concentric circles.
If $F_2 = 0$, (17) reduces to

$$ds^2 = d\bar{y}_1^2 + d\bar{y}_2^2$$

with $d\bar{y}_1 = H_1' dy_1 \qquad d\bar{y}_2 = G_2 dy_2$

In this case, the lines $\bar{y}_1$ = constant or y_1 = constant are parallel straight lines.

So, we may conclude that the Poiseuille flow of any simple fluid is only possible in a pipe of circular cross section, between two coaxial pipes of circular cross section of between two parallel planes. In these cases, the geometrical method considered here makes it possible to solve the problem and to get analytical solutions if an integrating factor can be obtained for a first order ordinary differential equation of the type

$$xdy = [L(y)dx + M(y)]dx \tag{20}$$

In the special case $c = 0$, $L = 0$ and the solution of (20) is trivial.

4. Poiseuille Flows for Special Simple Fluids

Now, we shall show that Poiseuille flows exist for which the curves of constant speed have constant curvature despite the fact that g_1 depends on y_1 and y_2.

Of course, in this case, the constitutive functions k_1 and k_2 must be such that the compatibility condition (7) doesn't depend explicitely on g_1 and the flows are no more radially symmetric.

Problem : find R', k_1, k_2 such that (7) does not depend explicitely on g_1 and that the curvature C_1 of the curves y_1 = constant only depends on y_1.

From $(4)_2$, the curvature C_1 is given by

$$C_1 = \frac{1}{g_1} (\eta + \nu R') \tag{21}$$

as $g_{1,2} \neq 0$, $C_{1,2} = 0$ if

$$(\eta' + \nu' R') g_1 - (\eta + \nu R') = 0 \tag{22}$$

If $c \neq 0$, there is only one possibility to satisfy the two conditions of the problem; that is

$$R' = k \qquad \eta = q g_1 \qquad \nu = p g_1 \tag{23}$$

In these expressions and in the following, latin small letters indicate arbitrary constants.

With (23), C_1 is a constant; all the curves y_1 = constant have the same constant curvature and the corresponding velocity profile is not physically admissible.

If $c = 0$ and $g_{1,2} \neq 0$, two cases arise giving Poiseuille flows for which the curves of constant speed have constant curvature.

I

$$\begin{aligned} R' &= \frac{1}{b y_1 + q} \qquad k_2 = \frac{1}{p} (g_1 - a k_1) \\ \mu &= - p g_1^2 + 2 b g_1 + s + w \sqrt{p g_1^2 - 2 b g_1 - s} \end{aligned} \tag{24}$$

II

$$\begin{aligned} &\tfrac{3}{2} r R''' + r^2 R' R'' - r s p R'^3 = - \left(\frac{R'''}{2R'}\right)' + 2 s p R' R'' \\ &k_2 = \frac{1}{p} (g_1 - a k_1) \qquad \mu = - p g_1^2 + r g_1 + s \end{aligned} \tag{25}$$

5. Geometrical Method of Resolution

There are special cases of (24) and (25) for which analytical solutions can be obtained for the corresponding Poiseuille problem. To apply the geometrical method of resolution presented here in each of these cases, we have first to solve the differential system (4) (6) in the components of the metric tensor g_{11} and g_{22}, second to find the coordinate transformation from the curvilinear coordinates $(y_1,$

$y_2)$ to isothermal one's (z_1,z_2) for which the two non zero components of the metric tensor are equal; finally, we have to search for the conformal mapping from (z_1,z_2) to cartesian coordinates (x_1,x_2). This last step can be performed in the following way :

$$\begin{cases} x_1 = X_1(y_1,y_2) \\ x_2 = X_2(y_1,y_2) \end{cases}$$

$$dx_1^2 + dx_2^2 = \sigma^2(y_1,y_2)(dy_1^2 + dy_2^2)$$

$$X_1 + iX_2 = f(w)$$

where f is a holomorphic function of $w = y_1 + iy_2$.

If we put $f'(w) = Re^{i\Theta}$, we have $R = \sigma(y_1,y_2)$.

As $f'(w)$ is a holomorphic function of w too, R and Θ have to verify Cauchy-Riemann equations which, in variables y_1, y_2 take the form

$$\begin{cases} \dfrac{\partial\Theta}{\partial y_1} = -\dfrac{1}{\sigma}\dfrac{\partial\sigma}{\partial y_2} \\ \dfrac{\partial\Theta}{\partial y_2} = \dfrac{1}{\sigma}\dfrac{\partial\sigma}{\partial y_1} \end{cases} \tag{26}$$

σ being known in terms of y_1, y_2, we get $\Theta(y_1,y_2)$ by solving this system.

$$f'(w) = \sigma e^{i\Theta} = \frac{\partial X_1}{\partial y_1} - i\,\frac{\partial X_1}{\partial y_2}$$

so that

$$\begin{cases} \dfrac{\partial X_1}{\partial y_1} = \sigma \cos\Theta \\ \dfrac{\partial X_1}{\partial y_2} = -\sigma \sin\Theta \end{cases} \tag{27}$$

We get $X_1(y_1,y_2)$ by integration.

6. Poiseuille Flows between Eccentric Circular Cylinders

The two special cases of (24) and (25) we shall consider are the following :

A. Case I with $s = w = 0$.

$$R' = \frac{1}{by_1 + q} \qquad \mu = -pg_1^2 + 2bg_1 \tag{28}$$

The corresponding constitutive functions are

$$k_1 = -\frac{mg_1}{pg_1 - 2b} \qquad k_2 = \frac{pg_1^2 + (am - 2b)g_1}{p^2g_1 - 2bp} \tag{29}$$

Solving equations (4) (6), it can be shown that the metric can be put in the isothermal form :

$$ds^2 = \frac{4}{(\bar{y}_1^2 + \bar{y}_2^2)^2} (d\bar{y}_1^2 + d\bar{y}_2^2) \tag{30}$$

where

$$\bar{y}_1 = - \frac{p}{by_1 + q}$$

Using (26) and (27), the conformal mapping from these coordinates to cartesian one's (x_1, x_2) can be obtained :

$$\begin{aligned} x_1 - B &= 2 \cos A \frac{\bar{y}_1}{\bar{y}_1^2 + \bar{y}_2^2} + 2 \sin A \frac{\bar{y}_2}{\bar{y}_1^2 + \bar{y}_2^2} \\ x_2 - D &= 2 \sin A \frac{\bar{y}_1}{\bar{y}_1^2 + \bar{y}_2^2} - 2 \cos A \frac{\bar{y}_2}{\bar{y}_1^2 + \bar{y}_2^2} \end{aligned} \tag{31}$$

These formulas, where A, B, D are arbitrary constants, can be inverted in order to get y_1 in term of x_1, x_2; this last expression gives the longitudinal velocity of the corresponding Poiseuille flow :

$$y_1 = \phi(x_1, x_2) = \frac{p}{b} \frac{(x_1 - B)^2 + (x_2 - D)^2}{\cos A(x_1 - B) - \sin A(x_2 - D)} - \frac{q}{b} \tag{32}$$

B. Case II with $r = s = 0$

$$\begin{aligned} \frac{R'''}{R'} &= h \qquad & \mu &= - pg^2_1 \\ k_1 &= k \qquad & k_2 &= \frac{1}{p} (g_1 - ak) \end{aligned} \tag{33}$$

1. $h > 0 : h = n^2$

From (33), it comes that

$$R' = L \, sh(ny_1 + m) \tag{34}$$

where L is an arbitrary constant.

System (4) (6) can be solved and gives rise to the following metric :

$$ds^2 = \frac{n^2/p^2L^2}{[\cos(ny_2 + u) - ch(ny_1 + m)]^2} (dy_1^2 + dy_2^2) \tag{35}$$

The conformal mapping to cartesian coordinates is given by

$$x_1 - B = - \frac{1}{pL} \frac{\cos A \, sh(ny_1 + m) + \sin A \sin(ny_2 + u)}{ch(ny_1 + m) - \cos(ny_2 + u)} \tag{36}$$

$$x_2 - D = \frac{1}{pL} \frac{- \sin A \, sh(ny_1 + m) + \cos A \sin(ny_2 + u)}{ch(ny_1 + m) - \cos(ny_2 + u)}$$

By inversion, we get the longitudinal velocity ϕ of the corresponding Poiseuille flow :

$$y_1 = - \frac{1}{n} \text{argth} \frac{2pL[\cos A(x_1 - B) + \sin A(x_2 - D)]}{1 + p^2L^2[(x_1 - B)^2 + (x_2 - D)^2]} - \frac{m}{n} \tag{37}$$

2. $h < 0 : h = - n^2$

From (33), it comes that

$$R' = L \sin(ny_1 + u) \tag{38}$$

In this case, we obtain the following metric

$$ds^2 = \frac{n^2/p^2L^2}{[ch(ny_2 + m) + \cos(ny_1 + u)]^2} (dy_1^2 + dy_2^2) \tag{39}$$

The change to cartesian coordinates is

$$x_1 - B = \frac{1}{pL} \frac{\cos A \sin(ny_1 + u) - \sin A \, sh(ny_2 + m)}{ch(ny_2 + m) + \cos(ny_1 + u)}$$

$$x_2 - D = \frac{1}{pL} \frac{\cos A \, sh(ny_2 + m) + \sin A \sin(ny_1 + u)}{ch(ny_2 + m) + \cos(ny_1 + u)} \tag{40}$$

By inversion, we get

$$y_1 = \frac{1}{n} \text{arctg} \frac{2pL[\cos A(x_1 - B) + \sin A(x_2 - D)]}{1 - p^2L^2[(x_1 - B)^2 + (x_2 - D)^2]} - \frac{u}{n} \tag{41}$$

In both cases B_1 and B_2, the coordinates (y_1, y_2) are nothing else than bipolar coordinates in the cross section of the tube so that the coordinate lines are non concentric orthogonal circles. The corresponding Poiseuille flows are rectilinear flows between two eccentric parallel circular cylinders. From (37), the curves y_1 = constant are circles whose centres have coordinates α, β given by

$$\alpha = B - \frac{1}{pL} \frac{ch(ny_1 + m)}{sh(ny_1 + m)} \cos A$$

$$\beta = D - \frac{1}{pL} \frac{ch(ny_1 + m)}{sh(ny_1 + m)} \sin A$$

and whose radii $R = \dfrac{1}{pL \, sh(ny_1 + m)}$

Considering a flow between two parallel circular cylinders whose axises do not coincide, let α_i, β_i, R_i and α_e, β_e, R_e the coordinates of the centres and the radii of the interior and exterior circles C_i and C_e respectively in the cross section of the flow. These data must be such that

$$d + R_i < R_e \tag{42}$$

where d is the distance between the centres of the circles given by

$$d^2 = (\alpha_i - \alpha_e)^2 + (\beta_i - \beta_e)^2$$

We shall suppose that the interior circle does not move and the exterior one moves with velocity v so that the equations of C_i and C_e in coordinates (y_1, y_2) are $y_1 = 0$ and $y_1 = v$.
We have thus six equations to determine the six constants m, n, B, D, A, L :

$$R_i = \frac{1}{pL \text{ sh } m} \qquad R_e = \frac{1}{pL \text{ sh}(nv + m)} \tag{43}$$

$$\alpha_i = B - \frac{1}{pL} \cos A \coth m$$
$$\beta_i = D - \frac{1}{pL} \sin A \coth m \tag{44}$$

$$\alpha_e = B - \frac{1}{pL} \cos A \coth(nv + m)$$
$$\beta_e = D - \frac{1}{pL} \sin A \coth(nv + m) \tag{45}$$

A, m, n are given by

$$\text{tg } A = \frac{\beta_i - \beta_e}{\alpha_i - \alpha_e} \tag{46}$$

$$\text{ch } m = \frac{1}{2R_i} \left[\frac{(R_e^2 - R_i^2)\cos A}{\alpha_i - \alpha_e} - \frac{\alpha_i - \alpha_e}{\cos A}\right] \tag{47}$$

$$\text{ch}(nv + m) = \frac{1}{2R_e}\left[\frac{(R_e^2 - R_i^2)\cos A}{\alpha_i - \alpha_e} + \frac{\alpha_i - \alpha_e}{\cos A}\right] \tag{48}$$

From (42), it follows that the second members of (47) and (48) are greater than 1. L, B, D are obtained from (43) and (44)

References

1. R.L. Fosdick and J. Serrin : Proc.R.Soc.Lond. A332 (1973), 311-333
2. G. Mayné, in "Trends in Appl.of pure Math.to Mech." E.Kröner and K.Kirchgässner, eds. - Lecture Notes in Physics 249, Springer, Berlin (1986)

Properly Posed Boundary Condition and the Existence Theorem for Supefluid He^4

Z. Peradzyński

Institute of Fundamental Technological Research,
Świętokrzyska 21, PL-00-049 Warsaw, Poland

1. Introduction

Superfluidity and superconductivity belong to the most striking manifestations of quantum effects on the macroscopic level. At low temperatures $T < 2.17$ K helium He^4 can be viewed as a two component fluid consisting of the normal fluid component which is a classical viscous fluid and the superfluid component which is an ideal fluid without viscosity. In addition, the superfluid component carries no entropy ($s = 0$) and as it follows from some general quantum mechanical considerations it should be irrotational. Vorticity is, however, possible in the form of quantized vortex lines when the region occupied by the superfluid becomes multiply connected, e.g. it contains a number of singular lines (diameter $\simeq$ inter-atomic distance). Any of such lines carries the single quantum of circulation. The quantum of circulation is $\varkappa = \frac{h}{m} \simeq 10^{-3}cm^2/sec$, where m is the atomic mass of He^4. Unlike classical fluids superfluid vortex lines do not move together with the fluid. The normal fluid interacts with the superfluid vortex lines through excitation quasiparticles, phonons and rotons. The difference between the velocity of superfluid vortex line and its mother fluid causes in turn the appearance of the Magnus force. Accordingly the local vortex lines velocity is assumed to be [1]

$$v_L = v_s + \beta'\rho_s(v_n - v_s) + \beta\rho_s\vec{\nu}_s \times (v_n - v_s), \tag{1}$$

where v_s is the superfluid velocity, ρ_s its density, v_n, ρ_n the velocity and density of the normal component, whereas $\vec{\nu}_s$ denotes the unit vector tangent to the vortex line. The coefficients β', β as well as ρ_s, ρ_n are temperature dependent but in the following we assume that they are constant, thus confining our further consideration to small variations of the temperature . We should emphasise also that by assuming ρ_s, ρ_n constant we exclude all acoustic phenomena such as the first and the second sound. The second sound arises from the possibility of converting the superfluid into the normal fluid and vice versa. Thus in this case ρ_n, ρ_s can vary, leaving the total density $\rho = \rho_s + \rho_n$ unchanged.

By excluding these interesting phenomena we want to concentrate on the vorticity dynamics in superfluid helium. Although the concept of an ideal fluid can be applied to the superfluid helium to a great extent, from the mathematical point of view, there is a fundamental difference between them. To explain this, let us consider a solid boundary in the flow of a classical fluid. In that case one assumes that the component of the velocity normal to the boundary is zero. For an Eulerian ideal fluid the vorticity is transported together with the fluid. In other words a fluid element which initially was at the boundary is transported all the time along the boundary. Therefore it does not matter if and how the vorticity is produced at the boundary. If it stays at the boundary never getting into the interior, it does not influence the flow inside. According to expression (1) the situation changes dramatically in the case of the superfluid. Even for superfluid velocities $\mathbf{v}_s$ remaining tangent to the boundary, the vortex lines can be transported into the interior of the flow, thus influencing the whole flow. For this reason one can not expect uniqueness of the initial boundary problem when $\mathbf{v}_s$ at $t = 0$ is given, $\mathbf{v}_s \cdot \vec{N} = 0$ is prescribed at the boundary and nothing is said about the vorticity flux through the boundary.

Because of apparent mathematical difficulties of working with a flow containing a great number of singular vortex lines we will use in the following the two-fluid model with smeared out superfluid vorticity. This procedure can be justified in the case of large densities of well ordered (not tangled) vortex lines. Let us note that for $\Omega = 1$ radian $\sec^{-1}$ this density is $2\cdot 10^3 \mathrm{cm}^{-2}$.

2. Equations and Boundary Conditions

According to Hall, Vinen, Bekarevich, Khalatnikov [1,2] the mean field equations for the superfluid component are

$$\frac{\partial \mathbf{v}_s}{\partial t} + \mathbf{v}_s \cdot \nabla \mathbf{v}_s + \nabla \mu = \vec{\omega} \times (\mathbf{v}_s - \mathbf{v}_L),$$
$$\operatorname{div} \mathbf{v}_s = 0, \qquad \vec{\omega} = \operatorname{rot} \mathbf{v}_s, \qquad \rho_s = \mathrm{const}, \tag{2}$$

where μ denotes the chemical potential. These equations must be supplemented by the Navier-Stokes type equations for the normal component

$$\frac{\partial v_n}{\partial t} + v_n \cdot \nabla v_n + \nabla p = -\frac{\rho_s}{\rho_n}\vec{\omega} \times (v_s - v_L) + \nu_n \Delta v_n,$$

$$\operatorname{div} v_n = 0, \qquad \rho_n = \text{const} \tag{3}$$

and by the expression for the vortex line velocity

$$v_L = v_s + \beta' \rho_s (v_n - v_s) + \beta\rho_s \frac{\vec{\omega}}{|\omega|} \times (v_n - v_s), \tag{4}$$

This equation will be considered in the four-dimensional domain Ω in the space-time E (Fig. 1) which is confined between two hyperplanes $t = 0$ and $t = \tau$. The side walls of Ω will be denoted by Σ $(= \bar{\Sigma})$. As one may expect the initial and the boundary conditions for the normal component are the same as in the case of Navier-Stokes equations

$$v_n|_{t=0} = \overset{\circ}{v}_n, \qquad v_n|_{\Sigma} = a, \tag{5}$$

whereas for the superfluid component apart from the initial condition

$$v_s|_{t=0} = \overset{\circ}{v}_s \tag{6}$$

one assumes that only the normal to the boundary component of v_s is given for $t \in [0,\tau]$

$$(v_s \cdot \vec{N})|_{\Sigma} = \varphi, \tag{7}$$

where $\vec{N} = \vec{N}(t,x)$ is the spatial three-dimensional unit normal to $\Sigma_t = \{t=\text{const}\} \cap \Sigma$. It is clear that a and φ must satisfy additional compatibility conditions coming from the fact that v_s, v_n are divergent free, i.e.for any $t \in [0,t]$

$$\int_{\Sigma_t} a \cdot \vec{N}\, d\sigma = 0 \qquad \text{and} \qquad \int_{\Sigma_t} \varphi\, d\sigma = 0, \tag{8}$$

As we have already noticed the conditions (5), (6), (7) do not guarantee the uniqueness of the solution.

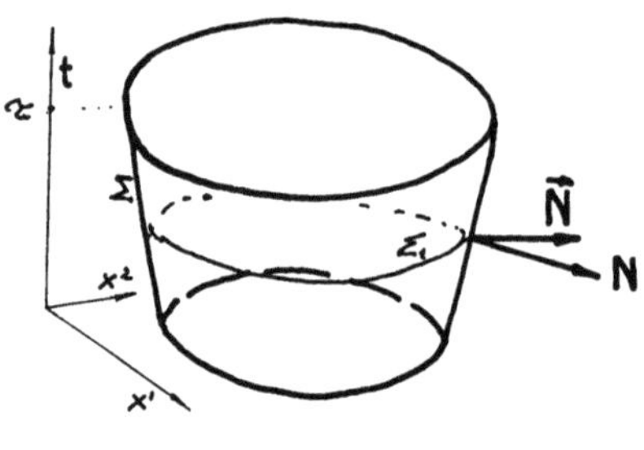

Fig. 1

In spite of lack of viscosity and as it follows from experiments the superfluid vortex lines are generated at the walls. Therefore we need another boundary condition expressing the rate of production of superfluid vorticity at the walls. This amounts to assume that the vorticity $\vec{\omega}$ at the boundary is given and forces us to use the vorticity equations

$$\frac{\partial \vec{\omega}}{\partial t} + \mathrm{rot}(\vec{\omega} \times \mathbf{v}_L) = 0, \qquad \mathrm{div}\, \vec{\omega} = 0. \tag{9}$$

The constraint $\mathrm{div}\, \vec{\omega} = 0$ for the initial condition is automatically satisfied if $\overset{o}{\omega} = \mathrm{rot}\, \overset{o}{\mathbf{v}}_s$. However, this equation leads also to some constraints for the boundary conditions. Similar problems were encountered in case of Euler equations [3,4]. As can be seen there, straightforward derivations of these constraints lead to lenghty and coordinate dependent expressions. To avoid it and to make the analysis more transparent we reformulate the problem in a geometrical setting.

3. Geometrical Content of the Superfluid Equations

In order to explain quantum properties of superfluid helium the Schrödinger equation has been used [5]. The wave function depending on t, x^1, x^2, x^3, which describes average properties of a simple particle of the fluid is called in this case "the order parameter". The role of the external potential is played by the chemical potential μ. It expresses the influence of other particles on the motion of the "average" particle. It seems therefore, that the proper object to start with is the action 1-form $A = -H_0 dt + \mathbf{p}\cdot d\mathbf{x}$, where $H_0 = m(\frac{1}{2} v_s^2 + \mu)$, $p = m\mathbf{v}_s$. Omitting for the convenience the atomic mass of He^4 we define

$$A = -(\tfrac{1}{2}v^2 + \mu)\, dt + \mathbf{v}_s \cdot d\mathbf{x}, \tag{10}$$

where $\mathbf{v}\cdot d\mathbf{x} = v_\alpha \cdot dx^\alpha$, $\alpha = 1, 2, 3$ (the summation convention is observed). As the action in quantum mechanics is quantized one comes to the conclusion that by integrating A along any closed contour C in the space-time one obtains the number of vortex lines contained within this contour.

$$\oint_C A = \frac{\hbar}{m} \times \text{number of vortex lines within } C,$$

where m is the atomic mass of He^4.

On the other hand the vortex line density can be measured by defining a certain 2-form, call it vorticity form, $J_{\mu\nu}dx^{\mu}\wedge dx^{\nu}$ in the space-time.

Definition. Given a point in the space-time, for any pair of infinitesimal vectors **X**, **Y** *we define*

$$J(\mathbf{X},\mathbf{Y}) = \frac{1}{2}\, J_{\mu\nu}(X^{\mu}Y^{\nu} - X^{\nu}Y^{\mu}), \qquad \mu,\nu = 0,1,2,3 \tag{11}$$

to be equal to $\frac{\hbar}{m}$ *multiplied by the number of vortex lines passing through the oriented surface element (parallelepiped) spanned on vectors* **X** *and* **Y**.

By definition for any two-dimensional surface S with a piecewise smooth ∂S we have

$$\int_S J = \oint_{\partial S} A.$$

Therefore J and A are related by the Stokes law

$$dA = J, \tag{12}$$

where d stands for the exterior derivative. By definition $dA = dA_{\nu} \wedge dx^{\nu} = A_{\nu,\mu}dx^{\mu} \wedge dx^{\nu}$. Applying once more exterior derivative to both sides of (12) one obtains

$$dJ = 0, \tag{13}$$

which expresses the conservation of vorticity.

To obtain further consequences of Eq.(12) the properties of J must be examined. Although we work in the Galillean space-time, it is convenient to introduce the four component velocity of vortex lines

$$\mathbf{v} = (1,\mathbf{v}_L)$$

with the first component equal to 1 as $\frac{dt}{dt} = 1$. For any k-form Q we define a scalar product $\mathbf{V} \lrcorner Q$ which is a (k-1)-form

$$(\mathbf{V} \lrcorner Q)\ (\underset{1}{\mathbf{X}},\underset{2}{\mathbf{X}},\ldots,\underset{k-1}{\mathbf{X}}) := Q(\mathbf{V},\underset{1}{\mathbf{X}},\ldots,\underset{k-1}{\mathbf{X}}).$$

Then we have

$$\mathbf{V} \lrcorner J = 0. \tag{14}$$

Eq.(14) expresses the fact that J represents the current of vorticity moving with velocity **V**. Consider in some inertial frame a vector **Y** representing a small segment of a line and let

$\mathbf{X} = \delta t(1,0,0,0)$ then according to the definition of J, $|J(\mathbf{X},\mathbf{Y})|$ is the number of vortex lines crossing the segment $\mathbf{Y}$ during the time interval δt. In particular, if this coordinate system is moving together with vortex lines the number $|J(\mathbf{X},\mathbf{Y})|$ is equal to zero - no lines are crossing $\mathbf{Y}$. On the other hand in this co-moving system $\mathbf{V} = (1,0,0,0)$ and therefore for any $\mathbf{Y}$ we have

$$J(\mathbf{V},\mathbf{Y}) \doteq 0. \tag{15}$$

This equality is, however, independent of the coordinate system which proves (14).

By inspection it can be seen from (15) that in the co-moving frame J is purely spatial form, i.e. $J_{0a} = -J_{a0} = 0$ and let us denote it by $\hat{\omega}$

$$J \doteq \hat{\omega}_{ab}\, dx^a \wedge dx^b, \qquad a,b = 1,2,3.$$

Transforming it back to the laboratory system, $dx^a \longrightarrow dx^a - v_L^a dt$, one obtains

$$J = \hat{\omega} + (\mathbf{v}_L \lrcorner\, \hat{\omega}) \wedge dt \tag{16}$$

which is the general form of J in any system of coordinates preserving the absolute time.

Let us emphasis that Eqs.(12)-(14) are coordinate independent. Decomposition (16) is also valid in any system of coordinates provided that t is the absolute time (not just the first coordinate). Let us note, that Galillean physics is endowed with absolute time and Euclidean metric on each hypersurface $t = \text{const}$ which in special (inertial) frames becomes constant in time and space. The contravariant counterpart of this metric is needed, for example, in order to define v_s^2 in curvilinear coordinates. As t, and thus also dt are given *a priori*, we may define the spatial exterior derivative $\bar{d}$ of any form ξ

$$\bar{d}\xi := d\xi - dt \wedge \dot{\xi}, \qquad \text{where } \dot{\xi} = \frac{\partial}{\partial t}\xi. \tag{17}$$

As one may expect the 2-form $\hat{\omega}$ is related by the standard duality in the Euclidean 3-space to the vorticity vector-field $\vec{\omega}$, $\omega^a = \epsilon^{abc}\hat{\omega}_{bc}$. Thus in the cartesian coordinates we have the following correspondences

$$\begin{aligned} &\hat{\omega} \longleftrightarrow \vec{\omega}, \qquad \mathbf{v}_L \lrcorner\, \hat{\omega} \longleftrightarrow \vec{\omega} \times \mathbf{v}_L, \\ &\hat{\omega}(\mathbf{X},\mathbf{Y}) \longleftrightarrow \det\,[\vec{\omega},\vec{\mathbf{X}},\vec{\mathbf{Y}}]. \end{aligned} \tag{18}$$

Similarly other useful correspondences can be derived

$$\bar{d}\hat{\omega} \longleftrightarrow \operatorname{div} \vec{\omega}, \qquad\qquad (19)$$

$$\bar{d}(\mathbf{v}_L \lrcorner \hat{\omega}) \longleftrightarrow \operatorname{rot}(\vec{\omega} \times \mathbf{v}_L).$$

Another important consequence of Eq.(13) and Eq.(14) can be easily obtained by using the notion of the Lie derivative. Given a vector-field on a manifold E, say $\mathbf{V}$, it generates a flow on E. One can think of V as of its velocity. This flow moves points of E and takes them into other places. Therefore it transforms functions into other functions. Consequently, the gradient of a function will be transformed into gradient of the transformed function. In this way differential forms as generated by gradients are transformed into other differential forms. In a similar way the flow carries vectors and tensors. Given any form on E (or a tensor-field), one defines its Lie derivative at a point p along the vector-field $\mathbf{V}$ by comparing the form defined at p with the one which was moved to p by the flow from some neighbouring point after time δt. The difference divided by δt gives the Lie derivative. The Lie derivative $\mathcal{L}_\mathbf{V}$ and exterior derivative d satisfy the useful identity [6,7]

$$\mathcal{L}_\mathbf{V} W = \mathbf{V} \lrcorner\, dW + d(\mathbf{V} \lrcorner\, W). \qquad (20)$$

Applying this to the vorticity form J and using (14) one arrives at

$$\mathcal{L}_\mathbf{V} J = 0, \qquad (21)$$

which expresses the local form of a Kelvin law of conservation of vorticity.

Summarizing, we have the definitions of the basic objects

$$A = -\left(\frac{1}{2}\, v^2 + \mu\right) dt + \mathbf{v}_s, \qquad J = \hat{\omega} + (\mathbf{v}_L \lrcorner\, \hat{\omega}) \wedge dt$$

where $\mathbf{v}_s = v_{sa} dx^a$ and $\hat{\omega} = \omega_{ab} dx^a \wedge dx^b$ are understood as spatial 1-form and 2-form, respectively. Then, there are equations for A and J

$$dA = J \longrightarrow dJ = 0 \longrightarrow \mathcal{L}_\mathbf{V} J = 0.$$

Developing the first equation of this sequence we have

$$dt \wedge \dot{\mathbf{v}}_s - \bar{d}\left(\frac{1}{2}\, v_s^2 + \mu\right) \wedge dt + \bar{d}\mathbf{v}_s = \hat{\omega} + (\mathbf{v}_L \wedge \hat{\omega}) \wedge dt.$$

Comparing terms with dt and using isomorphism (18) and (19) we obtain

$$\frac{\partial \mathbf{v}_s}{\partial t} + \nabla\left(\frac{1}{2}\, v_s^{\,2} + \mu\right) = - \vec{\omega} \times \mathbf{v}_L, \tag{22}$$

which by the identity $\nabla\frac{1}{2}\mathbf{v}^2 + \omega \times \mathbf{v} = \mathbf{v}\cdot\nabla\mathbf{v}$ is equivalent to Eq.(2) whereas the other terms give

$$\mathrm{rot}\ \mathbf{v}_s = \vec{\omega}. \tag{23}$$

In a similar way the equation $dJ = 0$ leads to

$$\bar{d}\omega + dt \wedge \dot{\hat{\omega}} + \bar{d}(\mathbf{v}_L \times \hat{\omega}) \wedge dt = 0,$$

which is equivalent to the following set of equations

$$\frac{\partial\vec{\omega}}{\partial t} + \mathrm{rot}(\vec{\omega} \times \mathbf{v}_L) = 0,$$

$$\mathrm{div}\ \vec{\omega} = 0. \tag{24}$$

The equation $\mathscr{L}_V J = V \lrcorner\, dJ = 0$ in an inertial frame corresponds to

$$\left[\frac{\partial}{\partial t} + \mathbf{v}_L\cdot\nabla\right]\vec{\omega} - (\vec{\omega}\cdot\nabla)\mathbf{v}_L - \vec{\omega}\ \mathrm{div}\ \mathbf{v}_L = 0 \tag{25}$$

which is an ordinary differential equation for $\dot{\vec{\omega}}$ along the trajectories of $V = (1, \mathbf{v}_L)$.

To close the system of superfluid equations we need the continuity equation and the state equation. They are

$$dq = 0, \qquad \rho_s = \mathrm{const},$$

where $q = \rho_s(dx^1 \wedge dx^2 \wedge dx^3 + v^1\, dt \wedge dx^2 \wedge dx^3 + v^2 dx^1 \wedge dt \wedge dx^3 + v^3 dx^1 \wedge dx^2 \wedge dt)$. Let us note that until this moment all our considerations were of purely kinematical nature. The only point where the dynamics must enter is the expression for $\mathbf{v}_L$, i. e. Eq.(4). To obtain it one considers forces acting on the vortex lines. Expression (4) becomes singular if $\vec{\omega} = 0$. Therefore either we will assume that in the region considered $\vec{\omega} \neq 0$ or we may regularize this expression, for instance, by assuming that

$$\mathbf{v}_L = \mathbf{v}_s + \beta'\rho_s(\mathbf{v}_n - \mathbf{v}_s) + \beta\rho_s \frac{\vec{\omega}}{\sqrt{\vec{\omega}^2 + \varepsilon^2}} \times (\mathbf{v}_n - \mathbf{v}_s). \tag{26}$$

To close this section let us note that contrary to the usual approach where vorticity is of secondary importance, in this approach the vorticity seems to be as important as the velocity.

4. The Boundary Conditions for the Vorticity

Any form defined on a manifold M can be transported (pulled back) on its submanifolds $\Sigma \subset M$. Suppose that Σ is given locally by $x^\nu = x^\nu(y^1,\dots,y^p)$, x being coordinates of the manifold and y - coordinates on the submanifold, then we have

$$dx^\nu \longrightarrow \frac{\partial x^\nu}{\partial y^\alpha}\, dy^\alpha .$$

Consequently, any form $\omega_{\nu..\mu} dx^\nu \wedge \dots \wedge dx^\mu$ is transformed into $\omega_{\nu..\mu} \frac{\partial x^\nu}{\partial y^\alpha} \dots \frac{\partial x^\nu}{\partial y^\alpha}\, dy^\alpha \wedge \dots \wedge dy^\beta$, which, indeed, is a form defined on the submanifold. Suppose $\partial\Omega$ is piecewise smooth, and suppose that all considered forms A, J, dJ,... are smooth in $\bar{\Omega}$ (i.e. they can be extended to some $\Omega' \supset \Omega$ in a smooth way). In such a case they can be pulled back on $\partial\Omega$ (except of singularities of $\partial\Omega$). Let A^*, J^*, $(dJ)^*$, ... denote these pull-back forms on $\partial\Omega$. Since the transport of forms commutes with exterior differentiation, we have

$$dJ = 0 \longrightarrow (dJ)^* = 0 \longrightarrow dJ^* = 0 .$$

Note, that in dJ d acts on Ω whereas in dJ^* it acts on $\partial\Omega$. Therefore the last equation is an intrinsic partial differential equation on $\partial\Omega$. In other words the coefficients $J_{\mu\nu}$ of the form J cannot be prescribed arbitrarily on $\partial\Omega$, they must satisfy

$$dJ^* = 0 .$$

For example, on parts of the boundary given by t = const the above equation is equaivalent to div $\vec{\omega} = 0$.

The equation $(dJ)^* = 0$ can be expressed also in terms of the coefficients of the form J at points of the boundary. $(dJ)^* = 0$ is equivalent to say that for any triple of vector-fields, say $\mathbf{X}$, $\mathbf{Y}$, $\mathbf{Z}$ on Ω which are tangent to $\partial\Omega$ we have

$$(dJ)\,(\mathbf{X},\mathbf{Y},\mathbf{Z}) = 0 \quad \text{at points of } \Omega. \tag{27}$$

The following identity is, however, valid for any 2-form J

$$(dJ)\,(\mathbf{X},\mathbf{Y},\mathbf{Z}) = \sum_{\text{cycl. permut. of } \mathbf{X},\mathbf{Y},\mathbf{Z}} \{\mathbf{X}\, J(\mathbf{X},\mathbf{Y}) - J([\mathbf{X},\mathbf{Y}],\mathbf{Z})\} \tag{28}$$

where $\mathbf{X}f$ denotes the differentiation of f along $\mathbf{X}$, i.e.

$$\mathbf{X}f = X^\nu \frac{\partial}{\partial x^\nu} f$$

and

$$[\mathbf{X},\mathbf{Y}] = \left(X^{\nu}\frac{\partial}{\partial x^{\nu}} Y^{\mu} - Y^{\nu}\frac{\partial}{\partial x^{\nu}} X^{\mu}\right)$$

defines a vector-field - the commutator of **X** and **Y**.

In the case of side boundary Σ (such, that it is nowhere tangent to any hypersurfaces $t = \text{const}$), at least one of **X**, **Y**, **Z** can be chosen to be time-like, i.e. $X^0 \neq 0$. Therefore we may assume $X = (1, X^1, X^2 X^3)$. The other two fields **Y**, **Z** can be chosen as spatial, Y^0, $Z^0 = 0$. In addition, they can be chosen in such a way that the whole triple is holonomic $[\mathbf{X},\mathbf{Y}] = [\mathbf{X},\mathbf{Z}] = [\mathbf{Y},\mathbf{Z}] = 0$. Then from (28) we have

$$\mathbf{X}\, J_{\mathbf{YZ}} + \mathbf{Z}\, J_{\mathbf{XY}} + \mathbf{Y}\, J_{\mathbf{ZX}} = 0 \tag{29}$$

where again $\mathbf{X} = \frac{\partial}{\partial t} + X^1\frac{\partial}{\partial x^1} + X^2\frac{\partial}{\partial x^2} + X^3\frac{\partial}{\partial x^3}$ and similarly $\mathbf{Z}f$, $\mathbf{Y}f$ denote differentiations along **Z** and **Y**. We obtained thus an evolution equation for $J_{\mathbf{YZ}}$. Let us note that as **Y**, **Z** are spatial (e.g. $\mathbf{Y} = (0, Y^1, Y^2, Y^3) = (0, \vec{Y})$) we have $J(\mathbf{Y},\mathbf{Z}) = \omega(\mathbf{Y},\mathbf{Z})$. Using the correspondence $\omega(\mathbf{Y},\mathbf{z}) = \det[\vec{\omega}, \vec{Y}, \vec{Z}] = \vec{\omega}\cdot(\vec{Y} \times \vec{Z})$ and noticing that $\vec{Y} \times \vec{Z}$ is perpendicular to Σ_t one observes that $J_{\mathbf{YZ}}$ can be identified with the normal component of vorticity field on $\partial\Omega$. Other components of J, i.e. $J_{\mathbf{XY}}$ and $J_{\mathbf{XZ}}$ can be identified as the combinations of $\vec{\omega}^{\perp}$ and tangent components of $\vec{\omega}$.

Conclusion. Suppose the tangent component $\vec{\omega}_{\parallel}$ of $\vec{\omega}$ is given at the side part Σ of the boundary of Ω. Let also $\omega^{\perp}$ be given on $\bar{\Sigma} \cap \{t=0\}$ then the evolution of $\omega^{\perp}$ is governed by Eq.(29).

Therefore only tangent component of $\vec{\omega}$ can be prescribed on the side boundary of Ω. For example, it can be assumed that

$$\vec{\omega}_{\parallel} = f(x, v_{bs}^2)\, \mathbf{v}_{bs} \times \vec{N}_t \tag{30}$$

where $\mathbf{v}_{bs} = \mathbf{v}_b - \mathbf{v}_s$ is the difference between the velocity of the boundary and the velocity of the superfluid. The function f can depend on the material properties of the boundary and $\vec{N}_t = (N^1(t,x), N^2(t,x), N^3(t,x))$ is the unit normal to the boundary. As we will see, however, $\vec{\omega}_{\parallel}$ must be given only at these points of Σ where the trajectories of the vector-field **V** are starting (for increasing t).

5. Existence Theorems

We begin with the linearized case. Suppose that $\mathbf{v}_L(t,x)$ is a given Lipschitz continuous function then we have

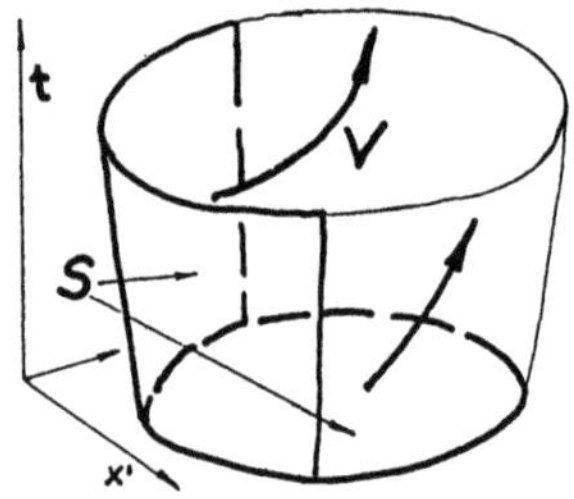

Fig. 2

Theorem 1. *Let* $S \subset \partial\Omega$ *be this part of* $\partial\Omega$ *on which the vector field* $V = (1, v_L)$ *is transversal to* $\partial\Omega$ *and* V *is pointing inward (i.e. its trajectories are starting from* S*). Let* $\hat{\omega}$ *(or* $\vec{\omega}$*) be given at points of* S *such that* $dJ^* = 0$. *Then, in the region covered by trajectories of* V *starting from* S, *there exists a unique solution of equations*

$$dJ = 0 \quad \Longleftrightarrow \quad \begin{cases} \dfrac{\partial\vec{\omega}}{\partial t} + \mathrm{rot}(\vec{\omega} \times v_L) = 0, \\ \mathrm{div}\, \vec{\omega} = 0. \end{cases}$$

To prove it, let us notice that having $\hat{\omega}$ and therefore also J on S we can solve the equation

$$\mathcal{L}_V J = 0$$

which is a system of linear and ordinary equations for coefficients of J along the trajectories of V, $\frac{dx}{dt} = V$. In this way we define J in the subset Ω_S of Ω which is covered by trajectories of V starting from S. By the construction of J out of V and $\hat{\omega}$ (Eqs(16)) the equation $V \lrcorner J$ is satisfied on S. We have, however,

$$\mathcal{L}_V(V \lrcorner J) = (\mathcal{L}_V V) \lrcorner J + V \lrcorner \mathcal{L}_V J = 0$$

as $\mathcal{L}_V V = 0$ and $\mathcal{L}_V J = 0$. Therefore $V \lrcorner J = 0$ everywhere in Ω_S. Similarly $dJ = 0$ at points of S. This follows from the fact that

a) $(dJ)(X,Y,Z) = 0$ for X, Y, Z tangent to S,

b) $V \lrcorner dJ = \mathcal{L}_V J = 0$,

but $V \lrcorner dJ = (dJ)(V,\cdot,\cdot,\cdot)$ and V is transversal to S, therefore at points of S, dJ vanishes for any set of vectors. Thus the coefficients of J are zero at S. We have

$$\mathcal{L}_V dJ = V \lrcorner d^2 J + d(V \lrcorner J) = 0$$

as $d^2 = 0$ and $V \lrcorner J = 0$. Therefore $dJ = 0$ in Ω_S which proves the theorem.

Let us consider now the side boundary Σ (i.e. this part which is nowhere tangent to any of hypersurfaces t = const). At given point $p \in \Sigma$ the vectorfield $\mathbf{V}$ can be situated with respect to Σ according to one of three possibilities: *a*) it may be pointing inward (into Ω), *b*) it can be tangent, *c*) it can be pointing outward. To avoid difficulties encountered in the case of tangent trajectories we assume now that $\mathbf{V}$ is nowhere tangent to Σ. (this may impose some constraints on the topology of the container). Thus, for any connected component of Σ, say Σ', only two possibilities can be encountered: *1*) trajectories of $\mathbf{V}$ are ending on Σ'; in this case no boundary conditions for $\vec{\omega}$ is needed, or *2*) trajectories of $\mathbf{V}$ are starting from Σ' and in this case, except of the initial condition for v_s, also the tangent component of vorticity must be given at Σ'.

Theorem 2. *Let* $\mathbf{V}_L$ *be given as in Theorem* 1 *and let at every point of* Σ, $\mathbf{V}$ *be transversal to* Σ. *Let* $\omega_{\parallel}$ *be given at this part of* Σ *on which* $\mathbf{V}$ *is pointing inward (in the case when* Σ *has many components) then there exists a unique solution for* J *at points of* Σ *satisfying the equations* $dJ^{*} = 0$.

To prove it take first the component of Σ on which $\mathbf{V}$ is pointing inward, call it Σ^{+}. Using the evolution equation (29) for $\omega^{\perp}$ and the initial condition for $\omega^{\perp}$ at $\Sigma^{+} \cap \{t=0\}$ one obtains coefficients of J defined on Σ^{+} and such that Eq.(29) which is equivalent to $dJ^{*} = 0$ is satisfied.

If Σ^{-} is a component of Σ on which $\mathbf{V}$ is pointing outward then, according to Theorem 1, J on Σ' can be obtained by solving equation $\mathcal{L}_{\mathbf{V}} J = 0$ with the starting values prescribed on Σ^{+} and $\partial\Omega \cap \{t=0\}$. As in this case $dJ = 0$ is satisfied then $dJ^{*} = 0$ on Σ^{-} follows automatically.

Concluding this paper let us comment on the fully nonlinear case. Using the well known results for Navier-Stokes system and combining them with our results for the superfluid equations, the local in time existence for the full system (2-4) can be proved. The dependence $\vec{\omega}_{\parallel} = \vec{F}(x, v_b - v_s, \vec{N})$ can be quite general provided that $\vec{F} \cdot \vec{N} \equiv 0$. However, $\vec{F}$ should not depend on ω as it leads to lost differentiability of certain mappings used in the proof of theorem 3, One must assume also that the expression for $\mathbf{V}$ is nonsingular at $t = 0$. If in certain points $\omega(t=0) = 0$ then expression (26) can be used.

Theorem 3. *Suppose that the initial conditions for* v_n *and* v_s *are such that* $\vec{V}|_{t=0}$ *is nowhere tangent to* Σ, $|\vec{V} \cdot \vec{N}| \geq \alpha > 0$. *Let on*

Σ^+ *the boundary conditions for* $\vec{\omega}_\parallel$ *be given in the following form* $\vec{\omega}_\parallel = \vec{F}(x, v_b - v_s, \vec{N})$, *where* $\vec{F} \cdot \vec{N} \equiv 0$. *Let* $v_s(0,x)$, $v_n(0,x) \in C^{k+1+\alpha}$, $0 < \alpha < 1$, $F \in C^{k+1+\alpha}$, *we assume also that approximate compatibility conditions between the boundary and initial conditions are satisfied.*

Then there exists a unique, local in time, solution for v_n *and* v_s, $\vec{\omega}$ *such that it satisfies Eqs* (2-4) *and the initial and boundary conditions* (5-7). *Moreover* v_s, $v_n \in C^{k+1+\alpha}$, $\omega \in C^{k+\alpha}$.

The proof will be published elsewhere.

This research was motivated to some extent by the papers [8,9,10] where the condition for the vorticity production at the wall of a container was used.

Acknowledgements.

This research was supported by Max-Planck-Institut fuer Stroemungsforschung at Goettingen and the Polish Academy of Sciences. The author likes to express his gratitude to Prof. E.A. Mueller for his kind interest in this work and to Prof. W. Fiszdon for many helpful discussions.

References

1. Hall, H. and W. Vinen. The rotation of liquid helium II. *Proceedings of the Royal Society, A238*, (1961) p.215.
2. **Бекаревич, И.Л., Халатников, И.М. Феноменологический вывод уравнений вихревого движеня** He II (A fenomenological deduction of vortex motion equations for He II). **ЖЭТФ**, 40, p.920 (*Soviet Phys. JETP*, 13), (1961) p.643.
3. Zajączkowski, W., Local stability of a nonstationary leakage problem for an ideal incompressible fluid, 3, Math. Meth. in the Appl. Sc. 4(1982), p.1.
4. **Антонцев, Ц.Н., Кажихов, А.В., Монахов, В.Н. Краевые задачи механики неоднородных жидкостей**, (The Boundary Value Problems in the Mechanics of Nonhomogeneous Fluids), (Nauka, Moscow, 1983).
5. Putterman, S.J. Superfluid Hydrodynamics, (North-Holland Publ. Comp., Amsterdam - London, 1974).
6. Sternberg, S. Lectures on Differential Geometry, (Prentice Hall, Inc. Englewood Cliffs, N.J., 1964)
7. Koszul, J.S. Lectures on Fibre Bundles and Differential Geometry, (Tata Institute of Fundamental Research, Bombay, 1960)
8. Fiszdon, W., Peradzyński, Z. Spin-up of He II as a Stefan problem, Ernst-Becker_Gedachtnis Kolloquium, Schriftreihe Wissenschaft und Technik, vol. 28 (T.H. Darmstadt 1986), p.57.

9. Peradzyński, Z,, Filipkowski, S., Fiszdon, W. Spin-up of He II in a cylindrical vessel of finite hight, (to be published).
10. Poppe, W., Schmidt, D.W., Theoretical investigation of the development of the normal fluid and superfluid velocity distributions during spin-up for He II-filled cylinder, MPI Strömungsforsch., Göttingen, Bericht 6/1987.

Mathematical Aspects of the Theory of Viscoelasticity

J. Brilla

Institute of Applied Mathematics and Computing Technique,
Comenius University, CZ-84215 Bratislava, Czechoslovakia

1. Introduction

It is very common to speak about applications of mathematics. We have the ISIMM symposia "Trends in Applications of Mathematics to Mechanics". We often forget about an influence and applications of other sciences to mathematics. Polish mathematician MAURIN /1/ writes : Analysis arised for "requirements" of mechanics and geometry. Many ideas of mathematical analysis originated from mechanics. Here it is necessary to mention variational principles-theorems and methods. The powerful Galerkin method is in fact the application of the principle of virtual work. Also modern numerical methods such as the finite element method and the boundary element method have been proposed by engineers. The basic idea of such a theoretical part of analysis, as the theory of distribution is has its origin in mechanics. In fact, in solutions of the influence of concentrated forces in mechanics using Fourier transform or an expansion into Fourier series there is an idea of application of functionals to such a "function". NADAI /2/ published in 1922 his solution of a circular plate supported at the boundary by concentrated forces expressed in the form of an expansion into Fourier series, what was incorrect from the point of view of mathematics of that time.

Mechanics and other sciences require from mathematics much more then mathematics can render them. Thus when dealing with such problems it is often necessary to introduce new mathematical ideas and get new mathematical results. Thus we arrive at the interaction of mechanics and mathematics or at applied mathematics.

In this paper we shall deal with mathematical aspects of the theory of viscoelasticity and show that it was necessary to develop convenient mathematical tools for this purpose. We have started to deal with a generalization of the finite element method using Laplace transform for the time variable /3/ and introduced in the sense of Laplace transform generalized variational principles and methods /4/. Then for error estimates it was convenient to introduce functional spaces of analytic functions valued in Sobolev spaces and for original form of problems weighted anisotropic Sobolev spaces /5-6/. Further it was also necessary to deal with spectral analysis of viscoelastic operators which are non-selfadjoint /7-8/. This theory has led also to interesting mathematical results. We have proved that viscoelastic operators are semisimple with exception of exceptional points which lay in the left halfplane of the transform parameter p and that the contraexample of Wilkinson /9/ for complex symmetric matrices is an exeptional point. The system of eigenvalues is biorthogonal to eigenvalues of the adjoint operator. It was given also a variational formulation of eigenvalue and eigenfunction problems for viscoelastic and other time dependent operators.

2. Constitutive Equations

The constitutive equation of an arbitrary linear viscoelastic material can be written in the form

$$t_{ij} = H_{ijkl}\, e_{kl} , \tag{1}$$

where H_{ijkl} represents a linear tensor operator, t_{ij} - a stress tensor and e_{kl} is an infinitisimal strain tensor. In tensor formulas we apply the summation rule with respect to indices which occur twice in an expression. On the basis of the assumptions of Onsager's theory for a linear rheological model, H_{ijkl} is a symmetrical tensor. In addition, it may be proved on the basis of the second law of thermodynamics /10/ that it is positive definite. The operator H_{ijkl} can have a differential or an integral form. Then according to /11/ the constitutive equation can be written in the form

$$H_{(r)}\, t_{ij} = H^{(s)}_{ijkl}\, e_{kl} \tag{2}$$

or

$$G_{(s)}\, e_{ij} = G^{(r)}_{ijkl}\, t_{kl} , \tag{3}$$

where

$$H^{(s)}_{ijkl} = \sum_{n=0}^{s} H^{(n)}_{ijkl} \frac{d^n}{dt^n} , \quad G^{(r)}_{ijkl} = \sum_{n=0}^{r} G^{(n)}_{ijkl} \frac{d^n}{dt^n} \tag{4}$$

are tensor operators and

$$H_{(r)} = \sum_{n=0}^{r} h_n \frac{d^n}{dt^n} , \quad G_{(s)} = \sum_{n=0}^{s} g_n \frac{d^n}{dt^n} \tag{5}$$

are scalar operators; h_n , g_n are constants. Tensor operators cannot be on both sides of (2-3) and $s = r$ or $s = r + 1$.

In the case of an integral form the constitutive equations can be written in the form

$$t_{ij} = \int_0^t G_{ijkl}(t - \tau)\frac{de_{kl}}{d\tau} d\tau \tag{6}$$

or

$$e_{ij} = \int_0^t J_{ijkl}(t - \tau)\frac{dt_{kl}}{d\tau} d\tau \quad , \tag{7}$$

where $G_{ijkl}(t)$ is a tensor of relaxation functions and $J_{ijkl}(t)$ a tensor of creep functions.

Applying Laplace transform, i.e, replacing the symbol d^n/dt^n by the transform parameter p, we have

$$\tilde{t}_{ij} = \tilde{H}_{ijkl}(p)\, \tilde{e}_{kl} \tag{8}$$

or

$$\sum_{n=0}^{r} h_n p^n \tilde{t}_{ij} = \sum_{n=0}^{s} H_{ijkl}^{(n)} p^n \tilde{e}_{kl} \quad , \tag{9}$$

and

$$\sum_{n=0}^{s} g_n p^n \tilde{e}_{ij} = \sum_{n=1}^{r} G_{ijkl}^{(n)} p^n \tilde{t}_{ij} \quad . \tag{10}$$

Similarly, application of the Laplace transform to (6-7) , yields

$$\tilde{t}_{ij} = p\, \tilde{G}_{ijkl}(p)\, \tilde{e}_{kl} \quad , \tag{11}$$

$$\tilde{e}_{ij} = p\, \tilde{J}_{ijkl}(p)\, \tilde{t}_{kl} \quad , \tag{12}$$

where tildes denote the Laplace transforms. In these formulas we have assumed zero initial values.

3. Generalized Variational Principles

Nonnegative work requirements are common in continuum mechanics. The work done to deform the material from stress free state must satisfy the positive work principle given by the relation

$$\int_0^t t_{ij}(\tau) \frac{de_{ij}(\tau)}{d\tau} d\tau \geq 0 \quad . \tag{13}$$

Now we consider

$$t_{ij}(t) = t_{ij}\, H(t) \quad , \tag{14}$$

where H(t) is the Heaviside function. Inserting (14) into (7) we arrive at

$$e_{ij}(t) = J_{ijkl}(t)\, t_{kl} \quad . \tag{15}$$

Then from the principle of positive work we get

$$\int_0^t t_{ij} \frac{dJ_{ijkl}(\tau)}{d\tau} t_{kl}\, d\tau = J_{ijkl}(t)\, t_{ij}\, t_{kl} \geq 0 \quad . \tag{16}$$

Applying Laplace transform to this inequality we find out that for real positive values of p it holds

$$\int_0^\infty J_{ijkl}(t)\, t_{ij}\, t_{kl}\, e^{-pt}\, d\tau = \tilde{J}_{ijkl}(p)\, t_{ij}\, t_{kl} \geq 0 \quad . \tag{17}$$

As for arbitrary t_{ij} the inequality (17) holds, we have for real positive values of p

$$p\, \tilde{J}_{ijkl}(p)\, \tilde{t}_{ij}(p)\, \tilde{t}_{kl}(p) \geq 0 \tag{18}$$

and hence also

$$\int_\Omega p\, \tilde{J}_{ijkl}(p)\, \tilde{t}_{ij}(p)\, \tilde{t}_{kl}(p)\, d\Omega =$$

$$= (p\, \tilde{J}_{ijkl}(p)\, \tilde{t}_{ij}(p), \bar{\tilde{t}}_{kl}(p)) \geq 0 \quad . \tag{19}$$

As the equality holds only for $\tilde{t}_{ij}(p)$ identically equal to zero, the bilinear form (18) is positive definite and for real positive values of $p > 0$ it holds

$$p\, \tilde{J}_{ijkl}(p)\, \tilde{t}_{ij}(p)\, \tilde{t}_{kl}(p) \geq \kappa^2 \sum_{i,j=0}^{3} \tilde{t}_{ij}^2(p) \quad . \tag{20}$$

As the matrix corresponding to (12) is positive definite the inverse matrix in (11) is also positive definite and for real $p > 0$ we have

$$p\, \tilde{G}_{ijkl}(p)\, \tilde{e}_{ij}(p)\, \tilde{e}_{kl}(p) \geq \gamma^2 \sum_{i,j=0}^{3} \tilde{e}_{ij}^2(p) \quad . \tag{21}$$

Now we can formulate the generalized variational principle for viscoelastic problems. Consider a viscoelastic problem which is given in the form of Laplace transform by the equation

$$A\tilde{u} = \tilde{f} \quad , \tag{22}$$

where A is an operator involving differential or integral viscoelastic tensor operators. Then we can construct the functional of the generalized potential energy

$$2\bar{V}(\bar{u}) = (A\bar{u},\bar{u}) - 2(\bar{f},\bar{u}) \tag{23}$$

and formulate the following variational theorem :

Theorem 1. The solution $\bar{u}$ of (22) makes the functional of the generalized potential energy (23) stationary and conversely the element which makes the generalized potential energy stationary gives the solution of (22) . Moreover for positive values of p the solution $\bar{u}$ of (22) minimizes the functional of the genralized potential energy.

The solution of the original problem is given by inverse Laplace transform.

By the inverse Laplace transform of (23) we arrive at a convolutional functional of potential energy

$$2V(u) = \int_0^{t} \left[A\, u(\tau) - 2f(\tau)\right] u(t-\tau)d\tau \tag{24}$$

introduced by GURTIN /12/. However when the operator A involves the tensor of relaxation functions $G_{ijkl}(t)$, the convolutional approach meets with difficuties as the complementary strain energy

$$\int_0^t \frac{dt_{ij}(\tau)}{d\tau} e_{ij}(\tau)\, d\tau \tag{25}$$

can assume also negative valües /13/.

Thus (21) cannot be replaced by (25) without further assumptions.

When applying proposed approach all variational principles derived for elastic problems can be generalized to viscoelastic problems by replacing stresses and strains by their Laplace transforms. Moreover also operator $B = A + \rho\, p^2$, which is an operator for dynamic problems, is positive definite for positive values of p and we can apply the theorem 1 also to dynamic problems.

4. Variational Methods of Solution

We shall look for the solution of (22) in the form

$$\bar{u}_n = \sum_{l=1}^{n} \bar{a}_l(p)\, v_l(x) \quad , \tag{26}$$

where $\{v_k\}^n$ are basis functions depending only on space variables and coefficients $\bar{a}_k(p)$ are functions of the transform parameter p.

We can apply in this way the finite element method. Then we shall consider global finite element basis functions, which are piecewise polynomials in space variables. Inserting (26) into (23) and making use of the usual variational procedure we arrive at

$$\frac{\partial \tilde{V}}{\partial \bar{a}_j} = \sum_{i=1}^{n} \bar{a}_i \, (Av_i \, , \, v_j) - (\tilde{f}, v_j) = 0 \quad . \tag{27}$$

Thus

$$\hat{a}_i(p) = \sum_{j=1}^{n} \frac{F_{ij}(p) \, (\tilde{f}, v_j)}{\Delta(p)} \quad , \tag{28}$$

where

$$\Delta(p) = |(Av_i \, , \, v_j)| \tag{29}$$

is the determinant and $F_{ij}(p)$ is the adjoint matrix of the matrix of the system. In the case of a constitutive equation of the differential type and also for many operators of integral type, $F_{ij}(p)/\ p$ is a rational function of p and the inverse transform can be achieved by the method of decomposition into partial fraction /14/.

For quasistatic problems it is very convenient to apply an approximate inverse Laplace transform. A very effective approximate method we have shown in /6/. It is realized by an expansion into Dirichlet series after an orthogonalization of exponential time functions. For the final solution it is necessary to solve n-problems for fixed real values of the transform parameter, what are elastic problems and combine them with a programme for an inverse transform.

5. Functional Analysis and Error Estimates

When dealing with a priori and error estimates we pass over to a weak formulation of our problem. We consider the following formulation

$$\left(\frac{1}{p^s} \tilde{A}(p) \, \tilde{u}, v \right) = \left(\frac{1}{p^s} \tilde{f}, v \right) \quad , \quad \forall \; v \in C_0^{\infty}(\Omega) \quad , \tag{30}$$

where in the case of viscoelastic problems the operator $A(p)$ is a strongly elliptic operator parametrically depending on the transform parameter p .

The operator $A(p)$ assumes in the case of a viscoelastic plate of a standard material the following form

$$\mathbf{A}(p) = \left(K_{ijkl} + \frac{K^1_{ijkl}}{p + \gamma} \right) (\cdot)_{,ijkl} + \rho \, p^2 \quad , \tag{31}$$

where subscripts preceded by a comma indicate differentiation with respect to corresponding spatial coordinates. K_{ijkl}, K^1_{ijkl} are positive definite tensors and γ is a positive constant. In the quasistatic case $\rho = 0$. Then we choose $s = 2$ in the dynamic case and $s = 0$ in the quasistatic case. Further we assume Dirichlet boundary conditions and nonhomogeneous initial conditions included by Laplace transform in $\tilde{f}$.

In order to analyse problem from mathematical point of view and to derive a priori and error estimates we introduce complex Sobolev spaces of functions $H^m(\Omega,\sigma)$ which are parametrically dependent on the transform parameter p and analytic with respect to p in the right hand halfplane $P_\sigma = \{p \mid \operatorname{Re} p \geq \sigma \geq 0\}$.

We define the norm in $H^m(\Omega,\sigma)$ as

$$||\tilde{u}||^2_{H^m(\Omega,\sigma)} = \int_\Omega \sum_{|\alpha|\leq m} |D^\alpha \tilde{u}|^2 \, d\Omega \quad , \tag{32}$$

where α is a multiindex.

Then for $p \in P_\sigma$, $\sigma > 0$ the bilinear form $(1/p^s\, A(p)\, \tilde{u}, \tilde{v})$, $\forall\, \tilde{v} \in H^m_0(\Omega, P_\sigma)$ fulfils the conditions of the Lax-Milgram theorem and we have

<u>Theorem 2.</u> For $\forall\, \tilde{f}/p^s \in H^m(\Omega,\sigma)$ $m \geq -2$ there exists a unique weak solution $\tilde{u} \in H^{m+4}(\Omega,\sigma)$ and such positive constant c , that

$$||\tilde{u}||_{H^{m+4}(\Omega,\sigma)} \leq c\, ||\tilde{f}/p^s||_{H^m(\Omega,\sigma)} \quad . \tag{33}$$

However this a priori estimates holds only for fixed values of p . When we want to derive global a priori estimates we have to introduce new functional spaces. We consider spaces of functions analytic in P_σ and valued in Sobolev spaces $K^{m,r}(\Omega,\sigma)$ endowed with the norm

$$||\tilde{f}||^2_{K^{m,r}(\Omega,\sigma)} = \sup_{p_1>\sigma} \frac{1}{2\pi} \sum_{k=0}^{r} \int_{-\infty}^{\infty} ||p^k\, \tilde{f}(p_1 + i\, p_2)||^2_{H^m(\Omega)}\, dp_2 \quad . \tag{34}$$

For $r = 0$, $K^{m,0}(\Omega,\sigma) = H_2(H^m,\sigma)$ what are Hardy spaces of functions valued in Sobolev spaces. We have delt with the corresponding a priori and error estimates in /3/. Thus the spaces $K^{m,r}(\Omega,\sigma)$ are a generalization of Hardy spaces.

Simultaneously we consider weighted anisotropic Sobolev spaces $H^{m,r}_{;0}(\Omega,R^+,\sigma)$ with the norm

$$\|f\|^2_{H^{m;r}_0(\Omega,R^+,\sigma)}$$

$$= \sum_{k=0}^{r} \int_0^\infty \|D_t^k f\|^2_{H^m(\Omega)} \exp(-2\sigma t)\, dt \quad , \tag{35}$$

where we have denoted $D_t^k = \partial^k/\partial t^k$ and assume that

$$D_t^k f(x,0) = 0 \quad , \quad k = 0,1,\dots,r-1 \quad . \tag{36}$$

Then it holds :

Theorem 3. The Laplace transform is an isometric isomorphic mapping of $H^{m;r}_0(\Omega,R^+,\sigma)$ onto $K^{m,r}(\Omega,\sigma)$.

Thus when we derive an a priori estimate for Laplace transform of the problem under consideration in $K^{m,r}(\Omega,\sigma)$ we get also the a priori estimate for original formulation of the problem.

When dealing with a priori estimates in these spaces we have to analyse estimates for initial conditions separately and now we assume homogeneous initial conditions. Then for a quasistatic problem $\rho = 0$, $s = 0$ we arrive at :

Theorem 4. Let $\tilde{f} \in K^{m,r}(\Omega,\sigma)$, $m \geq -2$, $r \geq 0$, then there exists a weak solution of (30,31, $\rho = 0$) $\tilde{u} \in K^{m+4,r}(\Omega,\sigma)$ and a positive constant c such that

$$\|\tilde{u}\|_{K^{m+4,r}(\Omega,\sigma)} \leq c\|\tilde{f}\|_{K^{m,r}(\Omega,\sigma)} \,. \tag{37}$$

When considering the dynamic problem with $\rho \neq 0$ in (30,31) we have:

Theorem 5. Let $f \in K^{m,r}(\Omega,\sigma)$ $m \geq -2$, $r \geq 0$, then there exists a weak solution of (30,31) $\tilde{u} \in K^{m+4,r+2}(\Omega,\sigma)$ and a positive constant c such that

$$\|\tilde{u}\|_{K^{m+4,r+2}(\Omega,\sigma)} \leq c\|\tilde{f}\|_{K^{m,r}(\Omega,\sigma)} \quad . \tag{38}$$

As spaces $K^{m,r}(\Omega,\sigma)$ and $H^{m;r}_0(\Omega,R^+,\sigma)$ are isometric and isomorphic we receive similar theorems and a priori estimates also for original formulation of the problem under consideration, i.e. in space-time variables. It is only necessary to replace in (37) and (38) Laplace transforms by original functions and spaces $K^{m,r}(\Omega,\sigma)$ by spaces $H^{m;r}_0(\Omega,R^+,\sigma)$.

Now we can deal with an error analysis for finite element solutions of viscoelastic problems.

We shall consider global finite element basis functions, which are piecewise polynomial in space variables. Since the set $\{v_k\}_1^n$ is linearly independent, it forms the basis of n-dimensional subspace $S_n(\Omega,p_\sigma)$ of $H^m(\Omega,p_\sigma)$ over the field of functions analytic in the right half-plane p_σ. It forms also with respect to space variables n-dimensional subspace of $K^{m,r}(\Omega,\sigma)$. Similarly we may introduce $S_n^{k,m}(\Omega,p_\sigma)$ families of finite element subspaces /15,16/, in which basis functions are polynomials in space variables of degree $k+1>m$ and form closed subspaces of $H^m(\Omega,p_\sigma)$.

In /3/ we have proved the generalization of finite element interpolation theorem. We denote the approximation error by

$$\tilde{e}_n = \tilde{u} - \tilde{u}_n \ , \tag{39}$$

where $\tilde{u}$ and $\tilde{u}_n$ are exact solution of (30) and its approximation in $H^m_{(n)}(\Omega,\sigma)$, respectively. Then applying Galerkin method in $S_n^{k,m}(\Omega,\sigma)$ families of subspaces and considering the finite element interpolation theorem, we arrive at

$$||\tilde{e}_n||_{H^j(\Omega,\sigma)} \leq c\ h^{s-j}\ ||\tilde{u}_n||_{H^s(\Omega,\sigma)} \ , \tag{40}$$

where h is the mesh parameter, c is a positive constant and $r \leq k+1$, $j<r$.

Then after passing over to norms of spaces $K^{m,r}(\Omega,\sigma)$ we have

$$||\tilde{e}_n||_{K^{j,r}(\Omega,\sigma)} \leq c\ h^{s-j}\ ||\tilde{u}_n||_{K^{s,r}(\Omega,\sigma)} \ . \tag{41}$$

Now in order to find the error estimate in terms of the data we apply the a priori estimate (37). Then for a quasistatic case we have:

Theorem 6. Let $S_n(\Omega,\sigma)$ belong to $S_n^{k,m}(\Omega,\sigma)$ families, where $k+1>m$. Then there exists a constant c independent of $\tilde{u}_n$ and h such that for a regular refinement and sufficiently small h it holds

$$||\tilde{e}_n||_{K^{m,r}(\Omega,\sigma)} \leq c\ h^{s-m}\ ||\tilde{f}||_{K^{s-4,r}(\Omega,\sigma)} \ . \tag{42}$$

As the Laplace transform is an isometric isomorphic mapping of $H^{m,r}_{;0}(\Omega, R^+,\sigma)$ onto $K^{m,r}(\Omega,\sigma)$ we get immediately

<u>Theorem 7.</u> Let $S_n(\Omega,\sigma)$ belong to $S^{k,m}(\Omega,\sigma)$ families where $k+1>m$. Then there exists a constant c independent of u and h

such that for a regular refinement and sufficiently small h it holds

$$||e_n||_{H^{m,r}_{;0}(\Omega,R^+,\sigma)} \leq c\, h^{s-m}\, ||f||_{H^{s-4,r}_{;0}(\Omega,R^+,\sigma)} \quad . \tag{43}$$

Similarly we can derive the error estimate in the dynamic case when we apply (38) .

Similar error analysis can be done also for other viscoelastic problems and for hyperbolic and parabolic equations and also for the boundary element method.

6. Spectral Analysis

Viscoelastic operators are complex symmetric. In the case of differential operators we arrive at following spectral problems

$$A(D^k)v = \sum_{k=0}^{n} A_k\, D^k v = 0 \quad , \tag{44}$$

where A_k are symmetric strongly elliptic operators.

When looking for the solution in the form $v = e \exp pt$ the spectral problem assumes the form

$$A(p)e = \sum_{k=0}^{n} p^k\, A^k\, e = 0 \quad , \tag{45}$$

When dealing with viscoelastic operators of integral type, we can approximate it by exponential kernels, and arrive to the equation

$$A(p)e = \left(A_0 + \sum_{k=1}^{n} \frac{A_k}{p + c_k} \right) e = 0 \quad . \tag{46}$$

Thus in both cases we have to analyse the spectral problem given by (45). Operators $A(p)$ are complex symmetric operators. It holds $A^*(p) = A(\overline{p})$ and

$$(A\, v,\, \overline{v}) = (v,\, \overline{A}\, \overline{v}) \quad . \tag{47}$$

When $A_i\, A_j \neq A_j\, A_i$, i.e. when operators A_k are noncommutative

$$A\, A^* \neq A^*\, A \tag{48}$$

and $A(p))$ is a nonnormal operator. Thus for the analysis it is not possible to apply the spectral theory of symmetric compact operators. However we have generalized some of it results.

We can deal with the spectral analysis of (45) from the point of view of a nonlinear eigenvalue problem /17/. However then valuable results of the linear spectral theory cannot be applied. Therefore we have defined a linear eigenvalue problem

$$A(p)\, e(p) = \sum_{k=0}^{n} p^k A_k\, e(p) = \lambda(p)\, e(p) , \qquad (49)$$

where $\lambda(p)$ for which solutions of (49) exist are eigenvalues and the corresponding solutions $e(p)$ are eigenvectors of (49). In general, eigenvalues and eigenvectors are functions of the parameter p. Nonlinear eigenvalues are then values of p, for which

$$\lambda(p) = 0 . \qquad (50)$$

For nonnegative real values of p $A(p)$ is a real symmetric strongly elliptic operator. Thus it has discrete spectrum and a complete pairwise orthogonal set of eigenvectors. Moreover we can prove:
Theorem 8. A complex symmetric elliptic operator $A(p)$ has a denumerable set of eigenvalues and eigenvectors, which make the functional

$$J(\lambda,e) = (A\, e, \overline{e}) - \lambda(e,\overline{e}) \qquad (51)$$

stationary.

This variational formulation is convenient for approximate solutions of eigenvalues and eigenvectors.

Points p, where it holds $(e(p),\overline{e(p)}) = 0$, we call exeptional points of the operator $A(p)$. Then it holds:
Theorem 9. Complex symmetric elliptic operators are semisimple with exception of exceptional points and their eigenvectors form with exception of exceptional points a complete system biorthogonal to eigenvectors of the adjoint operator.

It can be proved that in the case of viscoelastic operators there exists a right hand halfplane P_σ without exceptional points. Thus in P_σ a viscoelastic operator and its approximations can be diagonalized.

This theorem holds also for complex symmetric matrices. When the eigenvector e_n belonging to the eigenvalue λ of the multiplicity n fulfils condition $(e_n , \overline{e}_n) \neq 0$ the corresponding canonical form is diagonal and the matrix is simple. WILKINSON /9/ has shown an example of a complex symmetric matrix, which cannot be diagonalized. It is

$$A = \begin{bmatrix} 2i & i \\ 1 & 0 \end{bmatrix} . \tag{52}$$

This matrix has a two-fold eigenvalue $\lambda = i$ and the eigenvector $e_2 = [1 , -i]$, thus $(e_2 , \bar{e}_2) = 0$ and according to the above results it corresponds to the exceptional point and the matrix cannot be diagonalized.

Finally it is possible to prove basic theorems on analycity of eigenvalues and eigenvectors of A(p) and apply the theory for analysis of dynamic viscoelastic problems.

References

1. K. Maurin: Analiza., Vol.1(PWN, Warszawa 1971)
2. A. Nadai: Physikalische Zeitschrift 23 (1922)
3. J. Brilla: Numer. Math. 41, 1 (1983)
4. J. Brilla: In Thin Shell Theory New Trends and Applications, ed. by W. Olszak, CISM Courses and Lectures., Vol. 240 (Springer, Wien, New York 1980) p. 243
5. J. Brilla: In Equadiff 5, ed. by M. Greguš, Teubner-Texte zur Math., Vol. 47 (Teubner, Leipzig 1982) p. 64
6. J. Brilla: In Boundary Elements IX, Vol. 3 ed. by C.A. Brebbia, W.L. Wendland, G. Kuhn, Computational Mechanics Publications, Southampton, Boston (Springer, Berlin, Heidelberg, New York, London, Paris, Tokyo 1987) p.83
7. J. Brilla: In Equadiff 6, ed. by J. Vosmanský, M. Zlámal , Lecture Notes in Mathematics, No 1192 (Springer, Berlin, Heidelberg, New York, Tokyo 1986) p. 197
8. J. Brilla: In Trends in Applications of Pure Mathematics to Mechanics, ed. by E. Kröner, K. Kirchgässner, Lecture Notes in Physics., Vol. 249 (Springer, Berlin, Heidelberg, New York, Tokyo 1986) p. 65
9. J.H. Wilkinson: The Algebraic Eigenvalue Problem (Clarendon Press 1965)
10. W. Olszak, P. Perzyna: Ing.Arch. 28 (1959)
11. J. Brilla: In Proceedings of the Eleventh International Congress of Applied Mechanics, ed. by H. Görtler (Springer, Berlin, Heidelberg, New York 1964) p. 403
12. M.E. Gurtin: Arch. Ration. Mech. Anal. 13, 179 (1963)
13. R.M. Christensen: Theory of Viscoelasticity (Academic Press, New York, London 1971)

14. J. Brilla: In Mechnics of Visco-Elastic Media and Bodies, ed. J. Hult (Springer, Berlin, Heidelberg, New York 1975) p. 215

15. I. Babuška, A.K. Aziz: In The Mathematical Foundations of the Finite Element Method with Applications to Partial Differential Equations, ed. by A.K. Azis (Academic Press, New York 1972) p. 5

16. J.T. Oden, J.N. Reddy: An Introduction to the Mathematical Theory of Finite Elements (John Willey and Sons, New York, London, Sydney, Toronto 1976)

17. I.C. Gokhberg, M.G. Krein: Vvedenie v teoriju linejnych ne-samosoprjaznennych operatorov (Nauka, Moskva 1965) .

Part V

New Interactions of Mathematics and Mechanics

On Saint-Venant's Problem and Saint-Venant's Principle in Nonlinear Elasticity*

A. Mielke

Mathematisches Institut A, Universität Stuttgart,
D-7000 Stuttgart 80, Fed. Rep. of Germany

1. Introduction

The pioneering work of Saint-Venant in 1856 layed the basis for an important research field in elasticity, nowadays paraphrased by the names Saint-Venant's Principle and Saint-Venant's Problem. It is concerned with elastic deformations of a prismatic body whose elastic properties do not depend on the axial variable and which is loaded only near its end. Then, roughly spoken, the statement of Saint-Venant's Principle is that the deformation is essentially determined by the force and moment resultants (F,M) per cross-section. More precisely, the deformation is, up to perturbations decaying exponentially with the distance to the terminal surfaces, identical to a reference deformation having the same resultants (F,M). The characterization of these reference deformations is called Saint-Venant's Problem. Actually, Saint-Venant [17] gave the set of all reference solutions for the case of a homogeneous isotropic body in linearized elasticity.

While the linear theory for these subjects has progressed a great deal in the last twenty years [6,15,18,19], the results for the nonlinear case are considerably less complete [2,8,9,16]. They are mainly restricted to zero resultants in the treatment of Saint-Venant's Principle and to the so-called semi-inverse solutions [3,13] for Saint--Venant's Problem. Ericksen [4] pointed out that the set of all reference solutions is exactly the set of all deformations for an infinite prism having bounded strains along the whole body. Furthermore he conjectured that this set is closely related to the set of solutions of the rod equations. Muncaster [13,14] discusses extensively the underlying symmetry structure of the problem and, using a purely formal projection method, derives a corresponding rod model.

Here we want to present a mathematically rigorous approach which enables us to deal with the general nonlinear case, but which is restricted to deformations with sufficiently small strains along the whole body. The method is mainly based on viewing the governing partial differential equations as a formal differential equation (in a suitable function space over the cross-section of the body) with the axial variable taken as a time-like variable. As in the case of evolution equations, we then obtain all small bounded solution via reduction onto a finite-dimensional center

*) Research supported by ARO through MSI, Cornell University, by DFG (Bonn) under Ki 141/4-1, and by the European Community under ST2J- 0316-C (EDB).

manifold. In our case a adequate interpretation of the linear results show that the dimension of the center manifold is twelve, which corresponds to six rigid-body modes and six Saint-Venant's solutions. Furthermore, exploiting the symmetry structure reveals that the differential equation describing the flow on the center manifold has exactly the struture of the rod equations derived by Kirchhoff; yet the constitutive law of the rod will be nonlinear.

Moreover, a nonlinear version of Saint-Venant's Principle will be derived from the "normal hyperbolicity" of the center manifold. This property, well-known in the theory of dynamical systems, says that the solutions close to the center manifold tend either exponentially towards the center manifold or they move away exponentially. An appropriate interpretation of this phenomenon yields the result.

We restrict the present analysis to the case of a homogeneous, isotropic, and hyperelastic material. This is done only to give a brief introduction to the methods without too much involvement in technicalities.

2. The Equations

We denote the undeformed prism by $\Omega = \Sigma\times(-\ell,\ell)$, $\Sigma \subset \mathbb{R}^2$ being the cross-section. A mapping $\varphi = \varphi(x,t)$: $\Sigma\times(-\ell,\ell) \to \mathbb{R}^3$ is called a deformation of the prism, if it satisfies the equilibrium equations

$$\operatorname{div}[\frac{\partial W}{\partial F}(\nabla\varphi)] = 0 \text{ in } \Omega\,,\quad \frac{\partial W}{\partial F}(\nabla\varphi)n = 0 \text{ on } \partial\Sigma\times(-\ell,\ell). \tag{2.1}$$

Here $W = W(\nabla\varphi)$ is the stored energy function describing the material behavior. In particular, W does only depend on the symmetric strain tensor $E = (\nabla\varphi^t\nabla\varphi - I)/2$, i.e $W(\nabla\varphi) = \underline{W}(E)$. We define $S = S^t = S(E) = \partial\underline{W}/\partial E$ and assume $S(E) = \lambda(\operatorname{tr} E)I + 2\mu E + O(\|E\|^2)_{E\to 0}$, where the Lamé constants λ, μ are both fixed positive numbers.

Equation (2.1) is a system of three quasilinear partial differential equations of the second order for the three unknown functions φ_1, φ_2, and φ_3. Note that we do not prescribe boundary conditions at the terminal surfaces. This has several reasons. Firstly, we want to keep the invariance of (2.1) under rigid-body transformations, i.e. for every solution φ of (2.1) and every $(r,R) \in \mathbb{R}^3\times SO(3)$ the function $\psi = r + R\varphi$ is also a solution. Secondly, we only treat the existence problem for infinte prisms, where $\partial\Omega = \partial\Sigma\times\mathbb{R}$ (see Sect. 3+4). Thirdly, Saint-Venant's Principle for finite prisms (see Sect. 5) will be stated as an apriori estimate for given solutions of (2.1).

The symmetry under $\mathbb{R}^3\times SO(3)$ yields via Noether's Theorem six conservation laws. They are given as the resultants of the forces and the moments per cross-section

$$F = \int_\Sigma \nabla\varphi\, S(E)e_3\, dx = \text{const}, \quad M = \int_\Sigma \varphi\times\nabla\varphi\, S(E)e_3\, dx = \text{const}. \tag{2.3}$$

To retain the symmetry structure we use the following nonlinear ansatz:

$$\varphi(x,s) = \begin{bmatrix} r_1(s) \\ r_2(s) \\ r_3(s) \end{bmatrix} + R(s) \begin{bmatrix} x_1+u_1(x,s) \\ x_2+u_2(x,s) \\ u_3(x,s) \end{bmatrix}, \quad \dot{\varphi}(x,s) = \frac{\partial \varphi}{\partial s}(x,s) = R(s) \begin{bmatrix} v_1(x,s) \\ v_2(x,s) \\ 1+v_3(x,s) \end{bmatrix},$$

with $[u_i]_0 = 0$, $i = 1,2,3$, $[u_2]_1 - [u_1]_2 = [u_3]_1 = [u_3]_2 = 0$. (2.4)

(The brackets $[\cdot]_i$ are defined by $[f]_0 = \int_\Sigma f(x)dx/\sigma_0$, $[f]_i = \int_\Sigma x_i f(x)dx/\sigma_i$, where $\sigma_0 = \int_\Sigma dx$, $\sigma_i = \int_\Sigma x_i^2 dx$, $i = 1,2$.) Thus, $(r,R) \in \mathbb{R}^3 \times SO(3)$ give the mean rigid-body transformation of the cross-section $\Sigma \times \{s\}$, and (u,v) describe the internal displacements being responsible for the strains.

From (2.1) we obtain a differential equation for r, R, u and v, where (u,v) is an element of a suitable function space X over the cross-section Σ.

$$\dot{r} = R(e_3 + \gamma(u,v)), \quad \dot{R} = R\Gamma(u,v), \quad (\Gamma = -\Gamma^t) \tag{2.5a}$$

$$\frac{d}{ds}\begin{bmatrix} u \\ v \end{bmatrix} - L\begin{bmatrix} u \\ v \end{bmatrix} = N(u,v) \ \text{ in } \Sigma, \quad B\begin{bmatrix} u \\ v \end{bmatrix} = G(u,v) \ \text{ on } \partial\Sigma . \tag{2.5b,c}$$

Now fix a real number $s \in (0,1/2)$ and consider $w = (u,v)$ as an element of the Hilbert space $X = \{ (u,v) \in H^{s+1}(\Sigma) \times H^s(\Sigma) \mid [u]_0 = [u_2]_1 - [u_1]_2 = [u_3]_1 = [u_3]_2 = 0 \}$. (Here $H^s(\Sigma)$ denotes the Sobolev space of vector-valued functions $v: \Sigma \to \mathbb{R}^3$, each component v_i lying in $W^{s,2}(\Sigma)$.) The linear operators $L: D \to X$ and $B: D \to Y$ are given by

$$L\begin{bmatrix} u \\ v \end{bmatrix} = \begin{bmatrix} v_1 - [v_1]_0 + \frac{1}{2}([v_2]_1 - [v_1]_2)x_2 \\ v_2 - [v_2]_0 - \frac{1}{2}([v_2]_1 - [v_1]_2)x_1 \\ v_3 - [v_3]_0 - [v_3]_1 x_1 - [v_3]_2 x_2 \\ -u_{1,11} - u_{1,22} - \frac{\lambda+\mu}{\mu}(u_{1,11} + u_{2,12} + v_{3,1}) + [v_3]_1 \\ -u_{2,11} - u_{2,22} - \frac{\lambda+\mu}{\mu}(u_{1,12} + u_{2,22} + v_{3,2}) + [v_3]_2 \\ -\frac{\mu}{\lambda+2\mu}(u_{3,11} + u_{3,22}) - \frac{\lambda+\mu}{\lambda+2\mu}(v_{1,1} + v_{2,2}) \end{bmatrix} \tag{2.6a}$$

$$B\begin{bmatrix} u \\ v \end{bmatrix} = \begin{bmatrix} (\lambda+2\mu)u_{1,1} + \lambda(u_{2,2} + v_3) & \mu(u_{1,2} + u_{2,1}) \\ \mu(u_{1,2} + u_{2,1}) & (\lambda+2\mu)u_{2,2} + \lambda(u_{1,1} + v_3) \\ \mu(v_1 + u_{3,1}) & \mu(v_2 + u_{3,2}) \end{bmatrix} \begin{bmatrix} n_1 \\ n_2 \end{bmatrix} \tag{2.6b}$$

with $D = \{ w \in X \mid u \in H^{s+2}(\Sigma), \ v \in H^{s+1}(\Sigma) \}$ and $Y = H^{s+1/2}(\partial\Sigma)$.

Since $H^{s+1}(\Sigma)$ and $H^{s+1/2}(\partial\Sigma)$ are continuously embedded in $C^0(\Sigma,\mathbb{R})$ and $C^0(\partial\Sigma,\mathbb{R})$ resp. and since the trace mapping $\tau: H^{s+1}(\Sigma) \to H^{s+1/2}(\partial\Sigma)$; $f \to f|_{\partial\Sigma}$ is continuous, there is a zero-neighbourhood $V \subset D$ such that $N: V \to X$ and $G: V \to Y$ are smooth functions satisfying $\|N(w)\|_X + \|G(w)\|_Y = O(\|w\|^2_D)$ for $w \to 0$.

Moreover it is possible to find a smooth invertible mapping $z \to w = W(z) = z + O(\|z\|^2)$ such that the nonlinear boundary conditions (2.5c) transform into $Bz = 0$. Denoting $\underline{L}: D(\underline{L}) \to X;\ z \to Lz$, where $D(\underline{L}) = \{ z \in D \mid Bz = 0 \}$, equation (2.5) takes the form

$$\dot{r} = R(e_3 + \underline{\gamma}(z)),\ \dot{R} = R\underline{\Gamma}(z),\quad \dot{z} - \underline{L}z = \underline{N}(z). \tag{2.7a,b}$$

Accordingly the resultants F and M transform into

$$F = R\underline{n}(z),\ M = r \times F + R\underline{m}(z), \tag{2.8}$$

where $\underline{n}$ and $\underline{m}$ are integrals over Σ depending on $(z(.,s))$ only. The vectors $\underline{n}$ and $\underline{m}$ are called local contact force and local contact couple [1].

3. The Reduction

Here we show that all small bounded solutions of (2.7b) can be obtained by reducing the equation to a finite dimensional submanifold, the so-called center manifold. This well-known method in the theory of evolution equations was recently generalized to quasilinear elliptic problems [10]. To use these results we have to analyse the linear operator $\underline{L}: D(\underline{L}) \to X$ first. The following result is proved in [11].

Lemma 3.1

The operator $\underline{L}$ has a discrete spectrum, and zero is the only eigenvalue on the imaginary axis. The generalized kernel X_c of $\underline{L}$ is six-dimensional, in particular $X_c = \mathrm{span}\{z_1, z_2, z_3, z_4, z_5, z_6\}$ where $\underline{L}z_1 = \underline{L}z_2 = \underline{L}z_3 = \underline{L}z_4 = 0$, $\underline{L}z_5 = z_3$, $\underline{L}z_6 = z_4$. Moreover, $\underline{L}$ satisfies the resolvent estimate

$$\| (\underline{L} - i\xi)^{-1} \| = O(\tfrac{1}{|\xi|}) \ \text{ for } |\xi| \to \infty. \tag{3.1}$$

Remark:

The existence of the six reference solutions of the linear Saint-Venant's Problem [9,13] gives $z_1, \ldots, z_6$ explicitly; and hence $\dim X_c \geq 6$. On the other hand the linear Saint-Venant's Principle [15,19] can be used to show $\dim X_c \leq 6$.

Hence there is a $\underline{L}$-invariant splitting $X = X_c \oplus X_h$, where $X_c = PX$, $X_h = (I-P)X$, and P is a projection satisfying $\underline{L}P = P\underline{L}$. Denote $z = (z_c, z_h)$, $L_c = \underline{L}_{|X_c}$, and $L_h = \underline{L}_{|D(L)\cap X_h}$, then (2.7b) can be written as

$$\dot{z}_c - L_c z_c = P\underline{N}(z_c, z_h),\quad \dot{z}_h - L_h z_h = (I-P)\underline{N}(z_c, z_h). \tag{3.2}$$

For this system the center manifold theorem in [10] is applicable and we obtain

Theorem 3.2

There are neighborhoods $U_c \subset X_c$ and $U_h \subset D(L_h)$ of zero and a function $h = h(z_c) \in \mathcal{C}^1(U_c,U_h)$ with $h(0) = \frac{dh}{dz_c}(0) = 0$, such that $CM = \{ (z_c,h(z_c)) \in D(\underline{L}) \mid z_c \in U_c \}$ is a locally invariant manifold of (3.2). Moreover, every solution z of (3.2) with $(z_c(s),z_h(s)) \in U_c \times U_h$ for all $s \in \mathbb{R}$ lies on CM, i.e. $z_h(s) = h(z_c(s))$.

This shows that the problem of finding all small bounded solutions of (3.2) (Saint--Venant's Problem) is reduced to the discussion of an ordinary differential equation for z_c (remark $\dim X_c = 6$):

$$\dot{z}_c - L_c z_c = P\underline{N}(z_c,h(z_c)) . \tag{3.3}$$

Now we only have to solve (3.3), then, reversing all the transformations, we find $(u(.,s),v(.,s))$ and by solving (2.5a) the whole deformation $\varphi = \varphi(x,s)$ can be reconstructed via (2.4a).

4. Derivation of the Rod Equations

To analyse the form of the reduced equation (3.3) we use the integrals (F,M) from (2.8) and the symmetry structure of (2.7a) and (2.8).

For solutions on the center manifold we have to restrict z to $z_c + h(z_c)$. For (2.8) we obtain $n(z_c) = \underline{n}(z_c+h(z_c))$ and $m(z_c) = \underline{m}(z_c+h(z_c))$ with

$$(n(z_c),m(z_c)) = Tz_c + O(|z_c|^2) \quad \text{for } z_c \to 0. \tag{4.1}$$

Yet, $T: X_c \to \mathbb{R}^6$ can be computed from linearized elasticity, and the classical results [9,13] show that T is one-to-one. Hence (4.1) can be inverted to give $z_c = z_c(n,m)$. Inserting $z = z_c(n,m) + h(z_c(n,m))$ into (2.7a) and (2.8) comes down to

$$\dot{r} = R(e_3+\gamma(n,m)), \quad \dot{R} = R\Gamma(n,m), \tag{4.2a}$$

$$F = Rn = \text{const.}, \quad M = r\times F + Rm = \text{const.} \tag{4.2b}$$

However, by differentiating (4.2b) with respect to s and by multiplying with R^t we obtain, after substituting $\dot{r}$ and $\dot{R}$ via (4.2a),

$$\dot{n} = -\Gamma(n,m)n, \quad \dot{m} = -\Gamma(n,m)m - (e_3+\gamma(n,m))\times n. \tag{4.3}$$

But together with (4.2a) this constitutes a twelve-dimensional system of differential equations having exactly the form of Kirchhoff's rod equations [1,5,7,14]. Here γ

gives the shears and the extension, and $\Gamma = -\Gamma^t$ the bendings and the torsion of the rod as functions of the local contact forces and the local contact couples. In Kirchhoff's case γ and Γ were linear in (n,m); yet in our case we obtain general funtions containing more information on the prism (cf. [5]). (Kirchhoff's modell is gained back by the thin rod approximation which corresponds to introducing the scaling $s \to \epsilon \underline{s}$, $n \to \epsilon \underline{n}$, $m \to \epsilon \underline{m}$ and taking the limit $\epsilon \to 0$.)

Note that this method represents a mathematically rigorous way to link the three--dimensional beam theory to the one-dimensional rod theory. Furthermore we are able to calculate the material properties of the rod (i.e. γ and Γ) directly from the constitutive law of the material in the prism (i.e. $W = W(\nabla\varphi)$).

The equations (4.2a) and (4.3) are mostly discussed in the context of finite rods, then completed by appropriate boundary conditions for $s = \pm\ell$. For the solution of Saint--Venant's Problem we are rather interested in the qualitative behavior of the solutions of (4.3) as s tends to $\pm\infty$. In particular, if all solutions of (4.3) remain bounded then the set of reference solutions is simply obtained by solving (4.2a) and (4.3) with the initial data $(r(0), R(0), n(0), m(0)) = (0, I, F, M)$ for all the resultants $(F,M) \in \mathbb{R}^6$.

Some results on the solutions of (4.3) are given in [3,11,14], they concern mainly the class of semi-inverse solutions for which $(n,m) = \text{const}$. A more general treatment in [5] shows that (4.3), viewed as a dynamical system, exhibits chaotic solutions. Although those chaoticly deformed rods seem to be highly unstable and thus of no physical interest, it follows that a classification of the reference solutions is, in general, impossible.

5. On Saint-Venant's Principle

One natural formulation of Saint-Venant's Principle could be the following:

> There exist positive constants C and α, such that for every solution $\varphi: \Sigma\times(-\ell,\ell) \to \mathbb{R}^3$ of (2.1) (with sufficiently small strains) there exist a reference deformation ψ (i.e. ψ lies on the center manifold) which satisfies the estimate
>
> $$\| \varphi(\cdot,s) - \psi(\cdot,s) \|_{H^2(\Sigma)} \le C \cdot e^{-\alpha(\ell - |s|)} \quad \text{for } |s| \le \ell. \tag{5.1}$$

Here the important fact is that the constants C and α are independent of the lenght 2ℓ of the prism, i.e. the perturbations at the terminal surfaces have a fixed size. Thus the accordance between the actual deformation and the reference solution is the better the longer the prism is.

As it is more difficult to obtain pointwise (in s) estimates like (5.1), we derive only a weaker energy estimate. (A proof of (5.1) is given in [12].) Like Toupin [19] in the linear case, we want to relate the error in the whole prism $\Sigma\times(-\ell,\ell)$ to the error in a sub-prism $\Sigma\times(-\lambda,\lambda)$ with $0 < \lambda < \ell$. The result we obtain is

$$\int_{-\lambda}^{\lambda}\|e^{-\beta|s|}(\varphi(.,s)-\psi(.,s))\|^2_{H^2(\Sigma)}ds$$
$$\leq C_{\alpha,\beta}\cdot e^{-2(\alpha-\beta)(\ell-\lambda)}\cdot\int_{-\ell}^{\ell}\|e^{-\beta|s|}(\varphi(.,s)-\psi(.,s))\|^2_{H^2(\Sigma)}ds, \tag{5.2}$$

where $\beta\in(0,\alpha)$ can be chosen arbitarily, but $C_{\alpha,\beta}\to\infty$ for $\beta\to 0$.

We define the weighted L_2-spaces, $L_{2,\alpha}(Y)$, for arbitrary Hilbert spaces Y and positive constants α via

$$L_{2,\alpha}(Y)=\{\, f:\mathbb{R}\to Y \mid f \text{ measurable}, \ |f|_{\alpha,Y}<\infty\,\},\quad |f|^2_{\alpha,Y}=\int_{\mathbb{R}}\|e^{-\alpha|s|}f(s)\|^2ds.$$

It is shown in [10] that the linearized equation (3.2)

$$\dot{z}_c - L_c z_c = f_c(s),\ z_c(0)=\xi,\quad \dot{z}_h - L_h z_h = f_h(s) \tag{5.3}$$

has for each $(\xi,f_c,f_h)\in X_c\times L_{2,\alpha}(X_c\times X_h)$ a unique solution $z=(z_c,z_h)=K(\xi,f)\in L_{2,\alpha}(X_c\times D(L_h))$. Here $\alpha\in(0,\alpha_0)$ and $\alpha_0=\min\{|\mathrm{Re}\,\sigma| \mid \sigma \text{ eigenvalue of } L_h\}$. Hence, any bounded solution $z:\mathbb{R}\to D(\underline{L})$ of (3.2) satisfies the integral equation

$$z(s)=K(\xi,\underline{N}(z(.)))(s). \tag{5.4}$$

The solutions of (5.4) are uniquely determined by $z_c(0)=\xi$ and lie on the center manifold.

Let $y: (-\ell,\ell)\to D(\underline{L})$ be any given solution of (3.2) with $y(s)\in U_c\times U_h$ for $|s|<\ell$. From (5.4) we find a unique solution $\underline{z}$ with $\underline{z}_c(0)=y_c(0)$ and define

$$x(s)=\begin{cases} y(s) & \text{for } |s|\leq \ell-1,\\ y(s)+(|s|-\ell+1)[\underline{z}(s)-y(s)] & \text{for } \ell-1<|s|<\ell,\\ \underline{z}(s) & \text{else.}\end{cases}$$

and $\Delta(s)=x(s)-\underline{z}(s)$. Thus x satisfies (3.2) only up to a residuum r (i.e. $\dot{x}-\underline{L}x=\underline{N}(x)+r(s)$), where r is different from zero only if $\ell-1\leq|s|\leq\ell$. Accordingly x satisfies (5.4) up to the residuum $R=K(0,r)$. In particular, for $0<\beta<\alpha$, we have

$$|r|_{\alpha,X}\leq C\cdot e^{-(\alpha-\beta)\ell}|r|_{\beta,X},\quad |R|_{\alpha,D(L)}\leq C\cdot e^{-(\alpha-\beta)\ell}|R|_{\beta,D(L)}. \tag{5.5}$$

By subtracting the integral equations for $\underline{z}$ and x we obtain for Δ the equation

$$\Delta=K(0,\underline{N}(\Delta+x)-\underline{N}(x))-R. \tag{5.6}$$

Under appropriate restriction on x and Δ the operator on the right-hand side of (5.6) is a contraction in $\Delta \in L_{2,\alpha}(D(\underline{L}))$. Futhermore the unique solution Δ obeys the estimate

$$c_\alpha \cdot |R|_{\alpha,D(L)} \leq |\Delta|_{\alpha,D(L)} \leq C_\alpha \cdot |R|_{\alpha,D(L)}. \tag{5.7}$$

Using (5.5) and (5.7) allows the following estimate:

$$\begin{aligned}\int_{-\lambda}^{\lambda} \|e^{-\beta|s|}\Delta(s)\|^2 ds &\leq e^{2(\alpha-\beta)\lambda} \cdot \int_{-\lambda}^{\lambda} \|e^{-\alpha|s|}\Delta(s)\|^2 ds \leq e^{2(\alpha-\beta)\lambda} \cdot |\Delta|^2_{\alpha,D(L)} \\ &\leq e^{2(\alpha-\beta)\lambda} C_\alpha^2 \cdot |R|^2_{\alpha,D(L)} \leq e^{2(\alpha-\beta)\lambda} \cdot C_\alpha^2 \cdot e^{-2(\alpha-\beta)\ell} |R|_{\beta,D(L)} \\ &\leq C^2_{\alpha,\beta} \cdot e^{-2(\alpha-\beta)(\ell-\lambda)} |\Delta|^2_{\beta,D(L)}.\end{aligned}$$

However, this is exactly the desired estimate (5.2).

References

1. S.S. Antman: Large Lateral Buckling of Nonlinearly Elastic Beams, Arch. Rat. Mech. and Anal. 84, 293-305 (1984).
2. S. Breuer & J.J. Roseman: On SV's Principle in 3-dim. nonlinear elasticity, Arch. Rat. Mech. and Anal. 63, 191-203 (1977).
3. J.L. Ericksen: On the Formulation of Saint-Venant's Problem, in "Nonlinear Analysis and Mechanics: Heriot-Watt Symposium Vol. I", R.J. Knops ed, Pitman London - 1977, 158-186.
4. J.L. Ericksen: Problems for Infinite Elastic Prisms - St.-Venant's Problem for Elastic Prisms, in "Systems of Nonlinear Partial Differential Equations", J.M. Ball ed, NATO ASI Series C 111, Reidel Publ. Comp. 1983.
5. P.J. Holmes & A. Mielke: Complex Equilibria of Buckled Rods, Arch. Rat. Mech. Anal., to appear (1988).
6. K. Kirchgässner & J. Scheurle: Saint-Venant'Principle from a Dynamical Point of View, manuscript Universität Stuttgart (1986).
7. G. Kirchhoff: Über das Gleichgewicht und die Bewegung eines unendlich dünnen Stabes, J. für Mathematik (Crelle) 56, 285-313 (1859).
8. A.E.H. Love: A Treatise on the Mathematical Theory of Elasticity, Cambridge University Press (1927).
9. R.J. Knops & L.E. Payne: A Saint-Venant Principle for Nonlinear Elasticity, Arch. Rat. Mech. and Anal. 81, 1-12 (1983).
10. A. Mielke: Reduction of Quasilinear Elliptic Equations in Cylindrical Domains with Applications, Math. Meth. Appl. Sciences, to appear in Vol.10.
11. A. Mielke: On Saint-Venants Problem and Semi-Inverse Solutions in Nonlinear Elasticity, Arch. Rat. Mech. Anal., to appear (1988).
12. A. Mielke: On Saint-Venant's Principle for Beams in Nonlinear Three-dimensional Elasticity, manuscript Universität Stuttgart (1988).
13. R.G. Muncaster: Saint-Venant's Problem in Nonlinear Elasticity: A Study of Cross--Sections, in "Nonlinear Analysis and Mechanics: Heriot-Watt Symp. Vol. IV", R.J. Knops ed, Pitman London 1979, 17-75.

14. R.G. Muncaster: Saint-Venant's Problem for Slender Prisms, Utilitas mathematica 23, 75-101 (1983).

15. O.A. Oleinik & G.A. Yosifian: On the Asymptotic Behavior at Infinity of Solutions in Linear Elasticity, Arch. Rat. Mech. Anal. 78, 29-53 (1983).

16. J.J. Roseman: The Principle of Saint-Venant in Linear and Nonlinear Plane Elasticity, Arch. Rat. Mech. and Anal. 26, 142-162 (1967).

17. B. de Saint-Venant: Memoire sur la torsion des prismes, Memoires presentes par divers savant a l' academie de sciences de l' institut imperial de France, 2. Ser. 14, 233-560 (1856).

18. E. Sternberg & J.K. Knowles: Minimum Energy Characterizations of Saint-Venant's solution to the Relaxed Saint-Venant Problem, Arch. Rat. Mech. and Anal. 21, 89-107 (1966).

19. R.A. Toupin: Saint-Venant's Principle, Arch. Rat. Mech. Anal. 18, 83-96 (1965).

Hypermembrane Shells

J.L. Buchanan[1] *and R.P. Gilbert*[2]

[1]Department of Mathematics, U.S. Naval Academy
Annapolis, MD 21402, USA
[2]Department of Mathematical Sciences,
University of Delaware, Newark, DE 19716, USA

1. INTRODUCTION

The theme of this conference is the *application of pure mathematics to mechanics.* It seems fitting, therefore, to employ a branch of analysis, *generalized hyperanalytic function theory,* to solve the equations of elastic shell theory. Since the theory of generalized hyperanalytic functions is only a decade and a half old [4], [10], [11] it seems reasonable that we first give a short introduction to that topic. The reader wishing to learn more about the subject is directed to either the book by BUCHANAN and GILBERT [9], or the book by WENDLAND [15].

Generalized hyperanalytic function theory extends hyperanalytic function theory in exactly the same manner as generalized analytic function theory extends analytic function theory. BERS [3], HAACK [13], and VEKUA [14] showed that much of the analytic theory could be carried over to the weak solutions (in an L_p sense) of the differential equation

$$\frac{\partial w}{\partial \overline{z}} + a(x,y)\, w + b(x,y)\overline{w} = 0 \ . \tag{1.1}$$

The hyperanalytic functions are essentially the solution of the principal part of a real, first-order elliptic system. Whereas there is no *normal form* for elliptic equations of the first-order [15], under certain conditions such as continuity of the (complex) eigenvalues, and the invariance of the eigenvalue multiplicities are sufficient for a pseudo normal form to exit. This situation is valid for many physical systems including the equations of shell theory based on the DIKMEN [7], VEKUA [15] model. In this regard see an earlier paper by GILBERT and XU [12].

For expository reasons we give a short introduction to the theory corresponding to a real, first-order, elliptic system having a single, albeit multiple eigenvalue. DOUGLIS [8] showed in this case the principal part could be written concisely in terms of the hypercomplex algebra generated by 1, i, and the nilpotent e, where $e^{n+1}=0$, n is the multiplicity of the eigenvalue. Indeed, it has been shown that the principal part could be written as [8], [9], [10]

$$\partial w := \frac{\partial W}{\partial \bar{z}} + q(x,y)\frac{\partial W}{\partial z} , \tag{1.2}$$

where $q(x,y)$ is the nilpotent

$$q(x,y) = \sum_{k=1}^{n} q_k(x,y)e^k . \tag{1.3}$$

We make a few remarks about the hypercomplex algebra to give the reader a feeling for the function theory that we have at hand.

If $a := \sum_{k=0}^{n} a_k e^k$ is a hypercomplex number or order n, i.e. $e^n = 0$, then a_0 is called the complex part of a and $A := \sum_{k=1}^{r} a_k e^k$ is the nilpotent part. a_k is the kth component of a.

If $a_0 \neq 0$ then a unique multiplicative inverse a^{-1} or $1/a$ exists and is given by the formula

$$a^{-1} = \frac{1}{a_0}\sum_{k=0}^{r}(-1)^k\left[\frac{A}{a_0}\right]^k .$$

The norm of a hypercomplex number is given by

$$|a| := \sum_{k=0}^{r} |a_k| .$$

The logarithm of a hypercomplex number is

$$\log a := \log a_0 + \sum_{k=1}^{n}\frac{(-1)^{k-1}}{k}\left[\frac{A}{a_0}\right]^k .$$

The equation $\partial w = 0$ has a special solution called a generating solution, $t(z)$, such that $t(z) := z + T(z)$ where $T(z)$ is nilpotent. If $q(z) := q(x,y)$ is bounded and in $C^{\alpha}(\mathbf{R}^2)$ then there exists a *CAUCHY-POMPIEU* integral representation for the nonhomogeneous $\partial W = f(z) \in L_2(\mathbf{R}^2)$ of the form [8] [10]

$$w(z) = \frac{1}{2\pi i} \int_{\Gamma} \frac{w(\zeta) dt(\zeta)}{t(\zeta) - t(z)} - \frac{1}{\pi} \int_{D} \frac{t_{\zeta}(\zeta) f(\zeta)}{t(\zeta) - t(z)} d\zeta \, d\eta \tag{1.4}$$

where $\Gamma := \partial D$. Using the representation (1.4) much of the theory of analytic functions can be developed for the homogeneous equation $\partial W = 0$ [8]. For general first-order, elliptic systems much may be said using the hypercomplex notation; however, a rather simple example of a first-order elliptic system which covers a number of physical cases, and for which our theory [10] works admirably is that given by [9]

$$\partial W + A(x,y)W + B(x,y)\overline{W} = F(x,y) ,$$

where A, B, and F are hypercomplex valued functions in $L_{p,2}(\mathbf{R}^2)$, i.e.

$$A(x,y) := \sum_{k=0}^{n} a_k(x,y) e^k \in L_{p,2}(\mathbf{R}^2) \text{ , etc. .}$$

In this instance the VEKUA theory [14] can be recovered almost in its entirety [1], [9], [10]. The BERS theory of generating-pairs also has been developed by GILBERT, HILE [11] and BEGEHR, GILBERT [2].

In the present paper we shall be interested in a somewhat more complicated differential operator, namely

$$\overset{*}{D}W := \partial W + \overset{*}{q} \frac{\partial \overline{W}}{\partial \overline{z}} \text{ ,} \tag{1.5}$$

where ∂ is defined as in (1.2), and $\overset{*}{q}$ is a nilpotent function of the form (1.3). We are able to show that the study of first-order systems based on this principal part may be reduced to the generalized hyperanalytic equations thoroughly reported on in the book by BUCHANAN and GILBERT [9].

2. HIERARCHICAL SHELL THEORY

As a starting point we take the *hierarchical shell theory* developed by Dikmen in [7]. Details omitted in the following synopsis may be found in this work. Let **B** be a shell with surfaces $\omega^{\pm}$ and edge E. Let ω be a reference surface lying in **B**, typically but not essentially, the middle surface of the shell. Let θ^i , $i = 1,2,3,$ be a nondegenerate coordinate system for **B**. As the third coordinate has a distinguished role in a shell theory, ξ will often be used in lieu of θ^3. The equations of ω , $\omega^{\pm}$ are assume to be $\xi = 0$, $\xi = h^{\pm}(\theta^{\alpha})$ respectively. By convention, indices denoted by Greek letters range from 1 to 2, while indices denoted by Latin letters range from 1 to 3. The edge E of **B** is assumed to be generated by the ξ- lines.

Let $\mathbf{p}\left(\theta^{\alpha}, \xi\right)$ denote the position vector of a point in the shell **B** and $\mathbf{r}\left(\theta^{\alpha}\right) := \mathbf{p}\left(\theta^{\alpha}, 0\right)$ denote position on the reference surface ω. The vector $\mathbf{d}\left(\theta^{\alpha}, \xi\right) := \mathbf{p}\left(\theta^{\alpha}, \xi\right) - \mathbf{r}\left(\theta^{\alpha}\right)$ is assumed to have an expansion in powers of ξ, i.e.

$$\mathbf{p}\left(\theta^{\alpha}, \xi\right) = \mathbf{r}\left(\theta^{\alpha}\right) + \sum_{k=1}^{\infty} \xi^k \underset{k}{\mathbf{d}} \left(\theta^{\alpha}\right) . \tag{2.1}$$

Let

$$\{\mathbf{a}_1, \mathbf{a}_2, \mathbf{a}_3\} = \left\{ \mathbf{r}_{,1} , \mathbf{r}_{,2} \frac{\mathbf{a}_1 \times \mathbf{a}_2}{|\mathbf{a}_1 \times \mathbf{a}_2|} \right\}$$

with $\cdot_{,\alpha} := \dfrac{\partial}{\partial \theta^{\alpha}}$, be the local coordinate trihedron for ω. The coefficients of the first and second fundamental forms for ω are $a_{\alpha\beta}$ and $b_{\alpha\beta}$ respectively, i.e.

$$ds^2 = a_{\alpha\beta} d\theta^{\alpha} d\theta^{\beta} \quad \text{and} \quad -d\mathbf{r} \cdot d\mathbf{a}_3 = b_{\alpha\beta} d\theta^{\alpha} d\theta^{\beta}$$

with the usual summation convention. Finally $a := \det(a_{\alpha\beta})$ and $b := \det(b_{\alpha\beta})$.

For **B**

$$\left\{\mathbf{g}_i^*\right\} := \left\{\mathbf{p}_{,i}\right\}$$

denotes the local coordinate trihedron, g_{ij}^* the covariant metric tensor and $g^* := \det(g_{ij}^*)$.

If the vectors $\mathbf{T}^i(\theta^\alpha, \xi)$ are defined from the Cauchy stress tensor $\mathbf{T}^{ij}$ by $\mathbf{T}^i := \sqrt{g^*}T^{ij}\mathbf{g}_j^*$, then $\mathbf{T}^i\, d\theta^j\, d\theta^k (i \neq j \neq k)$ is the total force across an infinitesimal coordinate parallelogram generated by $d\theta^j$ and $d\theta^k$. An argument concerning the forces acting on an infintesimal coordinate parallelopiped yields the equations of balance for linear and angular momenta

$$\begin{aligned} \mathbf{T}^i_{,i} + \rho^* \sqrt{g^*}\mathbf{f}^* &= 0 , \\ \mathbf{g}_i^* \times \mathbf{T}^i &= 0 \ , \end{aligned} \tag{2.2}$$

where ρ^* is mass density in **B** and f^* is body force per unit mass.

Employing (2.1) it may be shown that the equations of balance imply

$$\underset{k}{\mathbf{M}}{}^{\alpha}{}_{|\alpha} + \rho\, \underset{k}{\mathbf{l}} - \underset{k}{\mathbf{m}} = 0 \ ,$$

$$\underset{k}{\mathbf{M}}{}^{\alpha} \times \mathbf{a}_\alpha + \sum_{j=1}^{\infty} \frac{j}{j+k} \underset{j+k}{\mathbf{m}} \times \underset{k}{\mathbf{d}} + \sum_{j=1}^{\infty} \underset{j+k}{\mathbf{M}}{}^{\alpha} \times \underset{k,\alpha}{\mathbf{d}} = 0 \ , \tag{2.3}$$

where ρ is mass density on ω, $\cdot_{|\alpha}$ is covariant differentiation and, for a shell in equilibrium and the *moment vectors* are given by

$$\sqrt{a}\, \underset{k}{\mathbf{M}}{}^{\alpha}(\theta^\beta) := \int_{h^-}^{h^+} \mathbf{T}^\alpha \xi^k \, d\xi \ ,$$

$$\sqrt{a}\, \underset{k}{\mathbf{m}}(\theta^\beta) := k \int_{h^-}^{h^+} \mathbf{T}^3 \xi^{k-1} d\xi \ , \tag{2.4}$$

$$\rho\sqrt{a}\, \underset{k}{\mathbf{l}}\, (\theta^\beta) := \int_{h^-}^{h^+} \rho^* \sqrt{g^*}\mathbf{f}^* \ \xi^k \, d\xi \quad +$$

$$\left\{ \xi^k \left[-h^+_{,\alpha} \mathbf{T}^\alpha + \mathbf{T}^3 \right] \right\}_{\xi = h^+} - \left\{ \xi^k \left[-h^-_{,\alpha} + T^3 \right] \right\}_{\xi = h^-} .$$

We make the special choice of coordinate system $\underset{1}{\mathbf{d}} = \mathbf{a}_3$, $\underset{k}{\mathbf{d}} = \mathbf{0}$ for $k > 1$, that is, take the ξ- lines to be normal to ω. In doing so the thinness condition

$$|h^+ - h^-| < \min\left(|R_1|, |R_2|\right) ,$$

where R_1 and R_2 are the principal radii of curvature must be imposed to ensure that the normals do not intersect within **B**. Also the edge of **B** is now assumed to consist of straight lines perpendicular to ω.

In this coordinate system Equation $(2.3)_2$ becomes

$$\underset{k}{\mathbf{M}}^{\alpha} \times \mathbf{a}_{\alpha} + \frac{1}{k+1} \underset{k+1}{\mathbf{m}} \times \mathbf{a}_3 - b_{\alpha}^{\gamma} \underset{k+1}{\mathbf{M}} \times \mathbf{a}_{\gamma} = 0 , \tag{2.5}$$

after an application of the Weingarten formulae. If in component form $\underset{k}{\mathbf{M}}^{\alpha} = \underset{k}{M}^{\alpha i} \mathbf{a}_i$, then from the Weingarten and Gauss formulae the following component form for Eqs. $(2.3)_1$ and (2.5) is obtained

$$\begin{aligned}
&\underset{k}{M}^{\alpha\beta}_{\ |\alpha} - b_{\alpha}^{\beta} \underset{k}{M}^{\alpha 3} + \rho \underset{k}{l}^{\beta} - \underset{k}{m}^{\beta} = 0 , \\
&\underset{k}{M}^{\alpha 3}_{\ |\alpha} - b_{\alpha\beta} \underset{k}{M}^{\alpha\beta} + \rho \underset{k}{l}^{3} - \underset{k}{m}^{3} = 0 , \\
&\underset{k}{M}^{21} - \underset{k}{M}^{12} = b_{\alpha}^{2} \underset{k+1}{M}^{\alpha 1} - b_{\alpha}^{1} \underset{k+1}{M}^{\alpha 2} , \\
&\underset{k}{M}^{\alpha 3} = b_{\beta}^{\alpha} \underset{k+1}{M}^{\beta 3} + \frac{1}{k+1} \underset{k+1}{m}^{\alpha} = 0 .
\end{aligned} \tag{2.6}$$

Definition of hypermembrane: *We shall say that a shell is a hypermembrane of order n if there is an n for which*

$$\underset{k}{\mathbf{M}}^{\alpha} = \mathbf{0} , \ \underset{k}{\mathbf{m}} = \mathbf{0} , \ \underset{k}{\mathbf{l}} = \mathbf{0} , \quad \text{for all} \quad k > n . \tag{2.7}$$

For $n = 0$ (2.7) is the usual defining condition of the membrane theory of shells. However for $n > 0$ it is a stronger condition on the moment vectors than is sometimes assumed in higher order bending theories in that the consequences of (2.7) and $(2.6)_3$, $(2.6)_4$ for $k = n$,

$$\underset{n}{M}^{21} - \underset{n}{M}^{12} = 0\,, \ \underset{n}{M}^{\alpha 3} = 0 \ ,$$

are not always assumed. See again DIKMEN [7].

We also assume that the moment vectors $\underset{k}{m}$ and $\underset{k}{l}$ are known. This is clearly reasonable in the latter case as $\underset{k}{l}$ involves only body and surface forces. However $\underset{k}{m}$ depends upon $\mathbf{T}^3$ in the interior of $\mathbf{B}$. Thus some further approximation is necessary here. For instance, if a linear relation

$$\mathbf{T}^3 = \mathbf{c}_0 + \mathbf{c}_1\,\xi$$

is assumed for $\mathbf{T}^3$ and ω is taken to be the middle surface of $\mathbf{B}$, then

$$\sqrt{a}\ \mathbf{T}^3 = \frac{1}{h}\underset{1}{\mathbf{m}} + \frac{6\xi}{h^3}\underset{2}{\mathbf{m}}\,, \tag{2.8}$$

with $h := h^+ - h^-$. Thus if $\mathbf{T}^3$ is known on the surfaces $\omega^\pm$, then (2.8) determines $\underset{1}{\mathbf{m}}$ and $\underset{2}{\mathbf{m}}$ and hence $\mathbf{T}^3$ is completely determined to this degree of approximation in the interior of $\mathbf{B}$ and on the edge E for $r \geq 2$ with $\underset{k}{\mathbf{m}} = 0$ for $k = 3 \cdots r$. For $n = 0$ no normal surface stresses are admissible, and for $n = 1$ only a stress couple can be specified, i.e. no normal surface shearing is permitted to this degree of approximation.

3. HYPERCOMPLEX FORMULATION

The hierarchial shell theory lends itself conveniently to the hypercomplex formulation. To this end, we introduce the *hypercomplex moments*

$$\mu^{\alpha i} := \sum_{k=0}^{r} \underset{r-k}{M}^{\alpha i} e^k \ , \tag{3.1}$$

$$l^i := \sum_{k=0}^{r} \underset{r-k}{l}{}^i e^k \quad ,$$

$$m^i := \sum_{k=0}^{r} \underset{r-k}{m}{}^i e^k \quad ,$$

$$\hat{m}^\alpha := \sum_{k=0}^{r-1} \frac{1}{r-k} \underset{r-k}{m}{}^\alpha e^k \quad .$$

The conservation of angular momentum (2.6) $_4$ becomes in hypercomplex form

$$\mu^{\alpha 3} = e b_\beta^\alpha \, \mu^{\beta 3} + e \hat{m}^\alpha \quad , \tag{3.2}$$

which may be solved to obtain

$$\mu^{13} = \frac{\hat{m}^1 \, e \left[1 - b_2^2 e\right] + \hat{m}^2 b_2^1 e^2}{\mathrm{D}} \quad ,$$

$$\mu^{23} = \frac{\hat{m}^2 \, e \left[1 - b_1^1 e\right] + \hat{m}^1 b_1^2 e^2}{\mathrm{D}} \quad , \tag{3.3}$$

where

$$\mathrm{D} = \left[1 - b_1^1 e\right] \left[1 - b_2^2 e\right] - b_2^1 b_1^2 e^2 \tag{3.4}$$

$$= 1 - 2He + Ke^2 \quad .$$

Here H is the *mean curvature* and K is the *Gaussian curvature*. From (3.3) $_1$, (3.3) $_2$ we realize that the $\mathrm{M}^{\alpha 3}$ are known providing the $\hat{m}^\alpha$ are known. The remaining equations of (2.6) become in hypercomplex form

$$\mu^{\alpha\beta}_{|\alpha} = b^\beta := b_\alpha^\beta \, \mu^{\alpha 3} - \rho l^\beta + m^\beta \quad , \tag{3.5}$$

$$b_{\alpha\beta} \mu^{\alpha\beta} = b^3 := -\mu^{\alpha 3}_{|\alpha} - \rho l^3 + m^3 \quad ,$$

$$\mu^{21} \mu^{12} = e \left[b_\alpha^2 \mu^{\alpha 1} - b_\alpha^1 \mu^{\alpha 2}\right] \quad , \tag{3.6}$$

where the functions b^i are known quantities under our assumptions.

Applying the formulae for covariant differentiation to $(3.5)_1$ yields

$$\mu^{11}_{,1}+\mu^{21}_{,2}+\left(2\Gamma^1_{11}+\Gamma^2_{12}\right)\Gamma_{11}+\Gamma^1_{12}\mu^{12}+\left(2\Gamma^1_{12}+\Gamma^2_{22}\right)\mu^{21}+\Gamma^1_{22}\mu^{22}=b^1 \ ,$$

$$\mu^{12}_{,1}+\mu^{22}_{,2}+\left(\Gamma^2_{11}\mu^{11}+\left(2\Gamma^2_{12}+\Gamma^1_{11}\right)+\mu^{12}+\Gamma^2_{12}+\Gamma_{12}\mu^{21}+\right. \tag{3.7}$$

$$\left.\left(2\Gamma^2_{22}+\Gamma^1_{12}\right)\mu^{22}=b^2 \ ,\right.$$

where the $\Gamma^\gamma_{\alpha\beta}$ are Christoffel symbols.

4. GENERALIZED HYPERANALYTIC FORM OF THE MOMENT EQUATIONS

If the Gaussian curvature of ω is everywhere positive then we may choose the coordinate system θ^α to be *isometric−conjugate*, i.e.

$$b_{11}=b_{22}=\hat{b},\ b_{12}=b_{21}=0 \ . \tag{4.1}$$

In an isometric-conjugate coordinate system $(3.5)_2$ takes on the simple form

$$\mu^{22}=-\mu^{11}+\hat{b}^3 \ , \tag{4.2}$$

where $\hat{b}^3:=\frac{b^3}{\hat{b}}$. After some manipulation the equations (3.7) with the use of (4.2) and (3.6) may be put into the form

$$\partial_{\bar{z}}w+\frac{1}{2}\,\mathrm{Re}\left(\pi\partial_z w\right)-\frac{1}{2}\,\mathrm{Re}\left(\pi\partial_{\bar{z}}w\right)=\frac{1}{2}\,h \tag{4.3}$$

where

$$\begin{aligned} w &:= \mu^{11}-i\mu^{12}\ ,\ z:=\theta^1+i\theta^2\ ,\\ p &:= p_1+ip_2\ ,\ h:=g-aw-bw\ , \end{aligned} \tag{4.4}$$

$$\hat{a}:=\frac{1}{2}\left\{\left(c_{11}+c_{22}\right)-\left(c_{21}-c_{12}\right)i\right\} ,$$

$$\hat{b} := \frac{1}{2}\left\{\left(c_{11}-c_{22}\right) - \left(c_{21}+c_{12}\right)i\right\} ,$$

$$g := g_1 - ig_2 , \qquad \text{and}$$

$$c_{11} := 2\Gamma^1_{11}+\Gamma^1_{12}-\Gamma^1_{22}+p_{1,2}+\left(2\Gamma^1_{12}+\Gamma^2_{22}\right)p_1 ,$$

$$c_{12} := 3\Gamma^1_{12}+\Gamma^2_{22}+p_{2,2}+\left(2\Gamma^1_{12}+\Gamma^2_{22}\right)p_2 , \tag{4.6}$$

$$c_{21} := \Gamma^2_{11}+2\Gamma^2_{22}-\Gamma^1_{12}+\Gamma^2_{12}p_1 ,$$

$$c_{22} := 3\Gamma^2_{12}+\Gamma^1_{11}+\Gamma^2_{12}p_2 ,$$

$$g_1 := b^1-\left(p_3\hat{b}^3\right)_{,2}-\Gamma^1_{22}b^3-p_3\left(2\Gamma^1_{12}+\Gamma^2_{22}\right)\hat{b}^3 , \tag{4.7}$$

$$g_2 := b^2-\hat{b}^3_{,2}-p_3\Gamma^2_{12}\hat{b}^3 - \left(2\Gamma^2_{22}+\Gamma^1_{12}\right)\hat{b}^3 .$$

$$p_1 := \frac{e\left(b^2_1+b^1_2\right)}{1-eb^2_2} , \; p_2 := \left(b^2_2-b^1_1\right)\sum_{k=1}^{r}\left(b^2_2\right)^{k-1}e^k , \; p_3 := \frac{-eb^1_2}{1-eb^2_2} . \tag{4.8}$$

We shall refer to solutions to an equation of the form

$$\overset{*}{D}w := \partial w + q_1\partial_z w + q_2\overline{\partial_z w} = 0 , \tag{4.9}$$

where q_1 and q_2 are hypercomplex and nilpotent, as *hyperanalytic functions*. Solutions to equations of the form

$$\overset{*}{D}w + aw + b\overline{w} = f , \tag{4.10}$$

where a, b, and f are hypercomplex are *generalized* hyperanalytic functions. For $q_2 \equiv 0$ these coincide with the hyperanalytic functions investigated by DOUGLIS [8], and the generalized hyperanalytic functions studied by GILBERT and HILE [10]. It is possible to obtain a Green's identity for the operator (4.9). Let us assume in what follows that G is a regular domain in C, and that $q_1, q_2 \in C^1_\alpha(\overline{G})$. If $w \in C^0_\alpha(\overline{G})$, and $\partial_z w$, $\partial_{\bar{z}} w \in L_p(\overline{G})$, then it may be shown that [6]

$$\iint_G \left[\partial_\zeta V_1 \overset{*}{D} w(\zeta) + \overline{\partial_\zeta V_2}\, \overline{\overset{*}{D} w(\zeta)}\right] d\sigma_\zeta = \frac{1}{2i} \int_{\partial G} w(\zeta) dV_1(\zeta) - \overline{w}(\zeta) d\overline{V}_2(\zeta) \quad (4.11)$$

for any $V_1, V_2 \in C^2_\alpha(\overline{G})$ satisfying

$$\partial_{\bar\zeta} V_1 + q_1 \partial_\zeta V_1 + \overline{q}_2\, \overline{\partial_\zeta V_2} = 0\,, \quad \partial_{\bar\zeta} V_2 + \overline{q}_2\, \overline{\partial_\zeta V_1} + q_1 \partial_\zeta V_2 = 0\,. \quad (4.12)$$

In analogy to the hyperanalytic case studied in [14] it is also possible to obtain a *genealized Cauchy-Pompieu representation* for the operator (4.9). It may be shown that for the preceding conditions on q_1, q_2 - etc. that

$$w(z) = \frac{1}{2\pi i} \int_{\partial G} w(\zeta) \left[\frac{d\zeta}{\zeta - z} + d\tau_1(\zeta,z)\right] - \overline{w}(\zeta) d\overline{\tau}_2(\zeta,z) \quad (4.13)$$

$$- \frac{1}{\pi} \iint_G \left\{\left[\frac{1}{\zeta - z} + \partial_\zeta \tau_1(\zeta,z)\right] \overset{*}{D} w(\zeta) + \overline{\partial_\zeta \tau_2(\zeta,z)}\, \overline{\overset{*}{D} w(\zeta)}\right] d\sigma_\zeta\,,$$

where $\tau_j(\zeta,z) := \sum_{k=1}^{r} \frac{(-1)^{k-1}}{k} \left[\frac{\psi_j(\zeta,z)}{\zeta - z}\right]^k$, and ψ_1, ψ_2 are solutions of the system

$$\partial_{\bar z} \psi_1 = - q_1 \partial_\zeta \psi_1 - \overline{q}_2 \frac{\psi_1}{\overline{\psi}_2}\, \overline{\partial_\zeta \psi_2} + \overline{q}_2 \frac{\psi_1}{\overline{\zeta} - \overline{z}}\,,$$

$$(4.14)$$

$$\partial_{\bar z} \psi_2 = - \overline{q}_2 \frac{\psi_2}{\overline{\psi}_1}\, \overline{\partial_\zeta \psi_1} + q_1 \psi_2 + q_1 \frac{\psi_2}{\zeta - z}\,.$$

Using the generalized Cauchy-Pompieu formula, and the imbedding properties of the integral operators appearing here, it is possible to solve the Riemann-Hilbert problem associated with differential equations of the type (4.3). Since numerous shell problems can be reformulated as Riemann-Hilbert problems this gives a useful application to our theory. This work will be reported on elsewhere [6].

The research of the second author was supported in part by the National Science Foundation through Grant DNS 8601318.

LITERATURE

[1] H. Begehr, R.P. Gilbert: in *Contemporary Mathematics* ed. by R.P. Gilbert, Vol. 11 (Amer. Math. Soc., Providence 1982) p.13.

[2] H. Begehr, R.P. Gilbert: Complex Vars (1988) (to appear).

[3] L. Bers: *Theory of Pseudo-Analytic Functions,* N.Y.U. Lecture Notes (Courant Inst., New York 1953).

[4] B.B. Bojarski: Annals. Polonici Math. *17* (1966), p. 281 (Russian).

[5] J.L. Buchanan, R.P. Gilbert: Appl. Anal. *11* (1981), p. 303.

[6] J. L. Buchanan, R.P. Gilbert: J. Elasticity (submitted)

[7] M. Dikmen: *Theory of Thin Elastic Shells,* Surv. Ref. Work in Math. *8* (Pitman, London 1982).

[8] A. Douglis: Comm. Pure Appl. Math. *6* (1953), p. 259.

[9] R.P. Gilbert and J.L. Buchanan: *First Order Elliptic Systems a Function Theoretic Approach,* Math. in Sci., Eng. *163* (Academic, New York 1983).

[10] R.P. Gilbert, G.N. Hile: Trans. Amer. Math. Soc. *195* (1974), p. 1.

[11] R.P. Gilbert, G.N. Hile: Math. Nachricten *72* (1976), p. 187.

[12] R.P. Gilbert, Y. Xu: J. Elasticity (1988) (to appear).

[13] W. Haack: Math. Nachr. *8* (1952), p. 123.

[14] I.N. Vekua: *Generalized Analytic Functions* (Pergamon, London 1962).

[15] I.N. Vekua: Matem. Sbornik. *31* (1952) p. 217 (Russian).

[16] W. Wendland: *Elliptic Systems in the Plane* (Pitman, London 1979).

On the Problem of Constructing the Mathematical Theory of Plates and Shells

T.S. Vashakmadze

Tbilisi, USSR

The formation of elasticity theory, as a scientific discipline (typically the field of continuum mechanics) is in its progressive mathematization. This process is characterized by the transition from the models, constructed by means of explicit hypotheses, to those that are an application of Noll-Truesdell's axiomatic theory and of mathematical structures of the Rykhlevski type on the structure of elastic material bodies, justifying phenomenological laws.

For the creation of a valuable rigorous theory it is necessary to overcome principle difficulties: along with the requirement of an adequate (within the exactness sufficient for practice at this stage) description of a real observable object (phenomenon), the model is to be realizable, i.e. it should permit the determination of the necessary characteristics.

In this direction the interest in the construction of valid approximate models and the development of general methods of investigation and estimation for problems of shell and plate theories has recently grown considerably, thanks to the studies of Goldenveizer, Friedrichs, Koiter, Vekua and Ciarlet.

Due to the great variety of important practical problems and the search for relevant convenient models, the theories of plates and shells were developed, their construction being based on the mechanical and geometric characteristics, assumed to be valid for bodies of small thickness. Therefore the development of general principles for constructing mathematical models free from such assumptions and based on the use of achievements of pure mathematics (with a possible estimate of inherent error of transition from three-dimensional problems to those of lower dimension) is, in our view, the central problem of continuum mechanics.

In this connection we shall report on some results, devoted to the study of problems of constructing the mathematical theory of elastic anisotopic non-homogeneous plates of variable thickness. The material presented announces and generalizes a number of results stated in [1–2].

1. The initial 3-dimensional problems of elasticity we write in the following form (see for example [3]):

$$\partial_j(\sigma_{ij} + \sigma_{kj}u_{i,k}) = f_i \quad , \quad x = (x_1, x_2, x_3) \in \Omega_h \quad , \tag{1}$$

$$l^{\pm}[u] = \sigma_{i3} + \sigma_{j3}u_{i,j} = g_i^{\pm} \quad , \quad x \in S^{\pm} \quad , \tag{2}$$

$$l(\partial_1, \partial_2, \partial_3)[u] = g \quad , \quad x \in S \quad , \tag{3}$$

$$\varepsilon_{ij} = \tfrac{1}{2}(u_{i,j} + u_{j,i} + u_{k,i}u_{k,j}) \quad , \sigma = B\varepsilon \quad , \varepsilon = A\sigma \quad , x \in \overline{\Omega}_h \quad ,$$
$$\Omega_h = \mathcal{D}(x_1, x_2)x]h^-(x_1, x_2) \quad , h^+(x_1, x_2)[\quad , S^{\pm} = \mathcal{D} \times h^{\pm} \quad ,$$
$$S = \partial\mathcal{D} \times [h^-, h^+] \quad , 2h = h^+ - h^- \quad , \hbar = \tfrac{1}{2}(h^- + h^+) \quad , B = \{b_{ij}\} \quad . \tag{4}$$

Here A and B depend on at most 13 independent elastic constants.

In the first part of this paper I shall recapitulate some results relating to the construction of refined theories and to the definition of new models without any simplifying hypotheses.

1. Let us introduce the average values with respect to plate thickness, both for the components of the displacement vector u and for the components of the stress tensor σ. Furthermore, we introduce the following averaging procedure, which we call the averaging according to Reissner:

$$u^*_\alpha = \frac{3}{2h^3}\int_{-h}^{h} tu\alpha(x_1, x_2, t)dt \quad , u^*_3 = \frac{3}{4h^3}\int (h^2 - t^2)u_3 dt \quad , \tag{5}$$

$$M_{\alpha\beta} = \int t(\sigma_{\alpha\beta} + \sigma_{\kappa\alpha}u_{\beta,\kappa})dt \quad , M_{\alpha\alpha} = M_\alpha \quad , \tag{6}$$

$$\psi_\alpha = \tfrac{1}{2}\int (h^2 - t^2)\sigma_{\alpha 3} dt \quad , Q_{ij} = \int (\sigma_{ij} + \sigma_{\kappa j}u_{i,\kappa})dt \quad , \tag{7}$$

$$\mathcal{F}^* = \frac{1}{2h}\int \mathcal{F}(x_1, x_2, t)dt \quad , \tag{8}$$

where

$$\sigma_{\alpha\beta} = (-1)^{\alpha+\beta}\left(\partial_{3-\alpha}\partial_{3-\beta}\mathcal{F} + \frac{1}{2h}\mathcal{P}_{\alpha\beta}(x_1, x_2)\right) \quad ,$$
$$\mathcal{P}_{\alpha\alpha,\alpha} - \mathcal{P}_{\alpha\beta,3-\alpha} = \int [f_\alpha - \partial_\beta(\sigma_{\beta\kappa}u_{\alpha,\kappa})]dt - (g^+_\alpha - g^-_\alpha) \quad ;$$

in (5–8) we have obtained from (1–4) the exigencies of the nonlocal representation (e.n.r) that in (11) [4, p. 490] were presented for the isotropic case.

When the generalized Hooke's law is represented in the form of (4), we have:

$$\begin{aligned}\frac{2h^3}{3}\mathcal{L}^0_4(\partial_1, \partial_2)u^*_3 = {} & \frac{2h^2}{3\delta}(1 + 2\gamma)\partial_\alpha S_{\alpha\beta}Q_{\beta 3} + 2h\mathcal{L}[u^*_3, \mathcal{F}^*] \\ & + h(g^+_{\alpha,\alpha} + g^-_{\alpha,\alpha}) + g^+_3 - g^-_3 - \frac{h^2}{2}(1 + 2\gamma) \\ & \times L_{33}(g^+_3 - g^-_3 + 4h\mathcal{L}[u^*_3, \mathcal{F}^*]) \\ & - \int [tf_{\alpha,\alpha} + (h^2 - t^2)(1 - L_{33})f_3]dt + r_3[u^*_3, \gamma] \quad ;\end{aligned}$$

$$\mathcal{L}_4^1(\partial_1,\partial_2)\mathcal{F}^* = -\frac{E}{2}\mathcal{L}[u_3^*,u_3^*] + (-1)^{\alpha+\beta} l_{\alpha\beta}\partial_{3-\alpha}\partial_{3-\beta}(g_3^+ + g_3^-) + r_{10}[\mathcal{F}^*] \quad ;$$

$$Q_{\alpha 3} - \frac{1}{3\delta}2h^2(1+2\gamma)S_{\alpha\beta}Q_{\beta 3} = -\frac{2h^3}{3}L_{\alpha\beta}u_3^* - \frac{h^2}{3}(1+2\gamma)L_{0\alpha}$$
$$\times (g_3^+ - g_3^- + 4h\mathcal{L}[u_3^*,\mathcal{F}^*]) + h(g_\alpha^+ + g_\alpha^-)$$
$$- \int [t f_\alpha - L_{0\alpha}(h^2 - t^2) f_3] dt + r_{7+\alpha}[Q_{\alpha 3},\gamma] \quad ,$$

where

$$\mathcal{L}_4^0 = c_{11}\partial_1^4 + 4c_{16}\partial_1^3\partial_2 + 2(c_{12}+2c_{66})\partial_1^2\partial_2^2 + 4c_{26}\partial_1\partial_1^3 + c_{22}\partial_2^4 \quad ,$$
$$\mathcal{L}_4^1 = a_{22}\partial_1^4 - 3a_{26}\partial_1^3\partial_2 + 2(a_{12}+a_{66})\partial_1^2\partial_2^2 - 3a_{16}\partial_1\partial_2^3 + a_{11}\partial_2^4 \quad ,$$
$$L_{\alpha 3} = c_{\alpha\alpha}\partial_\alpha^3 + 3c_{\alpha 6}\partial_\alpha^2\partial_{3-\alpha} + (c_{12}+2c_{66})\partial_\alpha\partial_{3-\alpha}^2 + c_{\alpha 3-\alpha}\partial_{3-\alpha}^3 \quad ,$$
$$L_{\alpha\alpha} = c_{\alpha\alpha}\partial_\alpha^2 + 2c_{\alpha 2}\partial_\alpha\partial_{3-\alpha} + c_{66}\partial_{3-\alpha}^2 \quad ,$$
$$L_{12} = c_{16}\partial_1^2 + (c_{12}+c_{66})\partial_{12} + c_{26}\partial_2^2 \quad ,$$
$$L_{33} = b_{33}^{-1}(b_{13}\partial_1^2 + 2b_{36}\partial_{12} + b_{23}\partial_2^2) \quad ,$$
$$S_{\alpha\beta} = b_{3+\beta,3+\beta}L_{\alpha\beta} - b_{45}L_{\alpha 3-\alpha} - \delta\partial_\beta L_{0\alpha} \quad ,$$
$$L_{0\alpha} = b_{33}^{-1}(b_{\alpha 3}\partial_\alpha + b_{36}\partial_{3-\alpha}) \quad , \quad \delta = b_{44}b_{55} - b_{45}^2 \quad ,$$
$$\mathcal{L}[u,v] = \partial_{11}u\partial_{22}v - 2\partial_{12}u\partial_{12}v + \partial_{11}v\partial_{22}u.$$

Numbers a, c and δ are expressed in terms of b_{ij}.

The rotation of the normals to the middle plane of the plate, u_α^*, and the bending and twisting moments, $M_{\alpha\beta}$, in accordance with the e.n.r., depend implicitly on the deflection u_3^*, the shear forces, $Q_{\alpha 3}$, the partial pressure, $g_3^+ - g_3^-$, and also the remainder terms, r.

Thus we have constructed a system of integro-differential equations which depend on arbitrary parameters $\gamma_i = \gamma_i(x_1,x_2)$, $i = \overline{1,8}$. If we denote

$$u^* = (u_1^*, u_2^*, u_3^*, M_1, M_2, M_{12}, M_{21}, Q_{13}, Q_{23}, \mathcal{F}^*)^T \quad ,$$
$$r = (r_1, r_2, \ldots, r_{10})^T \quad ,$$

then, for brevity and convenience, we can join them in the form of an operator equation like

$$\mathbb{P}_h(\partial_1,\partial_2;\gamma)U^* = \mathcal{G}(x_1,x_2;\gamma) + r \quad , \quad (x_1,x_2) \in \overline{\mathcal{D}} \quad .$$

We added here the necessary boundary conditions, averaged with respect to the plate's thickness resulting from the appropriate data on the lateral surfaces.

We shall rewrite the last system without r. Thus, we obtain the 2-dimensional parametric system of nonlinear equations:

$$\mathbb{P}_h(\partial_1,\partial_2;\gamma)u^h = \mathcal{G}(x_1,x_2;\gamma) \quad , \quad (x_1,x_2) \in \overline{\mathcal{D}} \quad , \quad (P_h)$$

where u^h denotes the approximate solution of the initial 3-dimensional problem (1–4).

The system has the structure typical of refined theories; a part of $(\mathbb{P}_h)$, corresponding to the functions u_3^*, $\mathcal{F}^*$ and $Q_{\alpha 3}$ (or u_3^h, $\mathcal{F}^h$, $Q_{\alpha 3}^h$), forms the closed elliptical nonlinear system of partial differential equations in D. The rest of the functions u_α^h, $M_{\alpha\beta}^h$ are expressed by means of obvious formulas.

As in the isotropic case [4], from scheme $(\mathbb{P}_h)$ we get refined theories and new models, which are equivalent to them with accuracy r.

2. Further, we remark that the method of construction of refined theories without hypotheses is applicable immediately to the nonhomogeneous and nonstationary cases. In these cases we get the same schemes and it is necessary to carry them through.

I call your attention to the problems connected with boundary conditions given on the lateral surface S and consider two simple examples.

Example 1. Assume that the problem (1–4) is linear, Ω_h is isotropic and homogeneous, $g \equiv 0$; the part of boundary conditions averaged along $\partial\mathcal{D}$ splits with respect to u_3^h and for definiteness takes the form:

$$u_3^h = \partial u_3^h / \partial n = 0 \quad . \tag{9}$$

Besides, let us assume $g_3^+ = q(x_1, x_2)$, $g_3^- = 0$.

Then the system of differential equations, corresponding to refined theories of plate bending takes the form

$$D\Delta^2 u_3^h = q - \frac{2-\nu}{3(1-\nu)}(1+2\gamma_1)h^2 \Delta q \quad , \tag{10}$$

$$\frac{h^2}{3}(1+2\gamma_2)\Delta Q_\alpha^h - Q_\alpha^h = Du_{3,\alpha}^h + \frac{h^2(1+2\gamma_3)}{3(1-\nu)}\partial_\alpha q \quad . \tag{11}$$

Let us choose the function q so that it is a nontrivial solution of Helmholtz's equation

$$\frac{2-\nu}{3(1-\nu)}h^2(1+2\gamma_1)\Delta q - q = 0 \quad .$$

Then, if a problem of the type (9, 10) has a unique solution, then $u_3^h \equiv 0$ in $\overline{D}$; and this result, of course, is not the solution of the initial problem (1–4) formulated for an arbitrary normal load $q(x_1, x_2)$.

Example 2. We deal with the same problem, $g \equiv 0$. Let us consider a sequence of functions $q_\kappa(x_1, x_2)$ contained in right-hand parts of (10), such as

$$\lim_{\kappa\to\infty} q_\kappa = \begin{cases} 0, & (x_1, x_2) \in \mathcal{D} \quad , \\ q^0(x_1, x_2) \equiv 0 \quad , & (x_1, x_2) \in \partial\mathcal{D} \end{cases} \quad .$$

Hence, if the equations (10–11) with corresponding boundary conditions averaged along $\partial\mathcal{D}$ and their right-hand parts being formed by means of q_κ, are solved subsequently, then sequences $(u_3^h)_\kappa$ and $(Q_\alpha^h)_\kappa$ will vanish in $\overline{D}$ in case of a unique solution, and it will contradict the solution of the initial problem (1–4).

Now we return to the problems of boundary conditions, connected with scheme $(I\!P_h)$. Let a lateral plane $S_1 c S$ be the part of the plane with normal Ox_1. The basic homogeneous boundary conditions given on S_1 are formulated as:

i)	$u_i = 0$	– rigidly clamped edge
ii)	$\sigma_{1i} = 0$	– a free edge
iii)	$\sigma_{11} = u_2 = u_3 = 0$	– a hinge-supported edge.

In the first case we have:

$$u_3^* = Q_2 = 0 \quad , \quad u_{3,1}^* = D_1^{-1}\psi_1 \quad , \quad D_1 = 2h^3\mu/3 \quad , \tag{12}$$

and

$$u_{3,11}^* = \partial_1(D_1^{-1}\psi_1) - D_3^{-1}\int_{-h}^{h} t\sigma_{33}(x_1,x_2,t)dt \quad , \quad D_3 = \ln\mu \quad . \tag{13}$$

Conditions (12, 13) will form four independent boundary conditions, corresponding to equations (10, 11).

Let us pay attention to the integral term from (13). It can be calculated by special quadrature formula with arbitrary parameter within the accuracy of refined theories. Just the same term appears in other basic averaged boundary conditions. If we compare the influence of interior stress state on the plate's edge, which is the main subject of many investigations with the integral terms of (13), we see that they are different.

3. The method of investigation above admits the generalization in case of plates with variable thickness, i.e. when $(h^- = h_1,\, h^+ = h_2)$:

$$\Omega_{\hbar} = \mathcal{D}(x_1,x_2) \times \,]h_1(x_1,x_2) \quad , \quad h_2(x_1,x_2)[\quad .$$

Let $\Omega_{\hbar}$ be an isotropic and homogeneous body and

$$v_\alpha(x_1,x_2,x_3) = u_{3,\alpha} - \mu^{-1}\sigma_{\alpha 3} \quad . \tag{14}$$

Then we have:

$$h^{-3}D\partial_\alpha[h^3\Delta u_{3,\alpha}(x_1,x_2,\hbar] = \sigma_{33}^+ - \sigma_{33}^- - \int_{h_1}^{h_2} f_3\,dt$$

$$+\partial_\alpha\left[h(\sigma_{\alpha 3}^+ + \sigma_{\alpha 3}^-) - \int_{h_1}^{h_2}(t-\hbar)f_\alpha\,dt\right] - (h_{2,\alpha}\sigma_{\alpha 3}^+ - h_{1,\alpha}\sigma_{\alpha 3}^-) - \frac{h}{1-\nu}$$

$$\times\,(\sigma_{33,\alpha}^+ h_{2,\alpha} + \sigma_{33,\alpha}^- h_{1,\alpha}) + \frac{1}{1-\nu}\int_{h_1}^{h_2}(h^2-(t-\hbar)^2)\Delta f_3\,dt$$

$$-\frac{1+\nu}{1-\nu}\int_{h_1}^{h_2}(h\partial_\alpha h+(t-\hbar)\partial_\alpha\hbar)f_{3,\alpha}\,dt-\frac{h^2(2-\nu)}{3(1-\nu)}1+2\gamma)\Delta(\sigma_{33}^{+}-\sigma_{33}^{-})+\frac{h^2}{3}$$
$$\times[\partial_\alpha(h+\hbar)\Delta\sigma_{23}^{+}+\partial_\alpha(h-\hbar)\Delta\sigma_{\alpha 3}^{-}]+\frac{4h^2}{3}\partial_\alpha h\Delta\sigma_{\alpha 3}(x_1,x_2,\hbar)+R_1 \quad , \tag{15}$$

$$\frac{4}{3}h\sigma_{\alpha 3}(x_1,x_2,\hbar)-\frac{2h^3}{3}\Delta\sigma_{\alpha 3}(x_1,x_2,\hbar)=-D\Delta u_{3,\alpha}(x_1,x_2,\hbar)+\frac{2}{3}h$$
$$\times(\sigma_{\alpha 3}^{+}+\sigma_{\alpha 3}^{-})-\int_{h_1}^{h_2}\left[(t-\hbar)f_\alpha-\frac{1+\nu}{2(1-\nu)}(h^2-(t-\hbar)^2)f_{3,\alpha}\right]dt$$
$$-\frac{1+2\gamma}{3(1-\nu)}h^2\partial_\alpha(\sigma_{33}^{+}-\sigma_{33}^{-})+R_{1+\alpha} \quad , \tag{16}$$

$$u_\alpha^{*}=-v_\alpha(x_1,x_2,\hbar)-\frac{3}{4h^3}\varrho_{Sm}[(h^2-(t-\hbar)^2)v_\alpha] \quad ,$$

$$M_\alpha= -D[v_{\alpha,\alpha}(x_1,x_2,\hbar)+\nu\partial_{3-\alpha}V_{3-\alpha}(x_1,x_2,\hbar)]$$
$$+\frac{h^2}{3(1-\nu)}(1+2\gamma)(\sigma_{33}^{+}-\sigma_{33}^{-})+R_{4+\alpha} \quad ,$$

$$M_{12}= -(1-\nu)D[v_{1,2}(x_1,x_2,\hbar)+v_{2,1}(x_1,x_2,\hbar)]+R_7\mu\Delta u_\alpha(x_1,x_2,\hbar)$$
$$+(\lambda^{*}+\mu)\partial_\alpha u_{\beta,\beta}(x_1,x_2,\hbar)=f_\alpha^{*}(x_1,x_2)+R_{7+\alpha} \quad , \tag{17}$$

$$f_\alpha^{*}(x_1,x_2)=\frac{1}{2h}\left(\int_{h_1}^{h_2}f_\alpha\,dt-\sigma_{\alpha 3}^{+}+\sigma_{\alpha 3}^{-}\right)-\frac{\lambda}{4(\lambda+2\mu)}\partial_\alpha[\sigma_{33}^{+}-\sigma_{33}^{-}]$$
$$-\mu[\Delta\hbar v_\alpha(x_1,x_2,\hbar)+2\partial_{\alpha\beta}\hbar v_\beta(x_1,x_2,\hbar)]-(\lambda^{*}+\mu)$$
$$\times[\partial_{\alpha\beta}\hbar v_\beta(x_1,x_2,\hbar)+2\partial_\beta\hbar v_\beta(x_1,x_2,\hbar)] \quad . \tag{18}$$

Here

$$D=\frac{2Eh^3}{3(1-\nu^2)} \quad , \quad \Delta=\partial_1^2+\partial_2^2 \quad , \quad \sigma^{\pm}=\sigma(x_1,x_2,h^{\pm}) \quad ,$$

$\varrho[\cdot]$ is the remainder term of Simpson's rule.

Thus in the case of a plate with variable thickness we arrive at the following conclusion:

The problem of the determination of the stress-strain state of the plate of variable thickness is split into two problems, which should be solved consecutively.

Problem I. We must solve the problem corresponding to equations (15–16), characterizing the deflection of the average surface, $u_3(x_1, x_2, \hbar)$, the shear forces Q_α, determined by means of $\sigma_{\alpha 3}(x_1, x_2, \hbar)$, the rotation angles with respect to the horizontal u^*_α, and the bending and twisting moments, $M_{\alpha\beta}$.

Problem II. Next we solve the problem of finding functions $u_\alpha(x_1, x_2, \hbar)$ based on (17) and taking into account conditions on S. Values f depend also on the values, defined from the first problem [see (14, 18)].

4. We use now Vekua's method [5], when

$$\Omega_h = \mathcal{D}(x_1, x_2) \times]0, \infty[\quad ,$$

and the problem of (1–4) is linear and isotropic simplicity, $h^- = 0$, $h^+ = +\infty$, $S = \partial\mathcal{D} \times]0, \infty[$, $S^\pm = \partial\Omega - S$.

For the sake of definiteness, we assume also that S^+ is free, and the vector of exterior stress acts on S^- and the Fournier-Lagger incomplete series expansion is true:

$$\sigma_3 = \exp(-\gamma z)\Big(g^- + \sum_\kappa \overset{\kappa}{\sigma}_3 (x_1, x_2)(L_\kappa(z) - L_{\kappa+l}(z)) \quad ,$$

$$\sigma_{\alpha\beta} = \exp(-\gamma z) \sum_\kappa \overset{\kappa}{\sigma}_{\alpha\beta}\, L_\kappa(z)\,,\; u = \exp(-\gamma z)\sum_\kappa \overset{\kappa}{u}\, L_\kappa(z) \quad ,$$

γ is some parameter.

Then the following theorem from approximation theory is valid.

Theorem. Let the series $\sum_\kappa \overset{\kappa}{u}_i$ and $\sum \operatorname{grad} \overset{\kappa}{u}_i$ be absolutely and uniformly convergent in D. Then for $\forall\varepsilon > 0$, $\exists N_0 = N_0(\varepsilon)$ such, that the reduced Galerkin-Vekua system approximates the initial boundary problem with ε accuracy when $N > N_0$.

The calculation of the truncated scheme is reduced to the consecutive inversion of two-dimensional differential problems in D for differential operators:

$$\Gamma_\alpha(\partial_1, \partial_2; \gamma) = \mu^\alpha(\lambda + 2\mu)(\Delta + \gamma^2 I_1)^{\alpha+2} \quad ,$$
$$\Gamma_3(\partial_1, \partial_2; \gamma) = -\mu \operatorname{rot}\operatorname{rot} + (\lambda + 2\mu)\operatorname{graddiv} + \mu\gamma^2 I_2$$

where I is the unit operator.

Literature

1. T.S. Vashakmadze: Rep. enlarged session Vekua Inst. Appl. Math., vol. **2**, N2 (Tbilisi, 1986) (Russian) pp. 29–31
2. T.S. Vashakmadze: Rep. enlarged session Vekua Inst. Appl. Math., vol. **2**, N3 (Tbilisi, 1986) (Russian), pp. 16–19
3. P.G. Ciarlet: Comp. mech. **1** (1986), pp. 177–202
4. T.S. Vashakmadze: *Trends Appl. Pure Math. to Mech.*, ed. by E. Kröner, K. Kirchgässner, Lect. N. in Phys., Vol. 249 (Springer, Berlin, Heidelberg 1986) pp. 487–491
5. I.N. Vekua: *Shell Theory: General Methods of Construction.* Pitman Advanc. Publ. Prog. Boston-London-Melbourne (1985), 287 p.

Application of Extremum Principles to Mixed Boundary-Value Problems in Non-linear Elasticity

J.B. Haddow[1] *and R.W. Ogden*[2]

[1]Department of Mechanical Engineering, University of Alberta, Edmonton, Alberta, Canada T6G 2G8

[2]Department of Mathematics, University of Glasgow, Glasgow G12 8QW, UK

1. Introduction

We consider the plane-strain compression of a rectangular block of elastic material bonded between parallel rigid plates with traction-free lateral surface. The framework for the discussion is the non-linear theory of elasticity, and the main objective is to determine the overall elastic properties of the bonded structure, expressed as a relationship between a measure e of the prescribed macroscopic strain (essentially the relative displacement of the plates) and the resulting force f per unit plate area. In the specialization to the linear theory this relationship takes the form

$$f = E_e e, \tag{1.1}$$

where E_e is the effective Young's modulus. The latter depends on the Young's modulus E of the material, Poisson's ratio ν and on the aspect ratio of the block, namely,

$$\eta = H/A, \tag{1.2}$$

where $2A$ is the width of the block and H is the distance between the plates prior to deformation. We define the macroscopic strain e as

$$e = 1 - h/H, \tag{1.3}$$

where h is the distance between the plates after deformation.

Let $U(e)$ denote the total stored elastic energy in the deformed body per unit reference volume. Then

$$f = U'(e), \quad U'(0) = 0, \tag{1.4}$$

and

$$E_e = U''(0), \tag{1.5}$$

where the prime denotes differentiation with respect to e.

Previous work on this problem relates mainly to the linear theory and, in particular, to determining expressions for E_e or bounds upon E_e. See, for example, GOODIER [4], GENT and LINDLEY [3], FLAVIN [2] and references contained in HADDOW and OGDEN [5]. In the non-linear context, attention has been restricted to the incompressible theory [6] with a special form of constitutive law.

The theory described in this paper is for a compressible material (although results for the incompressible theory can be deduced from those for the compressible theory by a limiting process) and, at least initially, no restriction is placed on the form of the strain-energy function.

2. Formulation of the Problem

Let Ω denote the in-plane cross-section of the body in its reference configuration, where

$$\Omega = \{(X_1,X_2) \mid -A \leqslant X_1 \leqslant A, \quad 0 \leqslant X_2 \leqslant H\},$$

and X_1,X_2 are rectangular Cartesian coordinates. The mapping $\chi = (\chi_1,\chi_2) : \Omega \to \mathbb{R}^2$ describes the deformation of Ω and the coordinates (x_1,x_2) of $(X_1,X_2) \in \Omega$ after deformation are given by

$$x_1 = \chi_1(X_1,X_2), \quad x_2 = \chi_2(X_1,X_2). \tag{2.1}$$

The deformation gradient matrix A has components $A_{ij} = \partial x_i/\partial X_j$, $i,j \in \{1,2\}$, is subject to the constraint

$$\det\mathsf{A} > 0 \tag{2.2}$$

and possesses the polar decomposition

$$\mathsf{A} = \mathsf{RU}, \tag{2.3}$$

where R is proper orthogonal and U is positive definite and symmetric.

The (in-plane) components of the nominal stress tensor are represented by the matrix S, and the stress-deformation relation for an elastic material is given in the form

$$\mathsf{S} = \partial W/\partial\mathsf{A}, \quad S_{ji} = \partial W/\partial A_{ij}, \tag{2.4}$$

where W is the strain-energy function per unit reference volume, which, for a homogeneous material, depends only on A. In the absence of body forces, the equilibrium equations are

$$\partial S_{ki}/\partial X_k = 0, \quad i \in \{1,2\}, \text{ in } \Omega, \tag{2.5}$$

with summation over k from 1 to 2.

For the problem in question, the boundary conditions are

$$x_1 = X_1 \quad \text{on} \quad X_2 = 0,H, \tag{2.6}$$

$$x_2 = 0 \quad \text{on} \quad X_2 = 0, \quad x_2 = h \quad \text{on} \quad X_2 = H, \tag{2.7}$$

$$S_{11} = S_{12} = 0 \quad \text{on} \quad X_1 = \pm A. \tag{2.8}$$

In what follows the boundaries $X_2 = 0$ and $X_2 = H$, with $-A \leqslant X_1 \leqslant A$, are jointly denoted by $\partial\Omega_x$, where $\partial\Omega$ is the boundary of Ω.

3. Energy and Stability Considerations

The energy functional

$$E\{\chi\} = \int_\Omega W(\mathbf{A})\mathrm{d}\Omega \tag{3.1}$$

involves no boundary integral, and provides the basis for a variational formulation of the problem (see [7] for a general treatment of variational principles in the context of the non-linear theory of elasticity).

We assume that W has sufficient regularity (as a function of $\mathbf{A}$) as is needed. We also assume that χ is (at least) twice continuously differentiable in Ω, and that it depends smoothly on the boundary data (i.e. on e). We write the associated one-parameter solution of the problem as χ_e. The total energy per unit reference volume may now be regarded as a (smooth) function of e, $U(e)$ say:

$$U(e) = V^{-1}E\{\chi_e\} = V^{-1}\int_\Omega W(\mathbf{A}_e)\mathrm{d}\Omega, \tag{3.2}$$

where $\mathbf{A}_e$ is the deformation gradient associated with χ_e and $V = 2AH$.

Let

$$f = -\frac{1}{2A}\int_{-A}^{A} S_{22}\,\mathrm{d}X_1 \tag{3.3}$$

denote the mean compressive force on the plane X_2 = const. (independent of X_2). It follows from (2.4) - (2.8) and the need for compatibility with the linear theory that (1.4) and (1.5) hold. It also follows that

$$U''(e) = V^{-1}\int_\Omega B_{kilj}\,A'_{ik}\,A'_{jl}\,\mathrm{d}\Omega, \tag{3.4}$$

a term involving $\mathbf{A}''$ vanishing by use of the divergence theorem and the boundary conditions, where the coefficients B_{kilj} are given by

$$B_{kilj} = \partial^2 W/\partial A_{ik}\,\partial A_{jl}. \tag{3.5}$$

In the linear theory χ_e depends linearly on e and (3.4) reduces to $U''(0)$, which is positive. Moreover, the solution will be symmetric about the lines $X_1 = 0$ and $X_2 = \frac{1}{2}H$. As e increases to values for which the use of a linear constitutive law is no longer appropriate this symmetry is maintained (at least up to some critical value of e) and, by continuity, the inequality

$$U''(e) > 0 \tag{3.6}$$

holds.

The symmetric solution is locally stable if

$$\int_\Omega B_{kilj}\,\dot{A}_{ik}\,\dot{A}_{jl}\,\mathrm{d}\Omega > 0 \tag{3.7}$$

for all incremental deformation fields $\dot{\chi}$ for which $\dot{\chi} = 0$ on $\partial\Omega_x$ and the associated gradient $\dot{A}$ does not vanish identically in Ω, B_{kilj} being evaluated for χ_e. Stability of χ_e implies that χ_e is a local minimizer of the energy functional(3.1) within an appropriate class of deformation fields satisfying (2.6) and (2.7), and, via (3.4) and (3.6), that f is a monotonic increasing function of e (with $f > 0$ corresponding to $e > 0$ in compression). Here, we are concerned with values of e for which the symmetric deformation is stable, so that, in particular, (3.6) holds. The smaller the value of the aspect ratio η, given by (1.2), the larger will be the value of e for which this stability holds. For large values of η buckling can be initiated at relatively small values of e, and we therefore concentrate attention on values of η for which buckling can realistically be excluded. We assume that (3.7), and hence (3.6), holds for $0 \leqslant e < 1$ in respect of the symmetric solution χ_e (which is assumed to exist for $0 \leqslant e < 1$). [Elsewhere we shall investigate criteria for failure of (3.7) and the incidence of bifurcation from the symmetric solution at some critical value of e, dependent on η.]

4. Complementary Energy

It follows from (3.6) that we may define a complementary energy function, $U_c(f)$ say, per unit reference volume (i.e. per unit area of Ω), by

$$U(e) + U_c(f) = ef, \tag{4.1}$$

where

$$e = U_c'(f), \quad U_c'(0) = 0, \tag{4.2}$$

$$U_c(f) = V^{-1}\int_\Omega \{\mathrm{tr}(\mathsf{SA}) - W(\mathsf{A})\}\,d\Omega - \mathrm{tr}(\bar{\mathsf{S}}) \tag{4.3}$$

and $\bar{\mathsf{S}}$ is the mean value of S over Ω.

In order to construct a complementary energy functional, dual to the energy functional (3.1), we use the fact that the stored energy must be independent of superimposed rigid-body rotations after deformation (the principle of objectivity), so that W depends on A only through the stretch matrix U appearing in (2.3), and define the stress matrix T (the Biot stress) conjugate to U as

$$\mathsf{T} = \partial W/\partial \mathsf{U}. \tag{4.4}$$

The identity

$$\mathrm{tr}(\mathsf{SA}) - W(\mathsf{A}) = \mathrm{tr}(\mathsf{TU}) - W(\mathsf{U}) \tag{4.5}$$

follows (see, for example, [7]). It is well known that W is not strictly convex as a function of A, and hence that the stress-strain law (2.4) is not uniquely invertible, so that a complementary energy density cannot be defined directly with respect to the conjugate variables (S,A). However, it is reasonable to assume, at least for deformations that are not too large, that W is a strictly convex function of U. Here we adopt this assumption, so that (4.4) is uniquely invertible, and define the complementary energy density function W_c by

$$W_c(\mathrm{T}) = \mathrm{tr}(\mathrm{TU}) - W(\mathrm{U}) \tag{4.6}$$

so that the inverse of (4.4) has the form

$$\mathrm{U} = \partial W_c/\partial \mathrm{T}. \tag{4.7}$$

The energy functional can now be expressed

$$E\{\chi\} = \int_{\partial\Omega_x} (S_{21}x_1 + S_{22}x_2)d\omega - \int_{\Omega} W_c(\mathrm{T})d\Omega, \tag{4.8}$$

where x_1 and x_2 are given by (2.6) and (2.7) respectively on $\partial\Omega_x$, and $d\omega = dX_1$ on $X_2 = H$, $-dX_1$ on $X_2 = 0$. In (4.8), T and S are calculated from χ through (2.3), (2.4) and (4.4).

Let λ_i, $i \in \{1,2\}$, denote the principal stretches, i.e. the principal values of U, and t_i the corresponding principal values of T. For an isotropic material W depends only on λ_1 and λ_2 and

$$W(\lambda_1,\lambda_2) = W(\lambda_2,\lambda_1) \tag{4.9}$$

and

$$t_i = \partial W/\partial\lambda_i \quad i \in \{1,2\}. \tag{4.10}$$

Equation (4.6) can then be expressed

$$W_c(t_1,t_2) = t_1\lambda_1 + t_2\lambda_2 - W(\lambda_1,\lambda_2), \tag{4.11}$$

$$W_c(t_1,t_2) = W_c(t_2,t_1), \tag{4.12}$$

and (4.7) becomes

$$\lambda_i = \partial W_c/\partial t_i \quad i \in \{1,2\}. \tag{4.13}$$

Moreover, we have the polar decomposition [7]

$$\mathrm{S} = \mathrm{TR}^T \tag{4.14}$$

analogous to (2.3). Unlike (2.3), however, (4.14) is not unique. Nevertheless, given S, T (and, if required, R) can be determined as follows: from (4.14), we have

$$\mathrm{SS}^T = \mathrm{T}^2, \quad \det\mathrm{S} = \det\mathrm{T} = t_1t_2, \tag{4.15}$$

from which can be calculated the principal axes of T (coaxial with T^2) and, to within an interchange of signs, the principal Biot stresses t_1,t_2. Thus, two pairs of values of (t_1,t_2) correspond to a given S, but only one of these pairs will be appropriate for the problem at hand; a further restriction on the values of (t_1, t_2) is derived below.

For an isotropic elastic material we can therefore regard the right-hand side of (4.8) as a functional of S, and we write this as

$$\mathrm{E}_c\{\mathrm{S}\} = \int_{\partial\Omega_x} (S_{21}x_1 + S_{22}x_2)\mathrm{d}\omega - \int_{\Omega} W_c(t_1, t_2)\mathrm{d}\Omega. \tag{4.16}$$

The connexion between this and (4.3) is

$$V^{-1}\mathrm{E}_c\{\mathrm{S}\} = ef - U_c(f), \tag{4.17}$$

and we also have

$$V^{-1}\int_{\Omega} \mathrm{tr}(\mathrm{SA})\mathrm{d}\Omega = ef + \mathrm{tr}(\bar{\mathrm{S}}). \tag{4.18}$$

Note that

$$\int_{-A}^{A} S_{21}\mathrm{d}X_1 = 0, \qquad 0 \leqslant X_2 \leqslant H \tag{4.19}$$

follows from symmetry and the boundary conditions.

Equations (4.18) and (4.19) also hold for non-isotropic materials, but (4.8) must then be regarded as a functional of S *and* R since then

$$\mathrm{T} = \tfrac{1}{2}(\mathrm{SR} + \mathrm{R}^T\mathrm{S}^T).$$

Henceforth, however, we confine attention to the isotropic case.

If S_f is the nominal stress associated with the stable symmetric solution χ_e then it follows that S_f is a local maximizer of the complementary energy functional (4.16) within an appropriate class of stress fields satisfying (2.5) and (2.8).

By considering the three different expressions for $U''(e)$, namely (3.4) and the corresponding expressions obtained by replacing $W(\mathrm{A})$ by $W(\mathrm{U})$ and $W(\lambda_1, \lambda_2)$, it can be shown that [5]

$$\int_{\Omega} (t_1\lambda_2 + t_2\lambda_1)\theta'^2\mathrm{d}\Omega = 0 \tag{4.20}$$

where θ' denotes the rate of rotation of the principal axes of U with respect to e. This provides information about the comparative signs of t_1 and t_2.

5. Extremum Principles

Let K denote the set of all twice continuously differentiable deformation fields which satisfy (2.6) and (2.7), depend smoothly on e, possess the symmetry imposed on χ_e and also satisfy (3.7), where B_{kilj} is calculated for the appropriate field. Let $\chi^* \in K$ and let S^* be calculated from the gradient A^* of χ^* through (2.4). By integrating along a path in K corresponding to the straight line joining S_f to S^* in the set obtained from K through the transformation (2.4), we then obtain

$$E\{\chi^*\} \geqslant E\{\chi_e\} \quad \text{for all } \chi^* \in K. \tag{5.1}$$

Thus, within the class K, χ_e is a global minimizer of the energy.

It is likely that the inequality (5.1) holds for a much wider class of deformation fields than K, given that W is a strictly convex function of U, but we do not attempt to characterize such a class here. We remark, however, that BALL [1] has established the existence of global minimizers of the energy functional for certain boundary-value problems under a number of hypotheses on W; his theory embraces the problem considered here. Certainly, (5.1) holds in the linear theory, and, by continuity, in the non-linear theory provided e is not too large.

Similarly, let Σ denote the set of all continuously differentiable stress fields, S^* say, which satisfy (2.5) and (2.8), possess the same symmetry that S_f inherits from χ_e, correspond, through (2.4), to locally stable deformation gradients (not necessarily the gradients of deformation fields), and depend smoothly on f^*, where f^* is defined by

$$f^* = -\frac{1}{2A}\int_{-A}^{A} S^*_{22}\,dX_1 \tag{5.2}$$

analogously to (3.3). Because of the required symmetry (4.19) holds for S^*_{21}.

By a similar argument to that leading to (5.1) we obtain

$$E_c\{S_f\} \geqslant E_c\{S^*\} \quad \text{for all } S^* \in \Sigma, \tag{5.3}$$

and, as with (5.1), the range of validity of this inequality extends beyond Σ although we do not consider such an extension here.

Let

$$U^*(e) = V^{-1}\int_\Omega W(A^*)\,d\Omega, \tag{5.4}$$

analogously to (3.2), and

$$U^*_c(f^*) = V^{-1}\int_\Omega W_c(T^*)\,d\Omega - \mathrm{tr}(\mathfrak{S}^*), \tag{5.5}$$

where T^* is the Biot stress associated with $S^* \in \Sigma$ with principal stresses t_1^*, t_2^* consistent with (4.20). Then $U^{*\prime\prime}(e) > 0$ and $U_c^{*\prime\prime}(f^*) > 0$, where, in general, (5.5) is not complementary to (5.4).

Taken together, the inequalities (5.1) and (5.3) can now be expressed in the form

$$U^*(e) \geqslant U(e) = ef - U_c(f) \geqslant ef^* - U_c^*(f^*), \tag{5.6}$$

which provides upper and lower bounds on the energy $U(e)$. The lower bound in (5.6) is maximised if f^* satisfies

$$e = U_c^{*\prime}(f^*). \tag{5.7}$$

The properties of $U_c^*(f^*)$ ensure that (5.7) is uniquely invertible to give f^* as a function of e, and the right-hand side of (5.6) is therefore a function of e, $U^{**}(e)$ say. We then have

$$U^*(e) \geqslant U(e) \geqslant U^{**}(e), \tag{5.8}$$

and the bounds are therefore expressed in terms of the parameter e.

Elementary bounds on $U(e)$ can be obtained by choosing χ^* to correspond to homogeneous uniaxial strain, with $\mathsf{A}^* = \mathsf{U}^* = \mathrm{diag}(1,\lambda_2)$, and $\mathsf{S}^* = \mathsf{T}^* = \mathrm{diag}(0, -f^*)$ to correspond to homogeneous uniaxial compression. Then (5.8) simplifies to

$$W(1,\lambda_2) \geqslant U(e) \geqslant W(\lambda_1,\lambda_2), \tag{5.9}$$

where $\lambda_2 = 1 - e$, and λ_1 is determined as a function of λ_2 from

$$\partial W(\lambda_1,\lambda_2)/\partial\lambda_1 = 0. \tag{5.10}$$

We note that strict convexity of W as a function of U is equivalent to

$$W(\lambda_1^*,\lambda_2^*) - W(\lambda_1,\lambda_2) - t_1(\lambda_1^* - \lambda_1) - t_2(\lambda_2^* - \lambda_2) > 0 \tag{5.11}$$

for all $(\lambda_1^*,\lambda_2^*) \neq (\lambda_1,\lambda_2)$, and this implies the outer inequality in (5.9). The local form of (5.11), namely

$$W_{11} > 0, \; W_{11}W_{22} - W_{12}^2 > 0, \tag{5.12}$$

where $W_{ij} = \partial^2 W/\partial\lambda_i\partial\lambda_j$, coupled with the inequality $W_{12} > 0$ ensures that (5.10) yields a unique value of $\lambda_1 > 0$ for each $\lambda_2 > 0$.

Application of the inequalities (5.8) for particular forms of strain-energy function $W(\lambda_1,\lambda_2)$ and specific choices of the fields χ^* and S^* has been considered by HADDOW and OGDEN [5]; in particular, some preliminary numerical results were given in [5].

6. Specialization to the Linear Theory

From the definition of $U^{**}(e)$ and (5.7), we obtain

$$U^{**\prime\prime}(e) = 1/U_c^{*\prime\prime}(f^*). \tag{6.1}$$

In the linear theory the chain of inequalities (5.8) is quadratic in e and, with the help of (6.1) and (1.5), can therefore be written as

$$U^{*''}(0) \geqslant E_e \geqslant 1/U_c^{*''}(0). \tag{6.2}$$

In terms of the components of the infinitesimal strain tensor e^*_{ij} (linear in e) and the stress tensor σ^*_{ij} (linear in f^*), we have

$$U^{*''}(0) = EV^{-1}\int_\Omega \{(1-2\nu)e'_{ij}e'_{ij} + \nu(e'_{ii})^2\}d\Omega/(1+\nu)(1-2\nu) \tag{6.3}$$

and

$$U_c^{*''}(0) = (1+\nu)V^{-1}\int_\Omega \{\sigma'_{ij}\sigma'_{ij} - \nu(\sigma'_{ii})^2\}d\Omega/E. \tag{6.4}$$

Use of (5.10) specialized to the linear theory shows that the elementary bounds (5.9) become

$$E(1-\nu)/(1+\nu)(1-2\nu) \geqslant E_e \geqslant E/(1-\nu^2). \tag{6.5}$$

The upper bound in (6.5) corresponds to the asymptotic value of E_e in the limit $\eta \to 0$ [2,4], and the lower bound to the asymptotic value of E_e as $\eta \to \infty$.

The bounds (6.2) are also applicable in the incompressible theory, and, by taking the limit $\nu \to \frac{1}{2}$, we obtain

$$U^{*''}(0) = \frac{4}{3}EV^{-1}\int_\Omega (e'_{11})^2 d\Omega \tag{6.6}$$

and

$$U_c^{*''}(0) = \frac{3}{4}E^{-1}V^{-1}\int_\Omega \{(\sigma'_{11} - \sigma'_{22})^2 + 4(\sigma'_{12})^2\}d\Omega. \tag{6.7}$$

We note that, in this case, $e_{11} + e_{22} = 0$ and the integrand in (6.7) is independent of the hydrostatic part of the stress. The upper bound in (6.5) becomes infinite and the lower bound becomes $4E/3$, which is consistent with the expression for E_e given by GENT and LINDLEY [3], namely

$$E_e = 4E(1+\eta^{-2})/3. \tag{6.8}$$

7. References

1. J.M. Ball: Arch. Rational Mech. Anal. **63**, 337 (1977)
2. J.N. Flavin: Int. J. Solids Structures **21**, 267 (1985)
3. A.N. Gent, P.B. Lindley: Proc. Inst. Mech. Engrs **173**, 111 (1959)
4. J.N. Goodier: J. Appl. Phys. **13**, 167 (1942)
5. J.B. Haddow, R.W. Ogden: to appear (1988)

6. W.W. Klingbeil, R.T. Shield: Z.A.M.P. 17, 281 (1966)

7. R.W. Ogden: *Non-linear Elastic Deformations* (Ellis Horwood, Chichester 1984)

Application of Bivariate B-Splines to a Plate Bending Problem

C.R. Traas

University of Twente, Faculty of Applied Mathematics,
P.O. Box 217, NL-7500 AE Enschede, The Netherlands

The tractability of bivariate quadratic B-splines in case of a plate bending problem has been investigated. It is shown that with a fair amount of automation these splines can be used very well. This result may open the way, in principle, to apply them for solving really complex problems from mechanics and from other disciplines.

1. Introduction

A significant new development in approximation theory is the concept of multivariate B-splines. A multivariate B-spline is a locally supported nonnegative piecewise polynomial function in more than one variable, which is defined by a set of knots in the Euclidean space.

Over the past few years a fairly complete theory of multivariate B-splines has been developed by many authors (see DAHMEN and MICCHELLI [1] for a survey). These splines are suitable for a wide range of applications, such as smooth surface fitting, solution of partial differential equations and computer aided design.

Numerical experience is rather limited. This is a consequence of the expected high computational cost of the multivariate B-spline, and the complexity of grid partitions in higher dimensions. However, by automatizing certain basic processes, which are required in the phase of preparing a computation with B-splines (knot placement), the weight of these objections can be reduced.

In particular, when choosing second degree piecewise polynomial splines in two variables only, which are of class C^1, there are no serious practical limitations with respect to applications.

To gain experience with the computational properties of bivariate quadratic B-splines and with the organizational matters around a computation with these splines (data bases, fundamental subroutines), the author has chosen a well-known and already in many ways solved problem from mechanics: the bending thin plate. If such a computation can be performed, and is succesful from various points of view, the way is open to apply these splines to less elementary problems where the C^1 property is required in view of conformity.

In this paper a description is given of multivariate B-splines, with a specialization to bivariate quadratic B-splines. It is shown that these splines can be applied very well, and that they give results comparable in accuracy with results of other methods based on piecewise quadratic polynomials.

2. Multivariate B-splines

The p-variate B-spline of degree d can be considered to be the "shadow image" of a (p+d)-simplex ($\subset R^{p+d}$) onto R^p. Since a (p+d)-simplex has p+d+1 vertices, the B-spline is nonzero in the convex hull of p+d+1 points in R^p. Any (hyper)plane passing through p of the vertices separates polynomial pieces within the support of the B-spline; if the knots are in "general position" (i.e. no p+1 knots are in one hyperplane), then these polynomial pieces match in C^{d-1}-sense.

The objective is, given a region $\Omega \subset R^p$, to construct a set of B-splines such that their linear span contains the space of all polynomials of degree d over the region Ω.

The construction of such a set proceeds as follows. Let the region $\Omega \subset R^p$ be partitioned into N p-simplices such that any two of these simplices are either disjunct, or have only a lower dimensional face (i.e. of dimension p-1 or lower) in common. Consider, for each of these p-simplices, the polyhedron of dimension p+d defined by the tensor product of the p-simplex with the standard d-simplex S_d. In each of these regions one constructs a collection of $\binom{p+d}{p}$ simplices of dimension p+d such, that the total collection of (p+d)-simplices, which are $\binom{p+d}{p}\cdot N$ in number, precisely fills the space $\Omega \times S_d$ in a proper way, i.e. any two of the (p+d)-simplices are either disjunct, or have only a lower dimensional simplex (a "face") in common.

Each of these (p+d)-simplices is a generator of a p-variate B-spline of degree d. In this way we have constucted a set of $\binom{p+d}{p}N$ B-splines over Ω. The supports of these B-splines are exactly, by construction (orthogonal projection from R^{p+d} onto R^p), the elements in the partition of Ω. We then have coinciding knots and thus discontinuities in the B-splines or in their derivatives.

To obtain global smoothness of class C^{d-1} the coinciding knots must be separated (pulled apart). This can be interpreted as projecting a slightly deformed space $\Omega \times S_d$ onto R^p. This deformation should be such that the p+d+1 knots for each B-spline are be in general position.

The practical computation of a B-spline is performed by using a recurrence relation:

$$M(\underline{x}|\underline{x}_0,\underline{x}_1,\cdots,\underline{x}_n) = \frac{n}{n-p}\sum_{j=0}^{n} \lambda_j M(\underline{x}|\underline{x}_0,\underline{x}_1,\cdots,\underline{x}_{j-1},\underline{x}_{j+1},\cdots,\underline{x}_n)$$

where M is the symbol for the B-spline, $n = p+d$,

$\underline{X}_i$ are the knots defining the B-spline, and $\underline{X} \in R^p$ is the position at which its value is computed.

The λ_j are local coordinates defined by

$$\underline{X} = \sum_{j=0}^{n} \lambda_j \underline{X}_j \quad , \quad \sum_{j=0}^{n} \lambda_j = 1.$$

The λ_j are not uniquely defined in this way; there is some freedom of choice. The start of the recursion is a set of lowest order B-splines:

$M(\underline{X}|\underline{X}_{i_0},\underline{X}_{i_1},\cdots,\underline{X}_{i_p})$, which are constants equal to $1/\mathrm{Vol}[\underline{X}_{i_0},\cdots,\underline{X}_{i_p}]$ within the convex hull of their knots (thus the inverse of the p-dimensional volume of this hull), and zero otherwise.

3. The Process of Pulling Apart

In the remaining sections we specialize to bivariate quadratic B-splines. These splines are defined by a set of 5 knots in R^2.

Let the region $\Omega \subset R^2$ be triangulated in a proper way. In the space defined by the tensor product of an arbitrary triangle from this triangulation with the standard 2-simplex S_2 a collection of six 4-simplices can be constructed along the lines described in section 2.

They generate six B-splines, all of which have the chosen triangle as their support, i.e. the vertices of the triangle are 3-fold knots.

We label the vertices of the triangle with the indices 0, 1 and 2, respectively, and the 3 (coinciding) knots in each of the vertices with a second index 0, 1 or 2. Then, as an example, 10 indicates, at vertex 1, the knot that will not be pulled, 11 the knot of type 1 that will be pulled from vertex 1, and 12 the knot of type 2 that also will be pulled from vertex 1.

The six 4-simplices generate the following six B-splines:

$M_1(x,y|00,01,02,12,22)$
$M_2(x,y|00,01,11,21,22)$
$M_3(x,y|00,01,11,12,22)$
$M_4(x,y|00,10,11,21,22)$
$M_5(x,y|00,10,11,12,22)$
$M_6(x,y|00,10,20,21,22)$.

The objective is to separate the coinciding knots in such a way that, for any of the six B-splines per triangle, the five defining knots are in general position, i.e. the B-splines belong to the class of functions which are continuously differentiable.

Assume, as an example, that the knots, 01,02,11 and 12 already have been pulled apart, and the next knot to be pulled is 21. In positioning this point certain straight lines ("forbidden lines") should be avoided. These lines are defined by the points (00,01), (00,10), (00,11), (00,20), (01,11), (10,11) and (10,20), respectively, as can be seen from the defining knot sets of the splines M_2, M_4 and M_6.
Next, in an analogous way, the knot 22 can be positioned.

In a process of pulling apart certain graphical aids can be very helpful. Projecting automatically the forbidden lines, associated with a certain triangle and the knots already available, onto the monitor screen, the new knots can be positioned very easily.

4. Boundary Conditions

It may occur that a function u(x,y), which has to be approximated, satisfies certain boundary conditions e.g. $u = 0$ and/or $\partial u/\partial n = 0$ along a boundary, or a condition of symmetry across a given line.
Conditions like $u = 0$ or $\partial u/\partial n = 0$ can be taken into account by a suitable placing of pulled knots. If both $u = 0$ and $\partial u/\partial n = 0$, all knots which are pulled from vertices located on the boundary, have to be pulled in the direction of the interior of the region Ω.
In the sequel we consider straight line boundary segments. If only $u = 0$ then one knot from a boundary vertex should be pulled to the interior of Ω, and the other one in the direction of the boudary. Then certain B-splines will have three of their knots on a boundary segment, resulting in discontinuous behaviour of the first derivative normal to this segment. Care has to be taken that the total collection of B-splines with 3 knots on the (straight) boundary generates a complete set of univariate piecewise linear basissplines for the normal derivative at this boundary.

If only $\partial u/\partial n = 0$, and u is arbitrary along a (straight) boundary (this is a condition of symmetry), then, from each boundary vertex, the two knots should be pulled both in the direction of the boundary. Then certain B-splines will have 4 knots on the boundary resulting in discontinuous behaviour of u(x,y) across this line. Care has to be taken, again, that the total collection of B-splines with 4 knots on the boundary generates a complete set of univariate piecewise quadratic basissplines for u along this boundary. However, without any further action the normal derivative $\partial u/\partial n$ would then be nonzero. By taking care that the total collection of B-splines with 3 points on the boundary generates a complete set of univariate piecewise linear basissplines for $\partial u/\partial n$, each B-spline with 4 points on the boundary can be linearly combined with one or more B-splines with 3 points on this line such, that new basisfunctions are created with the property that $\partial u/\partial n$ is zero at the boundary, while $u \neq 0$.

5. Surfaces Constrained by Satisfying a Differential Equation

Sometimes the problem at hand is to construct a function u(x,y) which not only satisfies given boundary conditions, but also satisfies a given partial differential equation in a region Ω. Adopting the Galerkin approach, or exploiting a variational principle (if it exists for the problem considered), a derived problem is to compute inner products, e.g. in case of a fourth order equation to compute integrals of the form

$$\iint_{\Omega} \frac{\partial^2 M_i}{\partial x^p \partial y^q} \cdot \frac{\partial^2 M_j}{\partial x^r \partial y^s} \, dxdy \quad , \quad p + q = r + s = 2.$$

By using recurrence relations, a second order partial derivative of a bivariate quadratic B-spline can be written as a linear combination of at most nine bivariate constant B-splines, each of these thus being defined by a set of 3 knots. The evaluation of the inner product therefore reduces to summing up at most 81 innerproducts of constant B-splines, each of them having the convex hull of three knots as its support. A problem, derived from this process, therefore is to determine the area of the intersection of two triangles, each having arbitrary orientation in the plane. This latter problem has been solved by applying linear programming methods. First, starting from a vertex of one of these triangles, a vertex of the boundary of the intersection of the two triangles can be found, using a few simplex steps, by requiring all of the six barycentric coordinates (there are two sets of 3 coordinates each) to be non-negative.

Next, the other vertices of the intersection are found, simply, by stepping from one vertex to the other one by pushing successively all barycentric coordinates to zero and keeping all of them non-negative. Once all vertices of the intersection are found, the area of the intersection is easily calculated.

6. Solving a Model Problem

The successful operation of the methods described is demonstrated using a model problem.

A thin, square plate, bending under its own weight, is considered. One side of the plate is clamped (boundary condition $u = 0$, $\partial u/\partial n = 0$) and the opposite side is simply supported (boundary condition $u = 0$). The two other sides are free. This configuration contains a line of symmetry. Therefore only one half of the plate needs to be analysed, and a symmetry condition is introduced.
The triangulation is sketched in figure 1, together with the pulled knots. The flaps at the left serve the purpose to match the symmetry condition and the free boundary condition, respectively, with the simply supported boundary condition. From each of these flaps three of the six B-splines are used. (See TRAAS [2]).

The results of the computation can be compared with the results obtained by GMELIG MEYLING [3] and the results reported in SCHWARZ [4]. These results are available in the points A through F, indicated in figure 1, and table 1 gives a survey.

The result from [3], given in table 1, has been obtained with a split-triangle quadratic element (HEINDL [5]), by partitioning the upper half of the thin plate (as given in figure 1, without the flaps) into 16 triangles (a uniform, unidiagonal "type I" grid). The fineness of this grid is comparable with that of the grid used in figure 1, and also the results are very well comparable.

The quoted results from [4] have been obtained with a bicubic rectangular polynomial conforming element and are considered, on the basis of convergence of the results with increasing grid fineness, to be correct in 4 digits.

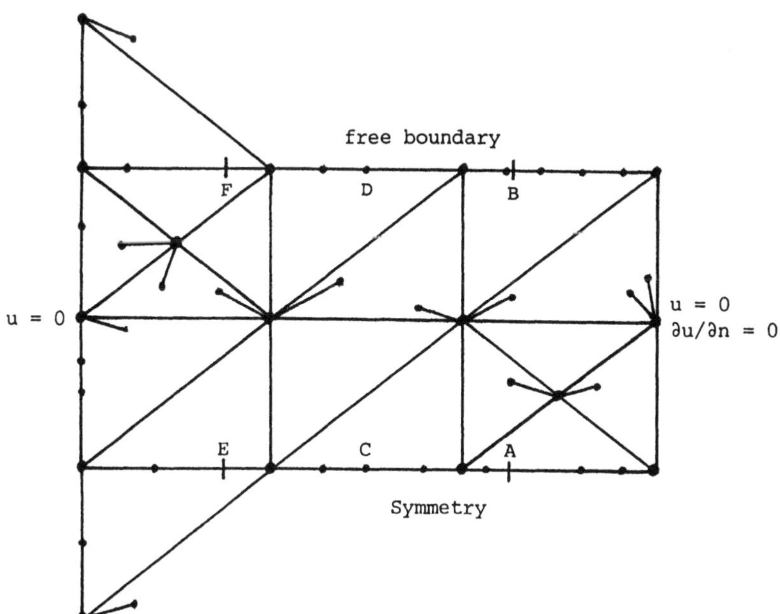

Fig 1.

Triangulation in modelproblem, and the pulling apart of knots.

Also in [3] convergence of the results is reported, while reducing mesh size; this convergence, however, is slow. Other results in [3] (not quoted in table 1), obtained with cubic polynomial conforming triangular elements (the Clough-Tocher element, and a representation in terms of Bézier-Bernstein, respectively) appear to be more accurate and to converge faster. The computation times for all of these methods are of comparable size.

7. Conclusions

It has been demonstrated that a computation with bivariate quadratic B-splines can be accomplished very well. Conditions for success are a well-designed data base, containing the necessary information about the grid and about the problem itself, and a set of subroutines to automatically prepare a major part of this information (e.g. the pulling apart, and special boundary conditions).
At this moment we have not yet obtained the maximum degree of automation, but once this aim will have been achieved, we expect that the method of bivariate, quadratic B-splines can be used as easily as other methods, however, with the additional advantage of very high flexibility.

8. References

1. W. Dahmen, C.A. Micchelli: In Surfaces in GAGD, ed. by R.E. Barnhill, W. Boehm (North-Holland Publishing Company, 1983), pp. 191-215.
2. C.R.Traas: In Approximation Theory V, ed. by C.K. Chui, L.L. Schumaker, J.D. Ward (Academic Press, 1986), pp. 595-598.
3. R.H.J. Gmelig Meyling: Private communication. To be published in Proceedings of Conference on Algorithms for Approximation, Shrivenham, UK (July 1988).
4. H.R. Schwarz: Methode der Finiten Elemente (Teubner Studienbücher, 1980).
5. G. Heindl: In Multivariate Approximation Theory, ed. by W. Schempp, K. Zeller (Birkhäuser Verlag, 1979), pp. 146-161.

Entropy and Dynamical Stability – a Method Due to Lagrange-Formalism as Applied to Thermodynamics of Irreversible Processes

K.-H. Anthony

Universität-Gesamthochschule-Paderborn, Theoretische Physik
D-4790 Paderborn, Fed. Rep. of Germany

1. Outline and motivation

Thermodynamics of irreversible processes can be included into Lagrange-formalism. Along the line of this formalism a methodical unification of thermodynamics, continuum mechanics and electromagnetism of matter may be obtained. The latter two topics have already been subsumed into Lagrange-formalism in the past.

Concerning thermodynamics I have shown so far, that the phenomena of heat transport, diffusion and chemical reactions can also be described within the framework of Lagrange-formalism [1,2,3]. So there is some good hope for a unified theory of phenomenological material behaviour. Especially the dynamics of point-like defects can be included by properly generalizing the concept of chemical reactions giving thus rise to a macroscopic description of disordered materials [3]. Keeping further in mind, that Lagrange-formalism allows for quite general dynamics even far away from equilibrium and that stability theory is an inherent topic of this theory, it gets apparent that Lagrange-formalism might also be useful for those instabilities and phase transitions of a material system which are due to a non-linear coupling of different degrees of freedom and which are traditionally subsumed into the concept of synergetics [4].

In Lagrange-formalism all information concerning the processes of a system is comprised into one function only, namely into the Lagrange-density-function. Thus the whole theory starts from the smallest possible entity. Using the well-established methods of Lagrange-formalism all characteristics of the system are developed by straightforward mathematics. For instance all balance and all constitutive equations are derived from the Lagrangian by means of Noether's theorem which is based on a set of universal invariance requirements relevant for all physics.

Thermodynamcis of irreversible processes can be subsumed into Lagrange-formalism by introducing a new set of complex-valued, fundamental field variables instead of the real valued traditional variables "temperature" and "mass densities" [1,2,3]. The complex valued fields, which will be called "field of thermal excitation" and "matter fields", give rise to an invariance of the Lagrangian with respect to common gauge transformations of these fields. It is a remarkable fact that the concept of entropy and the associated entropy balance equation (first part of the Second Law of Thermodynamics) is a quite natural outcome of Lagrange-formalism due to the gauge invariance of the Lagrangian. The entropy is a particular constitutive quantity which is completely derived from the Lagrangian. Its production rate is traditionally assumed to be non-negative for real physical systems (second part of the 2nd Law). This constitutive constraint is traditionally related with irreversibility of real thermodynamical processes and with stability of the processes. Of course a non-negative entropy production rate can also be incorporated into Lagrange theory. At a first glance, however, this seems to be a strange feature within Lagrange-formalism, which has no inherent support by the theory.

Stability theory is a part of Lagrange-formalism. To a certain extend it is possible to establish this theory in the sense of Ljapunov's direct method. Ljapunov-functionals are derived from the Lagrangian being thus intimately related with the dynamics of the system, i.e. they are defined by inherent structures of the system. This contrasts with the rather common fact in Ljapunov's stability theory, that the Ljapunov-functional is gained by ansatz without connection to the inherent structures of the system.

Of course a stability theory formulated in terms of Ljapunov's direct method and subsumed into Lagrange-formalism has its own worth. Especially the above-mentioned synergetical stabilities may be investigated. However, concerning thermodynamics it is a remarkable fact that - as a consequence of stability in Ljapunov's sense - a thermodynamical process is associated with a positive definite entropy production rate. Along this line the second part of the 2nd Law gets an inherent and quite natural structure of Lagrange-formalism, too. However, the stability concept is superior as compared with the positive definite entropy production rate. By this paper a short review of the relevant investigations will be given.

2. Some fundamental structures of Lagrange-formalism [5,6,1,2,3]

The real-valued Lagrange-density-function (Lagrangian)

$$\ell = \ell(\psi, \partial\psi) \tag{2.1}$$

contains all information concerning the processes of a physical system. ψ denotes the complete set of fundamental field variables:

$$\psi = \{ \psi_i(x,t) \ , \ i = 1,\dots,N \} \tag{2.2}$$

depending on space and time coordinates $x = (x^1, x^2, x^3)$ and t. Throughout this paper ∂ collects all first order derivatives:

$$\partial = \{ \partial_\alpha = \frac{\partial}{\partial x^\alpha} \ , \ \alpha = 1,\dots,3 \ , \ \partial_t = \frac{\partial}{\partial t} \} \ . \tag{2.3}$$

ℓ is the kernel of the action integral

$$J_1 = \int_{t_1}^{t_2} \int_{V(t)} \ell(\psi,\partial\psi)\, d^3x\, dt \tag{2.4}$$

which is taken over the actual volume V(t) of the system and the fixed time intervall $[t_1,t_2]$ between the beginning t_1 and the end t_2 of the process. Hamilton's variational principle is based on the action integral:

J_1 = extremum by free and independent variations $\delta\psi_i(x,t)$ of all fields $\psi_i(x,t)$ with fixed values at the beginning and at the end of the process: (2.5)

$$\partial\psi_i(x,t_{1,2}) = 0.$$

Variations at the boundary ∂V of the system are admissible depending, however, on the physical situation at the boundary.

Within the class of admissible processes the real processes are distinguished as solutions of Hamilton's principle.

Hamilton's principle is equivalent to the set of Euler-Lagrange-equations

$$\partial_t \frac{\partial \ell}{\partial(\partial_t\psi_i)} + \nabla\cdot\frac{\partial \ell}{\partial(\nabla\psi_i)} - \frac{\partial \ell}{\partial\psi_i} = 0 \quad , \quad i = 1,\dots,N, \tag{2.6}$$

which are the fundamental field equations of the system. These equations are supplemented by boundary conditions which in the special case of free variations $\delta\psi_i(x,t)$ at a sufficiently smooth boundary surface ∂V read as

$$\vec{n} \cdot \frac{\partial \ell}{\partial(\nabla\psi_i)} = 0 \quad , \quad i = 1,\dots,N \;. \tag{2.7}$$

This case is physically associated with a completely isolated system. $\vec{n}$ is the outer unit normal vector of the surface.

Equations (2.6) and (2.7) are solved by a real process $\psi(x,t)$ according to given initial values

$$\psi(x,t{=}0) = \psi_0(x) \quad . \tag{2.8}$$

Let me consider a one-parameter class of process perturbations which result in altered real processes $\psi(x,t,\vartheta)$. Of course these are again solutions of (2.6) and (2.7) with perturbed initial values

$$\psi(x,t{=}0,\vartheta) = \psi_0(x,\vartheta) \quad . \tag{2.9}$$

ϑ is the perturbation parameter. For $\vartheta = 0$ we come back to the unperturbed reference process

$$\psi(x,t,\vartheta{=}0) = \psi(x,t) \quad , \qquad \psi_0(x,\vartheta{=}0) = \psi_0(x) \quad . \tag{2.10}$$

The reader should keep in mind that I am not concerned with perturbations of the system but with perturbations of the processes only. This means that the Lagrangian (2.1) is completely preserved; for instance all material parameters are kept fixed. The perturbation of a process is physically realized by altering its initial state from the situation (2.8) to that one of (2.9).

The perturbation

$$\theta(x,t,\vartheta) \;=\; \psi(x,t,\vartheta) - \psi(x,t) \tag{2.11}$$

is completely ruled by the field equations (2.6) and the boundary conditions (2.7) for given initial values

$$\theta(x,t=0,\vartheta) = \psi_0(x,\vartheta) - \psi_0(x) \quad . \tag{2.12}$$

In this paper I restrict myself to the main linear part of the perturbation which to a linear approximation may replace the total perturbation θ:

$$\theta(x,t,\vartheta) \overset{\text{lin.}}{=} \eta(x,t) \cdot \vartheta \quad , \tag{2.13}$$

$$\eta(x,t) = \left.\frac{\partial\theta(x,t,\vartheta)}{\partial\vartheta}\right|_{\vartheta=0} = \left\{ \eta_i(x,t) = \left.\frac{\partial\, \psi_i(x,t,\vartheta)}{\partial\vartheta}\right|_{\vartheta=0} \;,\; i=1,\dots,N \right\} . \tag{2.14}$$

Subsequently $\eta(x,t)$ is also called perturbation of the given reference process $\psi(x,t)$. However, one should keep in mind throughout this paper that strictly speaking we are dealing with the main linear part of the total perturbation.

The set of functions $\eta = \{ \eta_i, i=1,\dots,N \}$ is a solution of Jacobi's equations

$$\partial_t \frac{\partial\Omega}{\partial(\partial_t\eta_i)} + \nabla\cdot\frac{\partial\Omega}{\partial(\nabla\eta_i)} - \frac{\partial\Omega}{\partial\eta_i} = 0 \;, \quad i=1,\dots,N \;, \tag{2.15}$$

and of the boundary conditions

$$\vec{n} \cdot \frac{\partial\Omega}{\partial(\nabla\eta_i)} = 0 \;, \qquad i=1,\dots,N \;, \tag{2.16}$$

for given initial values

$$\eta(x,t=0) \;=\; \eta_0(x) \;=\; \left.\frac{\partial\psi_0(x,\vartheta)}{\partial\vartheta}\right|_{\vartheta=0} . \tag{2.17}$$

- Eqs. (2.15) and (2.16) are also called variational or perturbational equations of the Euler-Lagrange-equations (2.6) and of the boundary conditions (2.7) respectively. They can easily be derived from (2.6) and (2.7) by inserting the solution $\psi(x,t,\vartheta)$ and applying the differentiation procedure (2.14). - The kernel Ω is derived from the Lagrangian:

$$2\Omega = 2\Omega(\; \psi,\partial\psi;\; \eta,\partial\eta \;)$$

$$= \sum_{i,j=1}^{N} \left(\frac{\partial^2\ell(\psi,\partial\psi)}{\partial\psi_i\,\partial\psi_j}\,\eta_i\,\eta_j + 2\sum_{\alpha=0}^{3} \frac{\partial^2\ell(\psi,\partial\psi)}{\partial\psi_i\,\partial(\partial_\alpha\psi)}\,\eta_i\,\partial_\alpha\eta_j \right.$$

$$+ \sum_{\alpha,\beta=0}^{3} \frac{\partial^2 \ell(\psi,\partial\psi)}{\partial(\partial_\alpha \psi_i)\,\partial(\partial_\beta \psi_j)}\, \partial_\alpha \eta_j\, \partial_\beta \eta_j \,) \tag{2.18}$$

- Here and subsequently I occasionally use the signature

$$x^0 = t \quad , \quad \partial_0 = \partial_t \tag{2.19}$$

for convenience. The factor 2 in 2Ω is conventional. -

Ω is a homogeneous quadratic form with respect to the perturbation η and its first derivatives. It further depends on the given reference process $\psi(x,t)$ via the second derivatives of the Lagrangian. As a consequence Jacobi's equations (2.15) are linear differential equations for the perturbations $\eta(x,t)$ of a given reference process $\psi(x,t)$, which itself enters into these equations. An analogous statement holds for the boundary conditions (2.16).

If a one-parameter class $\psi(x,t,\vartheta)$ of solutions of the fundamental eqs. (2.6) and (2.7) is given, which of course can be regarded as a one-parameter class of perturbations in the above-mentioned sense, then by definition (2.14) one gets immediately a solution η of (2.15) and (2.16). This statement is particularly important for the definition of the entropy (see ch. 4).

Jacobi's equations are fundamental for a stability theory based on the perturbations $\eta(x,t)$. Stability or instability of a given process $\psi(x,t)$ is decided to a linear approximation by the asymptotic behaviour of all its perturbations $\eta(x,t)$ for $t \to \infty$, i.e. stability properties are decided by the behaviour of the set of solutions $\eta(x,t)$ of (2.15) and (2.16) for the given reference process $\psi(x,t)$. All of these solutions differ by their respective initial values (2.17). If all solutions $\eta(x,t)$ tend to zero for increasing time the reference process $\psi(x,t)$ is stable. This question will be decided by means of Ljapunov's direct method (see ch. 5).

Applying Euler's formula to the quadratic form $\Omega(\ldots, \eta, \partial\eta)$ and taking further account of Jacobi's equations (2.15) we get a non-homogeneous balance equation

$$\partial_t\, \omega + \nabla \cdot \underset{(\omega)}{\vec{J}} = \underset{(\omega)}{\sigma} \tag{2.20}$$

which is fulfilled by each perturbation $\eta(x,t)$ of a given reference process $\psi(x,t)$. The constitutive quantities involved into (2.20) are defined as follows:

density :
$$\omega = \sum_{i=1}^{N} \frac{\partial\Omega(\psi,\partial\psi;\, \eta,\partial\eta)}{\partial(\partial_t \eta_i)}\, \eta_i \quad , \tag{2.21}$$

flux density:
$$\underset{(\omega)}{\vec{J}} = \sum_{i=1}^{N} \frac{\partial\Omega(\psi,\partial\psi;\, \eta,\partial\eta)}{\partial(\nabla \eta_i)}\, \eta_i \quad , \tag{2.22}$$

production rate density:
$$\underset{(\omega)}{\sigma} = 2\Omega(\psi,\partial\psi;\, \eta,\partial\eta) \quad . \tag{2.23}$$

All of these quantities are quadratic forms with respect to η and $\partial\eta$. They further depend on the given reference process $\psi(x,t)$. With regard to (2.18) it gets apparent that they are finally derived from the Lagrangian.

The formal structure of Jacobi's equations (2.15) and of the related boundary conditions (2.16) is the same as that one of (2.6) and (2.7). Therefore I introduce a "variational principle of the second kind" which is equivalent to eqs. (2.15) and (2.16):

$$J_2 = \int_{t_1}^{t_2} \int_{V(t)} 2\Omega(\psi,\partial\psi;\ \eta,\partial\eta)\ d^3x\ dt \quad , \tag{2.24}$$

J_2 = extremum by free and independent variations $\delta\eta_i(x,t)$ of all fields $\eta_i(x,t)$ with fixed values at the beginning and at the end of the perturbation: (2.25)

$$\delta\eta_i(x,t_{1,2}) = 0 \ .$$

Variations at the boundary ∂V are admissible.

The variables ψ, $\partial\psi$ indicate the given reference process $\psi(x,t)$ entering into this variational principle as a sort of "control parameters". They are not affected by the variational procedure. One could call (2.24, 25) a Hamilton's principle for the perturbations $\eta(x,t)$ of a given reference process $\psi(x,t)$. - The boundary conditions (2.16) are due to free variations at the boundary and they are again physically associated with a completely isolated system.

The universal methodical nature of Lagrange-formalism is essentially due to the group of universal "invariance transformations" which apply in the same manner to each physical process. An invariance transformation $\underline{T}\varepsilon_1,\ldots,\varepsilon_n$ transforms a <u>real</u> process into another <u>real</u> process:

$$(x',t',\psi',\eta') = \underline{T}\varepsilon_1\ldots\varepsilon_n\ (x,t,\psi,\eta) \quad . \tag{2.26}$$

The original real process $\psi(x,t)$ together with all of its perturbations $\eta(x,t)$ takes place at a particular region in space and time denoted in (2.26) by x and t. This process is dragged along by the physical operation $\underline{T}\varepsilon_1,\ldots,\varepsilon_n$ in order to produce a new real process $\psi'(x',t')$ together with all of its perturbations $\eta'(x',t')$ in a new region of space and time denoted by x' and t'. In (2.26) the symbols ε indicate the group parameters of an n-parameter Lie-group.

According to the relevant applications the invariance transformations (2.26) are assumed in the form

$$\left.\begin{aligned} x'^{\alpha} &= f^{\alpha}(x^{\beta},\ \varepsilon_{\kappa}) \\ \psi'_i &= F_i(x^{\beta},\ \psi_j,\ \varepsilon_{\kappa}) \\ \eta'_i &= G_i(x^{\beta},\ \psi_j,\ \eta_j,\ \varepsilon_{\kappa}) \end{aligned}\right\} \quad \left\{\begin{aligned} \alpha,\ \beta &= 0,\ 1,\ 2,\ 3 \\ \kappa &= 1,\ 2,\ldots,n \\ i,\ j &= 1,\ 2,\ldots,N\ . \end{aligned}\right. \tag{2.27}$$

The 1-operation of the invariance group is associated with $\varepsilon_{\kappa} = 0$, $\kappa = 1,\ldots,n$.

In order to get into agreement with definition (2.14) of the perturbations η the functions G_i have to be restricted to the special form

$$G_i = \sum_{k=1}^{N} \frac{\partial F_i(x^\beta, \psi_j, \varepsilon_\kappa)}{\partial \psi_k} \eta_k \quad . \tag{2.28}$$

In this paper the discussion is focused on the following two examples:

1. Time translations:

$$\begin{aligned} t' &= t + \varepsilon && \equiv f^0 \\ x'^\alpha &= x^\alpha && \equiv f^\alpha \ , && \alpha = 1,2,3 \ , \\ \psi'_i &= \psi_i && \equiv F_i \ , && i = 1,\ldots,N \ , \\ \eta'_i &= \eta_i && \equiv G_i \ , && i = 1,\ldots,N \ . \end{aligned} \tag{2.29}$$

   This one-parameter invariance group is physically associated with homogeneity of the time-continuum with respect to all physics or - in other words - with the methodical principle, that each process has to be reproducible at arbitrary times.

2. Gauge transformations:

$$\begin{aligned} t' &= t && \equiv f^0 \\ x'^\alpha &= x^\alpha && \equiv f^\alpha \ , && \alpha = 1,2,3 \ , \\ \psi'_k &= \psi_k \, e^{i\varepsilon} && \equiv F_k \ , && k = 1,\ldots,N \ , \\ \eta'_k &= \eta_k \, e^{i\varepsilon} && \equiv G_k \ , && k = 1,\ldots,N \ . \end{aligned} \tag{2.30}$$

   These transformations are due to the case of complex-valued field variables $\psi(x,t)$. The common gauge transformation corresponds with the fact that only phase differences of different fields ψ_k are physically relevant.

By definition the invariance transformations give rise to invariance properties of the Lagrangian ℓ as well as of the kernel Ω: Hamilton's principle (2.4,5) is simultaneously fulfilled by those different real processes $\psi(x,t)$ and $\psi'(x',t')$ which are connected by means of the invariance transformations (2.26,27). The same statement holds analogously for the 2nd kind variational principle (2.24,25) giving rise to the invariance criterion

$$2\Omega(\psi', \partial'\psi'; \eta', \partial'\eta') \underset{\psi,\eta,\varepsilon}{\equiv} 2\Omega(\psi, \partial\psi, \eta, \partial\eta) + \partial_t F(x, \psi, \eta, \varepsilon) \quad . \tag{2.31}$$

Here the arguments on the left hand side must be substituted by the formulae of the invariance transformations (2.27). In general the total time derivative $\partial_t F$ is a necessary term to fulfill the criterion for a given function Ω. It does not affect the variational procedure. The criterion is understood as an identity with respect to the field variables ψ and η and with respect to the group parameters ε. It delivers constraints for the analytical form of the kernel Ω and thus of the Lagrangian ℓ.

On the ground of criterion (2.31) and by means of Noether's theorem [5,6] each group parameter ε_κ of the invariance group (2.27) is associated with an inhomogeneous balance equation:

$$\varepsilon_\kappa \implies \partial_t a_\kappa + \nabla\cdot\vec{J}_\kappa = \sigma_\kappa \quad . \tag{2.32}$$

This equation is fulfilled for each perturbation $\eta(x,t)$ of a given refernce process $\psi(x,t)$. For the quantities involved into (2.32) the following constitutive equations are found by Noether's theorem:

At first for abbreviation the density and the flux density vector are joined together into a 4-vector $\vec{\lambda}$:

$$\vec{\lambda} = (\lambda^\alpha , \quad \alpha = 0,1,2,3) \quad , \tag{2.33}$$

$$a_\kappa = a_\kappa(x,t;\ \psi,\partial\psi;\ \eta,\partial\eta) = \lambda^0 \quad , \tag{2.34}$$

$$\vec{J}_\kappa = \vec{J}_\kappa(x,t;\ \psi,\partial\psi;\ \eta,\partial\eta) = \vec{\lambda} = (\lambda^1, \lambda^2, \lambda^3) \quad . \tag{2.35}$$

Then due to the general form (2.27,28) of the invariance transformations one obtains for $\alpha = 0, 1, 2, 3$:

$$\lambda^\alpha = \sum_{\beta=0}^{3} \Theta^\alpha_\beta \frac{\partial f^\beta}{\partial \varepsilon_\kappa} - \sum_{i=1}^{N} \left(\frac{\partial(2\Omega)}{\partial(\partial_\alpha \eta_i)} \frac{\partial G_i}{\partial \varepsilon_\kappa} + \delta^{\alpha 0} \frac{\partial F}{\partial \varepsilon_\kappa} \right) \Bigg|_{\varepsilon=0} \tag{2.36}$$

The production rate density is given by

$$\sigma_\kappa = - \sum_{i=1}^{N} \Bigg\{ \left(\frac{\partial(2\Omega)}{\partial \psi_i} - \sum_{\alpha=0}^{3} \partial_\alpha \frac{\partial(2\Omega)}{\partial(\partial_\alpha \psi_i)} \right) \left(\sum_{\beta=0}^{3} \partial_\beta \psi_i \frac{\partial f^\beta}{\partial \varepsilon_\kappa} - \frac{\partial F_i}{\partial \varepsilon_\kappa} \right)$$
$$- \sum_{\alpha=0}^{3} \partial_\alpha \left[\frac{\partial(2\Omega)}{\partial(\partial_\alpha \psi_i)} \left(\sum_{\beta=0}^{3} \partial_\beta \psi_i \frac{\partial f^\beta}{\partial \varepsilon_\kappa} - \frac{\partial F_i}{\partial \varepsilon_\kappa} \right) \right] \Bigg\} \Bigg|_{\varepsilon=0} \tag{2.37}$$

In (2.36) the abbreviation

$$\Theta^\alpha_\beta = \sum_{i=1}^{N} \frac{\partial(2\Omega)}{\partial(\partial_\alpha \eta_i)} \partial_\beta \eta_i - 2\Omega\, \delta^\alpha_\beta \tag{2.38}$$

is used. - This quantity, being the Legendre-transform of 2Ω, can be regarded as the "energy-momentum-tensor" of the perturbation η.

For the special case of time translational invariance (2.32-38) read as follows:

$$\text{time translation } \varepsilon \implies \partial_t h + \nabla\cdot\vec{J}_{(h)} = \sigma_{(h)} \ , \tag{2.39}$$

$$\text{density:} \quad h = h(\psi,\partial\psi;\ \eta,\partial\eta) = \Theta^0_0 = \sum_{i=1}^{N} \frac{\partial(2\Omega)}{\partial(\partial_t \eta_i)} \partial_t \eta_i - 2\Omega \ , \tag{2.40}$$

$$\text{flux density:} \quad \vec{J}_{(h)} = \vec{J}_{(h)}(\psi,\partial\psi;\ \eta,\partial\eta) = \sum_{i=1}^{N} \frac{\partial(2\Omega)}{\partial(\nabla \eta_i)} \partial_t \eta_i \ , \tag{2.41}$$

production rate density:

$$\sigma_{(h)} = \sigma_{(h)}(\psi,\partial\psi;\ \eta,\partial\eta) = -\sum_{i=1}^{N}\left[\frac{\partial(2\Omega)}{\partial\psi_i}\,\partial_t\psi_i + \sum_{\alpha=0}^{3}\frac{\partial(2\Omega)}{\partial(\partial_\alpha\psi_i)}\,\partial_t\partial_\alpha\psi_i\right]$$

$$= -\frac{\partial}{\partial t_{ex}}\,2\Omega(\psi,\partial\psi;\eta,\partial\eta) \tag{2.42}$$

where the operation $\partial/\partial t_{ex}$ acts on the arguments $\psi(x,t)$ and $\partial\psi(x,t)$ only.

For the gauge transformation (2.30) the balance (2.32) and the constitutive equations (2.33-38) are specialized in the following way:

phase translation ε ==> $\partial_t\pi + \nabla\cdot\vec{J}_{(\pi)} = \sigma_{(\pi)}$, (2.43)

density: $\pi = \pi(\psi,\partial\psi;\ \eta,\partial\eta) = -\,i\sum_k\left(\frac{\partial(2\Omega)}{\partial(\partial_t\eta_k)}\,\eta_k - c.c.\right)$ (2.44)

flux density: $\vec{J}_{(\pi)} = \vec{J}_{(\pi)}(\psi,\partial\psi;\ \eta,\partial\eta) = -\,i\sum_k\left(\frac{\partial(2\Omega)}{\partial(\nabla\eta_k)}\,\eta_k - c.c.\right)$ (2.45)

production rate density:

$$\sigma_{(\pi)} = \sigma_{(\pi)}(\psi,\partial\psi;\ \eta,\partial\eta) = i\sum_k\left(\frac{\partial(2\Omega)}{\partial\psi_k}\,\psi_k + \sum_{\alpha=0}^{3}\frac{\partial(2\Omega)}{\partial(\partial_\alpha\psi_k)}\,\partial_\alpha\psi_k - c.c.\right) \tag{2.46}$$

Here the summation k runs over the complex-valued fields η_k or ψ_k. The complex conjugated fields η_k^* or ψ_k^* are taken into account by "c.c." which indicates the complex conjugate of the respective preceding terms.

Due to the quadratic form (2.18) of the kernel 2Ω all of the constitutive quantities (2.33-38) and its specializations (2.40-42) and (2.44-46) are also quadratic forms with respect to η and $\partial\eta$. This fact will get important in ch. 5 with respect to stability theory.

2.1. Thermodynamics of irreversible processes within the framework of Lagrange-formalism. - Heat transport as an example

Heat transport as described by means of Lagrange-formalism is based on a complex-valued fundamental field variable $\psi(x,t)$ which I call "field of thermal excitation" [1,2]. The absolute temperature is defined from the excitation field as a secondary quantity by means of

$$T(x,t) = \psi\psi^* \geq 0 \quad . \tag{2.47}$$

It is a positive definite quantity by definition. The absolute zeropoint $T = 0$ is associated with a state of no thermal excitation: $\psi = 0$. On the other hand a temperature $T > 0$ is found whereever and whenever $\psi(x,t) \neq 0$. In (2.47) ψ^* is the complex conjugate of ψ.

The traditional linear theory of heat transport in a rigid material body [7,8] is completely comprised into the Lagrangian

$$\ell = - c\,\psi\psi^* - \frac{c}{\omega}\left[\frac{1}{2i}(\psi^*\partial_t\psi - \psi\partial_t\psi^*) + \frac{1}{2}\,\frac{\ln\frac{\psi\psi^*}{T_0}}{\frac{\psi\psi^*}{T_0}}\,\partial_t(\psi\psi^*)\right]$$

$$+ \sum_{\alpha,\beta=1}^{3} \frac{\lambda^{\alpha\beta}}{\omega}\left[\frac{1}{\psi\psi^*}\frac{1}{2i}\{(\psi^*\partial_\alpha\psi)(\psi^*\partial_\beta\psi) - (\psi\partial_\alpha\psi^*)(\psi\partial_\beta\psi^*)\}\right.$$

$$\left. + \frac{T_0}{2(\psi\psi^*)^2}\,\partial_\alpha(\psi\psi^*)\partial_\beta(\psi\psi^*)\right] . \tag{2.48}$$

The meaning of the symbols is as follows:

c: specific heat.- $\underline{\lambda} = (\lambda^{\alpha\beta})$: symmetric tensor of heat conductivity coefficients. For simplicity c and λ are assumed to be constant. - T_0: a reference temperature. - ω: a frequency which is necessary for dimensional reasons. It is not specified at this stage of the theory.

From the Lagrangian (2.48) all traditional balance and constitutive equations concerning heat transport can be derived. For subsequent use concerning entropy I especially want to focus the reader's attention on the invariance of the Lagrangian with respect to the gauge transformation (see (2.30))

$$\psi,\ \psi^* \quad ==> \quad \psi' = \psi e^{i\varepsilon}, \qquad \psi'^* = \psi^* e^{-i\varepsilon} . \tag{2.49}$$

Instead of ψ and ψ^* the Lagrangian (2.48) can more conveniently be based on the temperature T(x,t) and on the phase function $\phi(x,t)$ of the excitation field:

$$\psi(x,t) = \sqrt{T(x,t)}\; e^{i\phi(x,t)} . \tag{2.50}$$

Then the Lagrangian (2.48) takes the form

$$\ell = -\frac{1}{\omega}\left[\, cT\,\partial_t(\phi-\phi_0(t,T)) + (-\underline{\lambda}\cdot\nabla T)\cdot\nabla(\phi-\phi_0(t,T)) + \frac{1}{\omega}\,\partial_t\, G(T)\,\right] \tag{2.51}$$

with the abbreviations

$$\phi_0(t,T) = -\,\omega t + \frac{T_0}{2T} , \tag{2.52}$$

$$G(T) = + \frac{T_0}{2T}\int_{T_0/T}^{1} \frac{p(\xi T)}{\xi^2}\, d\xi . \tag{2.53}$$

The function p(T) entering into (2.53) is the hydrostatic pressure in the system which in the present case of a rigid material background is but a rigid body reaction. Its analytical form is not essential in the context of this paper.

A simple form of the Euler-Lagrange equations (2.6) is found from (2.51) by free variation of the field variables $\phi(x,t)$ and T(x,t). They read as

$$\delta\phi : \qquad c\,\partial_t T - \underline{\lambda} \cdot\cdot\, \nabla\nabla T = 0 , \tag{2.54}$$

$$\delta T : \qquad c\,\partial_t(\phi - \phi_0) + \underline{\lambda} \cdot\cdot\, \nabla\nabla(\phi - \phi_0) = 0 . \tag{2.55}$$

In (2.55) equation (2.54) has already been taken into account. These equations for the bulk are supplemented by the boundary conditions (2.7) of a thermally isolated system:

$$\delta\phi : \qquad \vec{n}\cdot(- \underline{\lambda}\cdot\nabla T) = 0 , \tag{2.56}$$

$$\delta T : \qquad \vec{n}\cdot[- \underline{\lambda}\cdot\nabla(\phi - \phi_0) + (- \underline{\lambda}\cdot\nabla T) \frac{T_0}{2T}] = 0 . \tag{2.57}$$

Obviously (2.54) is Fourier's law of heat conduction. The second equation (2.55) is solved on the ground of the solutions $T(x,t)$ of (2.54):

$$\phi(x,t) = \phi_0(t, T(x,t)) = - \omega t + \frac{T_0}{2T(x,t)} \tag{2.58}$$

This particular solution is associated with the traditional Principle of Local Equilibrium [7,8]. Due to this principle the traditional linear theory of irreversible processes is concerned only with those processes which locally and for all times run through equilibrium states. The Lagrangian (2.48,51) fits into this assumption. All consequences derived from this Lagrangian are essentially based on the use of the particular solution (2.58).

Due to the time translational invariance of the Lagrangian (2.48,51) one finds by Noether's theorem the heat flux

$$\vec{J} = - \underline{\lambda} \cdot \nabla T \quad . \tag{2.59}$$

Therefore due to the boundary condition (2.56) there is no heat flux across the boundary as it should be for an isolated system. The second condition (2.57) is fulfilled simultaneously with (2.56) on the ground of the solution (2.58).

2.2 Schrödinger's matter field as an example

As another example for the subsequent investigations I refer to Schrödinger's complex-valued matter field $\psi(x,t)$ which in quantum mechanics is associated with a sufficiently diluted flux of particles. If the particles suffer no external forces the quantumdynamics of the particle flux is comprised into the Lagrangian [9]

$$\ell = \frac{1}{2}\frac{\hbar}{i} (\psi\, \partial_t\psi^* - \psi^*\, \partial_t\psi) - \frac{\hbar^2}{2m} \nabla\psi\cdot\nabla\psi^* \quad . \tag{2.60}$$

$\hbar$ and m are Planck's constant and the particle mass respectively. From (2.60) one finds Schrödinger's equations as Euler-Lagrange-equations (2.6) by independent variations of ψ and ψ^*:

$$\delta\psi^*: \quad \frac{\hbar}{i}\, \partial_t\psi - \frac{\hbar^2}{2m}\, \Delta\psi = 0 , \tag{2.61}$$

$$\delta\psi : \quad - \frac{\hbar}{i}\, \partial_t\psi^* - \frac{\hbar^2}{2m}\, \Delta\psi^* = 0 . \tag{2.62}$$

The boundary conditions (2.7) for an isolated system are found in the form

$$\delta\psi^*: \qquad \vec{n} \cdot \nabla\psi = 0 \quad , \tag{2.63}$$

$$\delta\psi\ : \qquad \vec{n} \cdot \nabla\psi^* = 0 \quad . \tag{2.64}$$

These give rise to a vanishing matter flux across the boundary.

Of course Schrödinger's problem has nothing to do with thermodynamics. However, the subsequent formalism concerning stability works as well. It is also interesting to ask for the analogue of the entropy in the case of the matter field.

3. Stability theory - Ljapunov's direct method in a restricted sense

Let me come back to the perturbations $\Theta(x,t,\vartheta)$ of a given reference process $\psi(x,t)$. (See (2.11-14,17).) The process $\psi(x,t)$ is called asymptotically stable if for all sufficiently small initial perturbations $\Theta(x,t=0,\vartheta)$ one finds

$$\Theta(x,t,\vartheta) \to 0 \quad \text{for} \quad t \to \infty \quad . \tag{3.1}$$

Weak stability is found if the perturbation $\Theta(x,t,\vartheta)$ remains "sufficiently small" for $t > 0$, i.e. if it remains in a certain ε-neighbourhood of the reference process $\psi(x,t)$. If the perturbation is leaving each ε-neighbourhood the process $\psi(x,t)$ is called unstable [10,11].

To a linear approximation the stability of a given reference process $\psi(x,t)$ can be decided by the behaviour of the set of functions η (see (2.13,14) which solve Jacobi's equations (2.15) and the boundary conditions (2.16). As a necessary condition for asymptotic stability of the reference process $\psi(x,t)$ each solution $\eta(x,t)$ of (2.15,16) must be damped:

$$\eta(x,t) \to 0 \quad \text{for} \quad t \to \infty \quad . \tag{3.2}$$

Of course different solutions $\eta(x,t)$ of (2.15,16) belong to different initial perturbations (2.17).

Let me introduce a perturbation space which is spanned by the set of functions

$$P = \{\ \eta_i(x,t) \quad , \quad \partial_\alpha \eta_i(x,t) \quad ; \quad i = 1,\dots,N \quad ; \quad \alpha = 0,\dots,3\ \} \quad . \tag{3.3}$$

I further introduce a norm in perturbation space:

$$\left\| \eta(t) \right\| = \Big[\int_{V(t)} \sum_{i=1}^{N} (\eta_i^*(x,t)\eta_i(x,t) + \partial_t\eta_i^*(x,t)\ \partial_t\eta_i(x,t) + \nabla\eta_i^*(x,t)\cdot\nabla\eta_i(x,t))\ d^3x\ \Big]^{1/2} \tag{3.4}$$

Then the behaviour (3.2) will be understood in the sense of

$$\left\| \eta(t) \right\| \to 0 \quad \text{for} \quad t \to 0 \quad . \tag{3.5}$$

With regard to the three terms in (3.4) the relation (3.5) means that the perturbation η as well as its temporal and spatial fluctuations tend to zero.

For subsequent applications I also introduce a much weaker norm

$$\left\| \eta(t) \right\| = \Big[\int_{V(t)} \sum_{i=1}^{N} \eta_i^*(x,t)\eta_i(x,t)\, d^3x \Big]^{1/2} \quad , \tag{3.6}$$

by means of which the requirement (3.5) does not take care of the fluctuations.

The essential point of Ljapunov's direct method is to decide the behaviour (3.5) without solving of Jacobi's equations (2.15). This will be done by introducing in the perturbation space a real-valued Ljapunov-functional $\mathcal{L}[\eta]$ with the following properties in a full neighbourhood $\|\eta\| < R$ of $\eta = 0$ [10,11]:

1. $\mathcal{L}[\eta]$ positive definite, i.e. $\mathcal{L}[\eta] \geq 0$ for $\|\eta\| \geq 0$, (3.7)

2. $\mathcal{L}[\eta] \geq m(\|\eta\|)$ (3.8)

   with a monotonously increasing function m.

3. $\frac{d}{dt}\mathcal{L}[\eta] \leq 0$ (3.9)

   along the trajectories of all perturbations $\eta(x,t)$, i.e. of all solutions of Jacobi's equations (2.15,16).

If such a Ljapunov-functional can be found the reference process $\psi(x,t)$ will be stable (at least to a linear approximation). The relations "<" and "=" in (3.9) indicate asymptotic stability and weak stability respectively. As an essential point of Ljapunov's direct method the relation (3.9) has to be verified by means of Jacobi's equations (2.15) but without solving of these equations explicitly. The Ljapunov-functional $\mathcal{L}[\eta]$ is understood to depend on all variables (3.3) if the norm (3.4) is used. In the case of the weaker norm (3.6) it is assumed to depend on the functions $\eta_i(x,t)$ only.

Usually Ljapunov's direct method is applied to the full perturbation $\Theta(x,t)$. In my investigations, however, the method is applied to the main linear part $\eta(x,t)$ of the perturbation. This fact is expressed by the term "Ljapunov's direct method in a restricted sense".

In most cases a Ljapunov-functional is introduced into stability theory by assumption. In contrast to that I aim to define it completely from the dynamical structure of the system. In doing so it may happen that the functional $\mathcal{L}$ also depends on the reference process $\psi(x,t)$:

$$\mathcal{L}_t[\Psi,\eta] \quad . \tag{3.10}$$

This gives rise to an explicit time dependence of $\mathcal{L}$ due to the time variable t in the reference process $\psi(x,t)$. In (3.10) this fact is indicated by the index t. In order to avoid difficulties, which for instance may arise in those cases where the indicatrix in perturbation space

$$\mathcal{L}_t[\Psi,\eta] = 1 \tag{3.11}$$

is blowing up for increasing time, I have introduced the supplementary condition

(3.8). Of course this condition is not necessary for the case of an invariable indicatrix $[\eta] = 1$.

4. Gauge invariance and the concept of entropy - The Second Law of Thermodynamics

4.1 The general theory

All fundamental field variables $\psi_i(x,t)$ of the Lagrangian (2.1) are assumed to be complex-valued. Then the Lagrangian has to be invariant with respect to the group of common gauge transformations (2.30). This results in the fact that for a given real process $\psi(x,t) = \{ \psi_i(x,t) \;,\; i = 1,\ldots,N \}$ the set of functions

$$\psi'(x,t,\varepsilon) = \{ \psi_i'(x,t,\varepsilon) = \psi_i(x,t)e^{i\varepsilon} \quad , \quad i = 1,\ldots,N \} \tag{4.1}$$

defined by (2.30) describes again a real process for each choice of the group parameter ε.

I refer to the balance equation (2.20) which is based on the constitutive equations (2.21-23) and which is fulfilled by each perturbation $\eta(x,t)$ of a given reference process $\psi(x,t)$. (See (2.11-14).)

The one-parameter class of real processes $\psi'(x,t,\varepsilon)$ defined by (4.1) can be regarded as a special class of perturbations of the reference process $\psi(x,t)$. It is due to a common phase shifting ε of all field variables ψ_i. The group parameter ε of (2.30) is now interpreted as a perturbation parameter ϑ in the sense of (2.11). Then the associated main linear part $\eta(x,t)$ defined by (2.14) reads as

$$\eta(x,t) = \{ \eta_k(x,t) = \left.\frac{\partial\psi_k(x,t,\varepsilon)}{\partial\varepsilon}\right|_{\varepsilon=0} = i\psi_k(x,t) \quad , \quad k = 1,\ldots,N \} \,. \tag{4.2}$$

This set of functions solves Jacobi's equations (2.15) and the boundary conditions (2.16).

Joining together the preceding statements I get a balance equation (2.20) which for a given reference process $\psi(x,t)$ is fulfilled by the set $\eta(x,t)$ defined by (4.2). It is a remarkable fact that the set $\eta(x,t)$ is completely given by the reference process $\psi(x,t)$ except for the factors i. Thus I get a balance equation which exclusively operates in the state space. The perturbation space is eliminated. This balance is fulfilled by each real process $\psi(x,t)$. It is called the <u>entropy balance</u> of the system:

$$(2.20) \Longrightarrow \qquad \partial_t\, s + \nabla\cdot\vec{J}_{(s)} = \sigma_{(s)} \tag{4.3}$$

It is based on the constitutive equations for the <u>entropy density</u>

$$(2.21) : \quad \omega \Longrightarrow \quad s(\psi,\partial\psi) = \sum_{k=1}^{N} \left.\frac{\partial\Omega(\psi,\partial\psi;\,\eta,\partial\eta)}{\partial(\partial_t\eta_k)}\,\eta_k\right|_{\substack{\eta \,=\, i\psi \\ \eta^* \,=\, -i\psi^*}} , \tag{4.4}$$

for the <u>entropy flux density</u>

$$(2.22) : \vec{J}_{(\omega)} \Longrightarrow \vec{J}_{(s)}(\psi,\partial\psi) = \sum_{k=1}^{N} \frac{\partial\Omega(\psi,\partial\psi;\, \eta,\partial\eta)}{\partial(\nabla\eta_k)} \eta_k \Bigg|_{\substack{\eta \,=\, i\psi \\ \eta^* \,=\, -i\psi^*}} , \qquad (4.5)$$

and for the density of the entropy production rate

$$(2.23) : \sigma_{(\omega)} \Longrightarrow \sigma_{(s)}(\psi,\partial\psi) = 2\Omega(\psi,\partial\psi;\, \eta,\partial\eta) \Bigg|_{\substack{\eta \,=\, i\psi \\ \eta^* \,=\, -i\psi^*}} . \qquad (4.6)$$

In the case of thermodynamical processes eq. (4.3) is the first part of the Second Law of Thermodynamics. The positive definiteness of the entropy production rate, however, cannot be established at this stage of the theory. This property will come into play in a natural way by means of stability theory (ch. 5).

It is very important to keep in mind that the entropy concept is exclusively based on complex-valued field variables as fundamental variables of the Lagrangian. (See e.g. (2.48) and (2.60).) In principle one could switch over to a different set of variables in state space as e.g. is done in (2.51). However, the kernel 2Ω defined by (2.18) is not invariant with respect to a transformation of the state space variables $\psi_i(x,t)$. This is due to the second ordinary derivatives of the Lagrangian. The entropy concept, on the other hand, is understood as an invariant concept in state space. To take properly account of this requirement, one should use a covariant differential calculus in state space instead of the ordinary one which I have used in (2.18) and in the subsequent balance equation (2.20). The situation is analogous to that one of ordinary x-space where the ordinary differential calculus has to be replaced by the covariant Ricci-calculus, if a cartesian coordinate system is transformed into a curvilinear one. Following this analogy and looking as an example at the heat transport processes (ch. 2.1) the complex-valued thermal excitation field $\psi(x,t)$ and its complex conjugate $\psi^*(x,t)$ are, so to speak, the "cartesian coordinates" in state space, whereas the temperature $T(x,t)$ and the phase function $\phi(x,t)$ are "curvilinear coordinates".

4.2 Example: heat transport

Applying eqs. (4.3-6) to the Lagrangian (2.48) and using the particular solution (2.58) one gets the entropy balance for heat transport. Some lengthy calculations result in the following constitutive equations:

entropy density:

$$s = c \,\ln \frac{T}{T_0} , \qquad (4.7)$$

entropy flux density:

$$\vec{J}_{(s)} = \frac{-\,\underline{\lambda}\cdot\nabla T}{T} = \frac{\vec{J}}{T} , \qquad (4.8)$$

density of the entropy production rate:

$$\sigma_{(s)} = \frac{\nabla T \cdot \underline{\lambda} \cdot \nabla T}{T^2} = \vec{J} \cdot \nabla (\frac{1}{T}) \tag{4.9}$$

These expressions are well-known from traditional linear theory of heat transport [7,8]. $\vec{J}$ is the heat flux defined in (2.59). Due to the principle of local equilibrium the traditional thermodynamics of irreversible processes operates with the state functions of thermostatics. For the entropy density this procedure results in the expression (4.7). Concerning Lagrange-formalism this formula is closely related with the solution (2.58).

4.3 Example: Schrödinger's matter field

Looking at the linear Schrödinger equations (2.61, 62) it is quite clear that the associated Jacobi-equations (2.15) have the same form: ψ and ψ^* are substituted by η and η^*. Thus the kernel Ω has the same form as the Lagrangian (2.60):

$$\Omega(\eta,\eta^*,\partial\eta,\partial\eta^*) = \frac{1}{2}\frac{\hbar}{i}(\eta\, \partial_t \eta^* - \eta^*\, \partial_t \eta) - \frac{\hbar^2}{2m} \nabla\eta \cdot \nabla\eta^* \tag{4.10}$$

Applying (4.3-6) one gets a balance equation in state space

$$\partial_t\, s + \nabla \cdot \vec{J}_{(s)} = \sigma_{(s)} \tag{4.11}$$

which is fulfilled by the solutions of the field equations (2.61,62). The quantities entering into (4.11) are given by the constitutive equations

$$(4.4): \quad s = \left. \frac{\partial\Omega}{\partial(\partial_t \eta)}\eta + \frac{\partial\Omega}{\partial(\partial_t \eta^*)}\eta^* \right|_{\substack{\eta = i\psi \\ \eta^* = -i\psi^*}} = 0, \tag{4.12}$$

$$(4.5): \quad \vec{J}_{(s)} = \left. \frac{\partial\Omega}{\partial(\nabla\eta)}\eta + \frac{\partial\Omega}{\partial(\nabla\eta^*)}\eta^* \right|_{\substack{\eta = i\psi \\ \eta^* = -i\psi^*}} = -\frac{\hbar^2}{2m}\nabla(\psi\psi^*), \tag{4.13}$$

$$(4.6): \quad \sigma_{(s)} = \left. 2\Omega(\eta,\eta^*,\partial\eta,\partial\eta^*) \right|_{\substack{\eta = i\psi \\ \eta^* = -i\psi^*}}$$

$$= \frac{\hbar}{i}(\psi\partial_t\psi^* - \psi^*\partial_t\psi) - \frac{\hbar^2}{m}\nabla\psi \cdot \nabla\psi^* . \tag{4.14}$$

Of course we should not speak of an entropy balance in the context of Schrödinger's matter field. Nevertheless it is quite satisfactory that the entropy formalism of ch. 4.1 results in an identically vanishing state quantity s. The validity of the balance equation (4.11) can directly be confirmed by inserting Schrödinger's equations (2.61, 62) into (4.11,13,14).

5. Ljapunov-functionals designed by means of Lagrange-formalism

Let me consider the set of balance equations (2.32) and let me assume for simplicity a constant volume of the system: $V(t) = V_0$. Then (2.32) gives rise to a set of global balance equations associated with the group parameters ε_κ of the invariance group (2.27):

$$\varepsilon_\kappa \implies \frac{d}{dt} \underset{(\kappa)}{A}{}_t = \underset{(\kappa)}{\Sigma}{}_t - \underset{(\kappa)}{\Phi}{}_t \quad , \quad \kappa = 1,\ldots,n \quad . \tag{5.1}$$

The quantities involved in (5.1) are defined on the ground of eqs. (2.33-38):

$$\underset{(\kappa)}{A}{}_t [\Psi,\eta] = \int_{V_0} a_\kappa(\ \psi(x,t),\partial\psi(x,t);\ \eta(x,t),\partial\eta(x,t)\ )\ d^3x \quad , \tag{5.2}$$

$$\underset{(\kappa)}{\Sigma}{}_t [\Psi,\eta] = \int_{V_0} \sigma_\kappa(\ \psi(x,t),\partial\psi(x,t);\ \eta(x,t),\partial\eta(x,t)\ )\ d^3x \quad , \tag{5.3}$$

$$\underset{(\kappa)}{\Phi}{}_t [\Psi,\eta] = \int_{\partial V_0} \vec{n}\cdot\vec{J}(\ \psi(x,t),\partial\psi(x,t);\ \eta(x,t),\partial\eta(x,t)\ )\ d^2x \quad . \tag{5.4}$$

They are quadratic functionals of the perturbation field variables $\eta(x,t)$. In general they furthermore depend on the given reference process $\psi(x,t)$.

By definition the balance equations (5.1) are fulfilled by each perturbation $\eta(x,t)$ of a given reference process $\psi(x,t)$. Thus, these equations fit quite well into the scheme of Ljapunov's criterium of stability. Let me assume that in the set of functionals $\underset{(\kappa)}{A}{}_t$, $\kappa = 1,\ldots,n$ we can find at least one which can be taken as a Ljapunov-functional in a full neighbourhood of $\eta = 0$, i.e.

$$\underset{(\kappa)}{A}{}_t [\Psi,\eta] \geq 0 \quad \text{for} \quad \eta \geq 0 \quad , \tag{5.5}$$

$$\underset{(\kappa)}{A}{}_t [\Psi,\eta] \geq m(\ \eta\ ) \quad , \tag{5.6}$$

and

$$\frac{d}{dt} \underset{(\kappa)}{A}{}_t [\Psi,\eta] \leq 0 \tag{5.7}$$

along the trajectory of each perturbation $\eta(x,t)$ (see eqs. 3.7-9). Then a Ljapunov-functional is really completely designed by means of internal structures of Lagrange-formalism only, i.e. it is exclusively derived from the Lagrangian. Especially the requirement (5.7) can be decided in perturbation space by means of the associated balance equation (5.1) without solving of Jacobi's equations (2.15) explicitly. One has only to investigate the definiteness properties of the right hand side of (5.1).

Among the balance equations (5.1) that one associated with the time translational group (2.29) is of particular interest:

(2.39) ==>
$$\frac{d}{dt} \mathcal{H}_t = \underset{(h)}{\Sigma}{}_t - \underset{(h)}{\Phi}{}_t \quad , \tag{5.8}$$

(2.40),(5.2):
$$\mathcal{H}_t[\Psi,\eta] = \int_{V_0} h(\psi,\partial\psi;\ \eta,\partial\eta)\ d^3x \quad , \tag{5.9}$$

(2.42),(5.3): $$\underset{(h)}{\Sigma}{}_t[\Psi,\eta] = \int_{V_0} \sigma_{(h)}(\ldots\ldots\ldots)\, d^3x \quad , \tag{5.10}$$

(2.41),(5.4): $$\underset{(h)}{\Phi}{}_t[\Psi,\eta] = \int_{\partial V_0} \vec{n}\cdot\vec{J}_{(h)}(\ldots\ldots)\, d^2x \quad . \tag{5.11}$$

The functional $\mathcal{H}_t$ can be regarded as the Hamiltonian of the perturbations. In non-degenerated cases it is distinguished in so far as it contains all informations concerning the perturbations $\eta(x,t)$ of a given reference process $\psi(x,t)$. This is due to the fact that the perturbations $\eta(x,t)$ are completely defined by Jacobi's equations (2.15) and thus by the kernel 2Ω. The function h is the Legendre transform of 2Ω with respect to all perturbation velocities $\partial_t\eta_i$; it involves the same information as 2Ω. Thus $\mathcal{H}_t$ is an outstanding candidate for a Ljapunov-functional. - This statement holds except for those special cases, where 2Ω does not depend quadratically on <u>all</u> $\partial_t\eta_i$. In these cases the function h contains only reduced information as compared with 2Ω, i.e. Hamilton's formalism in perturbation space, which is based on the function h, gets degenerated in some way. Unfortunately the theory of heat transport as based on the Lagrangian (2.48,51) is such a degenerated case.

Because of its outstanding role with respect to the entropy concept let me further look at the gauge group (2.30), which by (5.1-5.4) and (2.43-2.46) results in another interesting candidate $\mathcal{N}_t$ for a Ljapunov-functional:

(2.43) ==> $$\frac{d}{dt}\mathcal{N}_t = \underset{(\pi)}{\Sigma}{}_t - \underset{(\pi)}{\Phi}{}_t \quad , \tag{5.12}$$

(2.44),(5.2): $$\mathcal{N}_t[\Psi,\eta] = \int_{V_0} \pi(\psi,\partial\psi;\ \eta,\partial\eta)\, d^3x \ , \tag{5.13}$$

(2.46),(5.3): $$\underset{(\pi)}{\Sigma}{}_t[\Psi,\eta] = \int_{V_0} \sigma_{(\pi)}(\ldots\ldots\ldots)\, d^3x \ , \tag{5.14}$$

(2.45),(5.4): $$\underset{(\pi)}{\Phi}{}_t[\Psi,\eta] = \int_{\partial V_0} \vec{n}\cdot\vec{J}_{(\pi)}(\ldots\ldots)\, d^2x \ . \tag{5.15}$$

There is one question left: How to investigate the definiteness properties of the functionals (5.2-4) or the special ones of (5.9-11, 5.13-15)? At first I would like to mention that the various functionals Φ_t, which are defined by surface integrals, allow for a stability analysis of open systems. If necessary these functionals can be transformed into volume integrals by means of the Gaussian theorem. Along this line they can be joined together with the functionals Σ_t in order to carry out the subsequent analysis.

Let

$$\mathcal{F}_t[\psi,\eta] = \int_{V_0} f(\ \psi(x,t),\partial\psi(x,t);\ \eta(x,t),\partial\eta(x,t)\ )\ d^3x \tag{5.16}$$

be one of the above-mentioned functionals A_t, $\mathcal{H}_t$, $\mathcal{N}_t$ or Σ_t. The kernel f is a homogeneous quadratic form with respect to η and $\partial\eta$. Thus the functional is positive definite, i.e.

$$\mathcal{F}_t[\Psi,\eta] \geq 0 \quad \text{for} \quad \|\eta\| \geq 0 \quad , \tag{5.17}$$

if and only if it has a proper minimum at $\eta = 0$ with respect to arbitrary variations of the functions $\eta_i(x,t)$. In order to realize this minimum two different cases must be distinguished:

1. The kernel f of (5.16) does not depend on any of the perturbation velocities $\partial_t\eta_i$, $i = 1,\dots,N$.
2. f depends at least on one of the velocities $\partial_t\eta_i$.

<u>ad 1:</u> The kernel f is of the form

$$f(\ \psi(x,t),\partial\psi(x,t);\ \eta_i(x,t),\nabla\eta_i(x,t),\ i = 1,\dots,N\ ) \quad . \tag{5.18}$$

Thus the integral (5.16) defines a variational problem of first order in three-dimensional space with respect to free variations of all functions $\eta_i(x,t)$. In order that the integral gets a minimum at $\eta(x,t) = 0$, which coincides with the positive definiteness (5.17), the condition of LEGENDRE and HADAMARD [12] has to be fulfilled as a necessary condition:

$$\sum_{\alpha,\beta=1}^{3} \sum_{i,j=1}^{N} \left.\frac{\partial^2 f(\psi,\partial\psi;\ \eta,\nabla\eta)}{\partial(\partial_\alpha\eta_i)\ \partial(\partial_\beta\eta_j)}\right|_{\eta=0} u_\alpha u_\beta\ v_i v_j \ \geq\ 0 \tag{5.19}$$

for arbitrary quantities u_α , $\alpha = 1,2,3$, and v_i , $i = 1,\dots,N$.

Ohne should keep in mind that only the derivatives of f with respect to spatial gradients $\nabla\eta$ are involved.

<u>ad 2:</u> The kernel f is of the form

$$f(\ \psi(x,t),\partial\psi(x,t);\ \eta_i(x,t),\nabla\eta_i(x,t),\underline{\underline{\partial_t\eta_i(x,t)}},\ i = 1,\dots,N) \ . \tag{5.20}$$

Here the variables $\partial_t\eta_i$ prevent the integral (5.16) from defining a variational problem immediately. However, the condition (5.17) is fulfilled if and only if

$$\int_{t_1}^{t_2} \mathcal{F}_t[\psi,\eta]\, dt \geq 0 \quad \text{for} \quad \|\eta\| \geq 0 \quad \text{and for } \underline{\text{fixed}} \text{ but nevertheless } \underline{\text{arbitrary}} \text{ time intervalls } [t_1,t_2]. \tag{5.21}$$

Together with the definition (5.16), this time integral gives rise to a four-dimensional variational problem of first order in space and time:

$$\int_{t_1}^{t_2} \int_{V_0} f(\ \psi,\partial\psi;\ \eta,\nabla\eta,\partial_t\eta\ )\ d^3x\ dt = \text{extremum by free variations of all functions } \eta_i(x,t). \tag{5.22}$$

In order that this integral gets a minimum at $\eta(x,t) = 0$, which via (5.21) again coincides with the positive definiteness (5.17), the Legendre-Hadamard condition has to be fulfilled:

$$\sum_{\alpha,\beta=0}^{3} \sum_{i,j=1}^{N} \frac{\partial^2 f(\psi,\partial\psi;\eta,\nabla\eta,\partial_t\eta)}{\partial(\partial_\alpha \eta_i)\partial(\partial_\beta \eta_j)}\bigg|_{\eta=0} u_\alpha u_\beta v_i v_j \geq 0 \qquad (5.23)$$

for arbitrary quantities u_α, $\alpha = 0,1,2,3,$ and v_i, $i = 1,\ldots,N$.
In contrast to (5.19) the derivatives of f with respect to spatial gradients $\nabla\eta$ as well as to velocities $\partial_t\eta$ are now involved.

Finally it should be mentioned, that the procedure of this chapter applies to the balance equation (2.20) as well.

5.1 Example: Schrödinger's matter field

The investigations are based on the function Ω of eq. (4.10) which does not depend on the reference process $\psi(x,t)$. This results in the fact, that each functional $A_{(\kappa)}{}^t$ (see (5.2)) does not depend explicitly on time, i.e. the associated indicatrices (3.11) in perturbation space are invariant in time.

The functional $\mathcal{H}_t$ associated with the time translational invariance is based on the function

$$h = \frac{\hbar^2}{m} \nabla\eta^* \cdot \nabla\eta \geq 0 \quad . \qquad (5.24)$$

(See (2.40, (4.10), (5.9).) However, this function is but a truncated quadratic form with respect to all variables η, $\nabla\eta$ and $\partial_t\eta$. Thus for the functional $\mathcal{H}$ we find the property

$$\mathcal{H}_t[\eta] = \int_{V_0} h \, d^3x \begin{cases} \geq 0 \\ = 0 \end{cases} \quad \text{for } \|\eta\| \begin{cases} > 0 \\ = 0 \end{cases} \qquad (5.25)$$

which does not fit into the scheme (5.5) and (5.6), i.e. it defines no Ljapunov-functional. (Of course is referred to the norm (3.4)). This lack is due to the special form of the kernel Ω (eq. 4.10) and thus to the Lagrangian (2.60).

For the case of Schrödinger's matter field we are more successful with the functional (5.13) associated with gauge invariance. The function π is taken from (2.44) and (4.10):

$$\pi = 2\hbar\, \eta\eta^* \qquad (5.26)$$

Thus, because of $\quad \hbar > 0 \qquad (5.27)$

the functional $\mathcal{N}[\eta]$ fits into the requirements (5.5) and (5.6):

$$\mathcal{N}_t[\eta] = \int_{V_0} \pi \, d^3x \geq 0 \quad \text{for} \quad \|\eta\| \geq 0 \quad , \qquad (5.28)$$

$$\mathcal{N}_t[\eta] \geq m(\|\eta\|) \qquad (5.29)$$

where the monotonous function m can be chosen as

$$m(\|\eta\|) = 2\hbar \|\eta\| \quad . \qquad (5.30)$$

Here the norm $\| \eta \|$ in perturbation space is understood in the sense of (3.6).

Due to (4.10) one gets for the flux density (2.45) and for the production rate (2.46)

$$\underset{(\pi)}{\vec{J}} = i \frac{\hbar^2}{m} (\eta \nabla\eta^* - \eta^* \nabla\eta) \quad , \tag{5.31}$$

$$\underset{(\pi)}{\sigma} = 0 \quad . \tag{5.32}$$

This results in the following equations for the functionals $\underset{(\pi)}{\Sigma}$ and $\underset{(\pi)}{\Phi}$:

$$\underset{(\pi)}{\Sigma}{}_t[\eta] = 0 \tag{5.33}$$

$$\underset{(\pi)}{\Phi}{}_t[\eta] = \int_{\partial V_0} \vec{n} \cdot \underset{(\pi)}{\vec{J}} \, d^2x = 0 \quad . \tag{5.34}$$

In (5.34) I have used $\underset{(\pi)}{\vec{J}} = 0$ (5.35)

which follows from the boundary conditions for the perturbations η :

$$\vec{n} \cdot \nabla\eta = 0 \quad , \tag{5.36}$$

$$\vec{n} \cdot \nabla\eta^* = 0 \quad . \tag{5.37}$$

These boundary conditions are the variational equations of (2.63,64). They belong to a materially isolated system.

Using (5.1,33,34) I finally get

$$\frac{d}{dt} \mathfrak{N}_t[\eta] = 0 \tag{5.38}$$

along the trajectories of each perturbations $\eta(x,t)$. This equation fits into the third part (5.7) of Ljapunov's criterion. Thus the functional $\mathfrak{N}_t[\eta]$ has been realized as a Ljapunov-functional which guarantees weak stability of each reference process $\psi(x,t)$. "Weak stability" means in this case that the perturbation $\eta(x,t)$ remains in the Ljapunov-surface

$$\mathfrak{N}_t[\eta] = 2\hbar \int_{V_0} \eta\eta^* \, d^3x = 2\hbar \| \eta \|^2 = \text{constant} \tag{5.39}$$

for all times,i.e. in linear approximation the perturbed process $\psi(x,t,\vartheta)$ remains at the initially given "distance" from the reference process $\psi(x,t)$.

5.2 Example: Heat transport

The functionals $\underset{(\kappa)}{A}{}_t$ defined by (5.2) now depend on the reference process $\psi(x,t)$, i.e. the associated indicatices (3.11) depend on time and therefore the condition (5.6) of Ljapunov's criterion gets essential.

As in the case of Schrödinger's matter field the functional $\mathcal{G}_t[\Psi,\eta]$ of (5.9) does not define a Ljapunov-functional because of the special form of the Lagrangian

(2.48). One should keep in mind that this Lagrangian is associated with linear theory of heat transport within the framework of the principle of local equilibrium; e.g. inertia terms are missing. Thus from this point of view the Lagrangian (2.48) is but a truncated one. Generalizing this Lagrangian in order to take also into account those processes which run outside local equilibrium [13] the situation concerning the use of $\mathcal{G}_t$ as a Ljapunov-functional may be improved.

Let me consider the functional (5.13). Using (2.48,18,44) one finds after some lengthy calculations

$$\pi = \frac{c}{\omega}\left[2\eta\eta^* + i\left(\frac{\eta^2}{\psi^2} - \frac{\eta^{*2}}{\psi^{*2}} \right) T_0\left(1 - \ln\frac{T}{T_0} \right) \right] . \tag{5.40}$$

We have now to keep in mind that we are dealing with processes running through local equilibrium states. These processes are associated with the particular solution (2.58) which gives rise to a constraint for the process trajectories in state space and which subsequently results in the constraint

$$\eta_\phi = - \frac{T_0}{2T(x,t)^2}\, \eta_T \tag{5.41}$$

for the trajectories in perturbation space. In (5.41) η_ϕ and η_T are the perturbations associated with the fields $\phi(x,t)$ and $T(x,t)$ respectively. (See (2.14) as applied to ϕ and T.) Taking further account of (2.50) the perturbations η, η^*, η_T and η_ϕ are related by

$$\eta = \frac{\psi}{2T}\eta_T + i\,\psi\,\eta_\phi \quad , \tag{5.42}$$

$$\eta^* = \frac{\psi^*}{2T}\eta_T - i\,\psi^*\,\eta_\phi \quad . \tag{5.43}$$

Inserting (5.41-43) into (5.40) one finally gets

$$\pi = \frac{c}{2\omega T_0}\left[z\left(1 + 3z^2 + \frac{1}{2} z^2 \ln z \right)\Big|_{z = T_0/T} \right] \eta_T^2 \tag{5.44}$$

where $$[\ldots\ldots]\Big|_{z = T_0/T} > K > 0 \quad \text{for} \quad 0 \le T < \infty \tag{5.45}$$

as can easily be shown. Together with

$$c\ ,\ \omega\ ,\ T_0 > 0 \tag{5.46}$$

we thus get a functional $\mathfrak{M}_t[\eta]$ which fits into the conditions (5.5,6) and which is constrained to those processes running through local equilibrium states:

$$\mathfrak{M}_t[\Psi,\eta] = \int_{V_0} \pi\, d^3x \ \ge\ 0 \quad \text{for} \quad \|\eta\| \ge 0 \quad , \tag{5.47}$$

$$\mathfrak{M}_t[\Psi,\eta] \;\geq\; m(\|\eta\|) \;=\; \frac{c}{2\omega T_0} K \|\eta\| \; . \tag{5.48}$$

Here the norm (3.6) is used.

For a thermally isolated system the functional $\underset{(\pi)}{\Phi}{}_t$ of (5.15) allows for analogous arguments as in the preceding chapter:

$$\underset{(\pi)}{\Phi}{}_t[\psi,\eta] \;=\; 0 \tag{5.49}$$

Finally the functional $\underset{(\pi)}{\Sigma}{}_t[\Psi,\eta]$ of (5.14) is based on the function $\sigma_{(\pi)}$ the essential part of which is found to be

$$\sigma_{(\pi)}(\psi,\partial\psi;\, \eta,\partial\eta) \;=\; -\frac{1}{T_0}\, [\, z(1+z^2)\,]\Big|_{z\,=\,T_0/T} (\nabla\eta_T\cdot\underline{\lambda}\cdot\nabla\eta_T) + \ldots \tag{5.50}$$

where
$$[\ldots\ldots]\Big|_{z\,=\,T_0/T} > K > 0 \tag{5.51}$$

within each finite temperature range. The formula (5.50) is again due to the procedure (5.41-43) which has already been applied to (5.40-44). Obviously the kernel $\sigma_{(\pi)}$ fits into the scheme (5.16,18,19). Thus by (5.19) the relation

$$\underset{(\pi)}{\Sigma}{}_t[\Psi,\eta] \;=\; \int_{V_0} \sigma_{(\pi)}\, d^3x \;\leq\; 0 \tag{5.52}$$

holds within a neighbourhood of $\eta = 0$, if the

tensor $\underline{\lambda}$ of heat conductivity is positive definite . (5.53)

From (5.1,49,52) I finally get the relation

$$\frac{d}{dt}\mathfrak{M}_t\,[\Psi,\eta] \;\leq\; 0 \tag{5.54}$$

which holds along the trajectories of the perturbations $\eta(x,t)$ of a given reference process $\psi(x,t)$. This relation fits into Ljapunov's criterion (5.7). Thus the functional $\mathfrak{M}_t[\Psi,\eta]$ defined by (5.13,44) is a Ljapunov-functional which within the class of local equilibrium processes manifests the stability of each heat transport process, provided (5.53) holds.

6. Concluding remarks: Entropy and stability

Within Lagrange-formalism entropy and stability are separate concepts. Nevertheless the stability condition (5.53) concerning the positive definite tensor $\underline{\lambda}$ of heat conductivity gives rise to a positive definite entropy production rate (4.9). For heat conduction this is the second part of the 2nd Law of Thermodynamics. On the other hand a positive definite entropy production rate (4.9) results in (5.53) and thus in stability of heat conduction processes. However, this one-to-one correspondence might be accidental due to the special form of the Lagrangian (2.48).

I expect that in general stability is superior as compared with the positive entropy production rate. Further investigations towards this statement will be done.

7. Literature

1. K.- H. Anthony: In Continuum Models of Discrete Systems 4, ed. by O. Brulin and R.K.T. Hsieh (North Holland Publ. comp., Amsterdam 1981) p.481
2. K.- H. Anthony: In Disequilibrium and Self-Organisation, ed. by C.W. Kilmister (Reidel Publ. Comp., Dordrecht (The Netherlands) 1986) p. 75
3. K.- H. Anthony: Archives of Mechanics, Warsaw, (1988), in press
4. H. Haken: Synergetics, 2nd ed., (Springer, Berlin 1978)
5. E. Schmutzer: Symmetrien und Erhaltungssätze der Physik (Akademie-Verlag, Berlin 1972)
6. E.M. Corson: Introduction to Tensors, spinors and Relativistic Wave Equations (Blackie and Son, London 1957)
7. S.R. DeGroot and P. Mazur: Non-Equilibrium Thermodynamics (North Holland Publ. Comp., Amsterdam 1969)
8. J. Meixner and H.G. Reik: Thermodynamik der irreversiblen Prozesse, in "Handbuch der Physik" III/2, ed. by S. Flügge, (Springer, Berlin 1959)
9. E.Fick: Einführung in die Grundlagen der Quantenmechanik (Akademische Verlagsgesellschaft, Leipzig 1968)
10. V.I. Zubov: Methods of A.M. Lyapunov and their Applications (Nordhoff, Groningen (The Netherlands) 1964)
11. W. Hahn: Theorie und Anwendungen der direkten Methode von Ljapunov, Springer Ser. "Ergebnisse der Mathematik und ihrer Grenzgebiete", (Springer, Berlin 1959)
12. R. Klötzler: Mehrdimensionale Variationsrechnung (Birkhäuser, Basel 1970)
13. K. H. Anthony und H. Knoppe: In Kinetic Theory and Extended Thermodynamics, ed. by I. Müller and T. Ruggeri (Pitagora editrice, Bologna 1987) p.15

A Thermodynamic Theory for Turbulence

J. Verhas

Technical University Budapest, H-1521 Budapest, Hungary

Introduction

Phenomenological thermodynamics is often regarded as a science dealing with averages over chaotic molecular motions, while its basic principles are free from stochastic considerations. This fact leads to the idea that thermodynamics should be able to deal with other averages as well, even with averages over turbulent domains.

The outlines of a turbulence model based on thermodynamics can be conceived. From the success and failures of the attempts made so far, one can see that neither the linearity of Onsager's eqations nor the hypothesis of local equilibrium is tenable for turbulence models. Turbulent fluid is not a simple fluid since the viscometric functions and Reynolds' similarity conditions are incompatible except for the laminar case.

Some new concepts recently introduced in non-equilibrium thermodynamics enable us to construct a new turbulence model avoiding the weak points of the old theories. Introducing dynamic state variables considering also their transport and using non-linear constitutive equations instead of Onsager's linear ones, a more efficient model might be obtained.

Departing from linearity causes some difficulties since everything depending anyhow on anything is a too slippery ground to work on. So we had to look for further physical principles to determine the constitutive equations. Such a guide has been found in Reynolds' similarity conditions and another in the idea of Reynolds' turbulent stress tensor.

Encouraged by the success of the thermodynamic method in rheology, the turbulent Navier-Stokes fluid is approximated by a body which has one dynamic variable. For simplicity, isochoric motions are assumed only, body forces are neglected.

1. The Thermodynamic Model

The balance equations for the average motion are rather similar to the usual ones. The balance equation for the momentum reads

$$\varrho\frac{d\underline{v}}{dt} + \text{grad } p = \text{Div } \underline{\underline{t}} \qquad (1)$$

Here $\underline{v}$ is the average velocity. The scalar quantity p consists of two parts: the average of the scalar pressure and the trace of Reynolds' tensor. The tensor $\underline{\underline{t}}$ includes the average of the deviatoric part of Cauchy's stress and that of Reynolds' tensor.

The balance of the mass

$$\text{div } \underline{v} = 0, \qquad (2)$$

is well known for incompressible fluids.

The internal energy balance reads

$$\varrho\frac{du}{dt} + \text{div } \underline{J}_u = \underline{\underline{t}}:\underline{\underline{d}} \qquad (3)$$

where $\underline{\underline{d}}$ stands for the symmetric part of the average velocity gradient.

The specific entropy (entropy per unit mass) depends on the specific internal energy and on a dynamic variable ξ. Its physical meaning and tensorial order are to be specified later. According to the Morse lemma and the second law, the dynamic variable can always be chosen so that the form of the entropy function be

$$s = s^e(u - \frac{\gamma}{2}\,\underline{\xi}\cdot\underline{\xi}) \qquad (4)$$

where $s^e(u)$ is the equilibrium entropy function and γ is an arbitrary number as yet. The general form of the entropy balance is

$$\varrho\frac{ds}{dt} + \text{div } \underline{J}_s = \sigma_s >= 0 \qquad (5)$$

Here the entropy current $\underline{J}_s$ consists of the classical term and the one belonging to the transport of the dynamic variable:

$$\underline{J}_s = \frac{1}{T}\underline{J}_u - \frac{\gamma}{T}\,\underline{\xi}:\underline{\underline{J}}_\xi \qquad (6)$$

The current density $\underline{\underline{J}}_\xi$ is that of the dynamic variable. Its last index refers to the direction of the transport.

The dynamic variable can be a symmetric second order tensor with zero trace or a vector. The former choice seems more natural than the latter while the latter furnishes a model easier to handle.

2. Constitutive Equations

The entropy production density for the present case is obtained from eqs. (3)-(6)

$$\sigma_s = \frac{1}{T}\Big[t:d-\gamma\xi:(e\overset{\circ}{\xi}+\mathrm{Div}J_\xi)-(J_u-\gamma\xi:J_\xi)\frac{\mathrm{grad}T}{T}-\gamma J_\xi:\mathrm{Grad}\,\xi\Big] \tag{7}$$

Here we introduced the corrotating time derivative $\overset{\circ}{\xi}$ of the dynamic variable ξ ($\overset{\circ}{\xi} = \frac{d\xi}{dt} + \xi\omega - \omega\xi$ for a tensor or $\overset{\circ}{\xi} = \frac{d\xi}{dt} - \omega\xi$ for a vector) and applied

$$\xi:(\xi\omega - \omega\xi) = 0 \qquad \text{or} \qquad \xi(\omega\xi) = 0 \tag{8}$$

since ω is the antisymmetric part of the average velocity gradient. The constitutive equations for isothermal motions have the form

$$\begin{aligned} t &= L_{11}d - L_{12}\gamma\xi - \gamma L_{13}\mathrm{Grad}\xi \\ e\overset{\circ}{\xi} + \mathrm{Div}\,J_\xi &= L_{21}d - L_{22}\gamma\xi - \gamma L_{23}\mathrm{Grad}\xi \\ J_\xi &= L_{31}d - L_{32}\gamma\xi - \gamma L_{33}\mathrm{Grad}\xi \end{aligned} \tag{9}$$

The L_{ik} coefficients are higher order tensors. If Onsager's linear laws held they should be material properties (constants) but in our case they are functions. To specify these functions we turn to Reynolds' similarity conditions. It means, we believe in the validity of the Navier-Stokes equations, but we are not able to get their solutions for turbulent flows. Consequently, we look for other methods, at present for thermodynamic ones. At the same time we presume these differential equations (i.e (1) and (9)) obey the same similarity conditions as the Navier-Stokes equations do. So, from similarity considerations, the coefficients L_{11}, L_{13}, L_{31} and L_{33} turn out to be constant, L_{12}, L_{21}, L_{23} and L_{32} depend linearly on the dynamic variable ξ. The coefficient L_{22} can contain terms linear on d, ω or $\mathrm{Grad}\xi$ and one quadratic in ξ. The details are rather lengthy. The sketched argumentation leads to the alternative sets of constitutive equations.

If the dynamic variable is a second order tensor we have

$$\underline{\underline{t}} = 2\nu\underline{\underline{d}} - A_1\left[\underline{\underline{\xi}}^2 - \frac{1}{3}(\underline{\underline{\xi}}:\underline{\underline{\xi}})\underline{\underline{\delta}}\right]$$

$$\varrho\dot{\underline{\underline{\xi}}} + \mathrm{Div}\,\underline{\underline{J}}_\xi = A\left[\frac{1}{2}(\underline{\underline{\xi}}\,\underline{\underline{d}}+\underline{\underline{d}}\,\underline{\underline{\xi}}) - \frac{1}{3}(\underline{\underline{\xi}}:\underline{\underline{d}})\underline{\underline{\delta}}\right] - \frac{\varrho^2}{\nu}B(\underline{\underline{\xi}}:\underline{\underline{\xi}})\underline{\underline{\xi}} \qquad (10)_a$$

$$J_{\xi ijk} = -c_1\nu\xi_{ij,k} - c_2\nu\delta_{ij}\xi_{kl,l} - c_3\nu[\delta_{ik}\xi_{jl,l} + \delta_{jk}\xi_{il,l}] - c_4\nu(\xi_{ik,j} + \xi_{jk,i})$$

with $3C_2+2(C_3+C_4)=0$ as the trace of $\underline{\underline{\xi}}$ is assumed to be zero here. The proper choise of the quantity in (4) leads to $A_1=1$ and A, B, C_1, C_2 and C_3 are universal constants, similar to Karman's one but in a more general sense.

If a vector is chosen for the dynamic variable the constitutive equations read

$$\underline{\underline{t}}^V = 2\nu\underline{\underline{d}} - \varrho(\underline{\xi}\circ\underline{\xi} - \frac{1}{3}\xi^2\underline{\underline{\delta}})$$

$$\varrho\frac{d\underline{\xi}}{dt} + \mathrm{Div}\,\underline{\underline{J}}_\xi = -\varrho L^1_{22}\omega\underline{\xi} - L^2_{22}\underline{\underline{d}}\,\underline{\xi} - \varrho L^3_{22}(\underline{\xi}\,\mathrm{grad})\underline{\xi} - \frac{\varrho}{2}L^4_{22}\,\mathrm{grad}\,\underline{\xi}^2 -$$

$$- \varrho L^5_{22}\underline{\xi}\,\mathrm{div}\,\underline{\xi} - \frac{\varrho^2}{\nu}L^6_{22}\xi^2\underline{\xi} \qquad (10)_b$$

$$\underline{\underline{J}}_\xi = -\varrho L^*_{32}\underline{\xi}\circ\underline{\xi} - \nu L^1_{33}\,\mathrm{Grad}\,\underline{\xi} - \nu L^2_{33}(\mathrm{Grad}\,\underline{\xi})^T - \nu L^3_{33}\,\mathrm{div}\,\underline{\xi}\,\underline{\underline{\delta}}$$

Here $L_{13}=\pm L_{31}=0$ as a consequence of the Onsager-Casimir reciprocal relations and of the hypothesis that a laminar solution always exists.

3. Discussion

The next task is to try the constitutive equations (10). The differential equations deriving from them for different experimental situations have to be solved and the outcomes are to be compared to the experimental results. This work has just started and some preliminary computations show that the model given here is not far from reality.

As a rough first testing, some computations with tensorial dynamic variables have been done with the help of a microcomputer for stationary flows in smooth walled tubes and between smooth parallel walls. The values C_2, C_3 and C_4 were taken zero for comfort. Blasius' formula for both cases was reproduced, nevertheless the accuracy enables us to estimate the parameters cautiouosly.

The computations for a vector dynamic variable are a bit easier. Taking L^3_{22}, L^4_{22}, L^5_{22}, L^6_{22} and L_{32} zero for comfort, we have obtained rather considerable solutions. The radial and the axial components of the vector are plotted against the distance from the axis in Figs. 1 and 2 respectively. Figure 3 shows the velocity distribution at several Reynolds numbers. The drag coefficient against the Reynolds number is plotted in Fig. 4. Here the solid line is Blasius' one verified experimentally a large number of times.

However, the numerical problems seem to need methods more adapted to their peculiarities than the methods available for more or less general purposes. On the other hand, finding the correct place of the dynamic variable $\underline{\underline{\xi}}$ in a consistent picture can also be of great help as it might give further restrictions reducing the number of independent parameters.

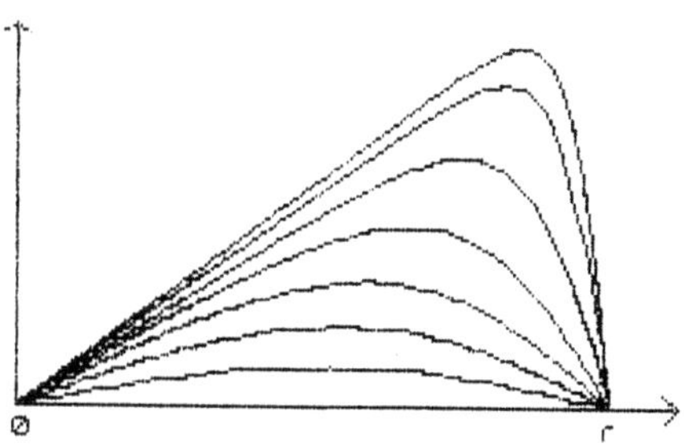

Fig. 1.

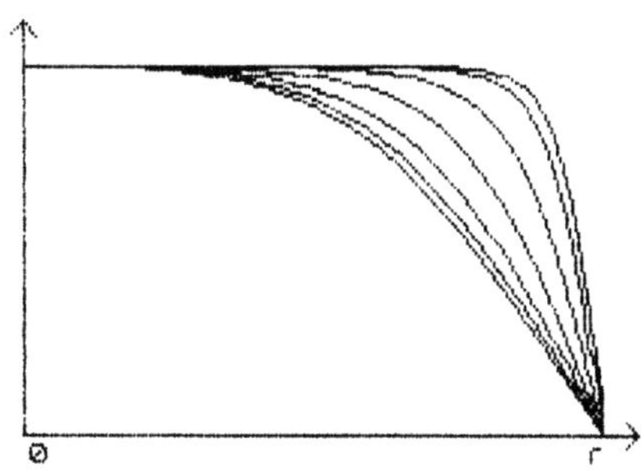

Fig. 2.

The radial component / The axial component
of the dynamic variable.

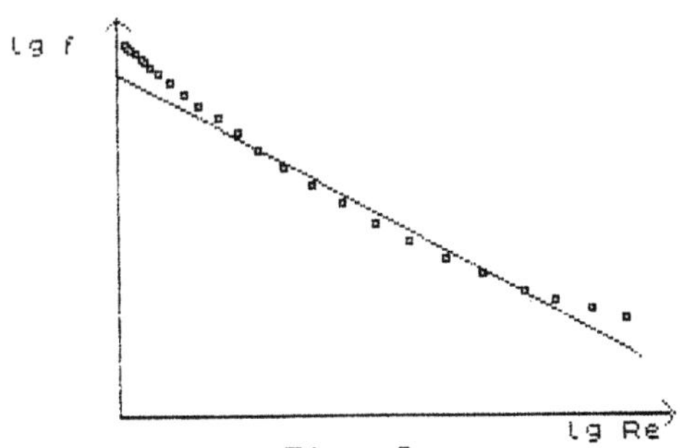

Fig. 3.

The logarithm of the drag coefficient plotted against the logarithm of Reynolds' number.

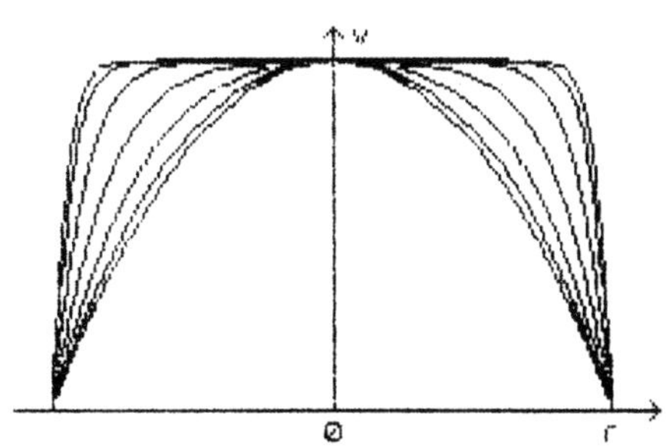

Fig. 4.

The velocity plotted against the distance from the axis.

Appendix: Reynold's Similarity Conditions

The Navier-Stokes equations for an incompressible fluid read

$$\rho\frac{\partial \underline{v}}{\partial t} + \rho(\underline{v}\ \mathrm{grad})\underline{v} + \mathrm{grad}\ p = \eta \Delta \underline{v} \qquad \text{(A1)}$$

$$\mathrm{div}\ v = 0.$$

Here $\underline{v}$ is the velocity, p is the pressure and Δ is the Laplaceian. The boundary conditions are assumed homogeneous, i.e. the velocity is zero at the walls. For two geometrically similar problems, one introduces the transformations

$$\begin{aligned} \underline{v} &= v_o \underline{v}' \\ p &= p_o p' \\ \underline{r} &= D\ \underline{r}' \\ t &= t_o t' \end{aligned} \qquad \text{(A2)}$$

where D is a characteristic length of the geometry and v_o, p_o and t_o are numbers to be chosen properly. If we make use of these transformations we get for the Navier-Stokes equations

$$\frac{\rho D^2}{t_o \eta}\frac{\partial \underline{v}'}{\partial t'} + \frac{\rho v_o D}{\eta}(\underline{v}'\mathrm{grad}')\underline{v}' + \frac{p_o D}{\eta v_o}\mathrm{grad}' p' = \Delta' \underline{v}' \qquad \text{(A3)}$$

$$\mathrm{div}' \underline{v}' = 0$$

and the boundary conditions remain unchanged.

Choosing

$$t_o = \frac{\rho\ D^2}{\eta}; \quad v_o = \frac{\eta}{\rho\ D}; \quad p = \frac{\eta^2}{\rho\ D^2}, \qquad \text{(A4)}$$

we get

$$\frac{\partial \underline{v}'}{\partial t'} + (\underline{v}'\mathrm{grad}')\underline{v}' + \mathrm{grad}' p' = \Delta' \underline{v}' \qquad \text{(A5)}$$

$$\mathrm{div}' \underline{v}' = 0$$

which is universal and does not contain any criterion number. It is worth mentioning that Reynolds' number is a parameter of a solution and not in the equations (A5).

It is demanded that the system of differential equations consisting of (1) and (9) get a universal form after the transformations (A2) and (A4). This leads to the following transformation rules:

$$d = \frac{\eta}{\rho D^2} d'; \quad \omega = \frac{\eta^2}{\rho D^2}\omega'; \quad t = \frac{\eta}{\rho D^2} t'; \quad \xi = \frac{\eta^2}{\rho D}\xi'; \quad J_\xi = \frac{\eta}{\rho D^2} J'_\xi \qquad \text{(A6)}.$$

which bring (1) to the needed universal form and transform (9) into

$$\underline{\underline{t}}' = \frac{L_{11}}{\varrho}\underline{\underline{d}}' - \frac{L_{12}D\gamma}{\varrho}\underline{\underline{\xi}}' - \frac{L_{13}\gamma}{\varrho}\mathrm{Grad}'\underline{\underline{\xi}}'$$

$$\dot{\underline{\underline{\xi}}}' + \mathrm{Div}'\underline{\underline{J}}_\xi' = \frac{L_{21}D}{\varrho}\underline{\underline{d}}' - \frac{L_{22}\gamma D^2}{\varrho}\underline{\underline{\xi}}' - \frac{L_{23}\gamma D}{\varrho}\mathrm{Grad}'\underline{\underline{\xi}}' \qquad \text{(A7)}$$

$$\underline{\underline{J}}_\xi' = \frac{L_{31}}{\varrho}\underline{\underline{d}}' - \frac{L_{32}\gamma D}{\varrho}\underline{\underline{\xi}}' - \frac{L_{33}\gamma}{\varrho}\mathrm{Grad}'\underline{\underline{\xi}}'$$

The coefficients have to be universal, so L_{12}, L_{21}, L_{23} and L_{32} are homogeneouos linear functions of $\underline{\underline{\xi}}$. The coefficients L_{11}, L_{13}, L_{31} and L_{33} have to be constants. At last L_{22} contains a term quadratic in $\underline{\underline{\xi}}$ and three terms linear in Grad $\underline{\underline{\xi}}$, $\underline{\underline{d}}$ and $\underline{\underline{\omega}}$ respectively. As L_{21} and L_{23} have their counterparts in L_{22}, they can be dropped. Moreover, the supposition that a laminar solution always exists takes the coefficient L_{31} zero. From the Onsager-Casimir relation $L_{13}=\pm L_{31}$ we infer that $L_{13}=0$. From here, making use of the theory of isotropic functions, we get the possible forms of the constitutive equations.

References

M.Morse: Trans.Am.Mat.Soc. 27 345 (1925)

J.Rotta: Z. fur Phys. 129 547 (1951); Ibid 131 51 (1951)

J.O.Hinze: Turbulence. Mc Graw-Hill. New York 1959.

S.R.De Groot-P.Mazur: Non-Equilibrium Thermodynamics. North-Holland Publ. Co. Amsterdam 1962.

I.Gyarmati: Non-Equilibrium Thermodynamics. Springer, Berlin, 1970.

I.Muller: Arch.Rat.Mech.Anal. 26 118 (1967); Ibid 40 1 (1971); 41 319 (1971)

B.E.Launder-D.B.Spalding: Mathematical Models of Turbulence. Acad.Press. London 1972.

I.Gyarmati: Non-Equilib. Thermodyn. 2 233 (1977)

M.Frias: Thermodynamically Structured Flows. Rheology III. 749. Plenum Press. New York (1980)

G.Lebon: Int.J. Engng. Sci. 18 727 (1980)

G.A.Kluitenberg: Physica 87A 302 (1977); 109A 91 (1981)

W.Ebeling: Ann.d.Phys. 40(7) 25 (1983); Ibid 40(7) 51 (1983)

G.A.Maugin: Int.J.Engng.Sci. 21 705 (1983)

J.Verhas: J.Non-Equilib. Thermodyn. 8 201 (1983)

J.Verhas: Ann.d.Phys. 40(7) 189 (1983)

D.Jou, C.Perez-Garcia and J.Casas-Vazquez: J.Phys.A. 17 2799 (1984)

J.Verhas: Acta Phys.Hung. 55 275 (1984)

J.Verhas: Acta Mech. (Vienna) 53 125 (1984)

M.T.Landahl-E.Mollo-Christensen: Turbulence. Cambridge Univ.Press. Cambridge, London 1986.

B.Nyiri: Acta Phys.Hung. 60 245 (1986)

J.Verhas: Int.J.Heat Mass Transfer 30 1001 (1987)

Approach to Lower Dimensional Dynamical Systems: Thermodynamical Aspects

M. Grmela

École Polytechnique de Montréal, P.O. Box 6079, Station A, Montréal, Québec, Canada H3C 3A7

Thermodynamics arises as a new information obtained by combining models formulated on different levels of description.

1. Introduction

Let us consider one of the well studied physical systems that exhibit hydrodynamic instabilities (e.g. Rayleigh-Bénard cell - see e.g. CHANDRASEKHAR [1]). We shall model this system on two levels of description. The first level is the level of hydrodynamics (i.e. hydrodynamic fields are chosen to describe states of the system and hydrodynamic equations - see e.g. Refs. [1],[2] - are the governing equations of the model). On the second level, the quantities that are chosen to describe states are more microscopic than hydrodynamic fields. This means that to one state on level 1 corresponds, in general, a whole family of different states on level 2. For example, we can choose one particle distribution functions as the level 2 state variables. Governing equations on the level 2 are introduced by expressing the insight into the physical processes involved in terms of the level 2 state variables. Experience indicates that if our interest is limited to the information contained in results of hydrodynamic measurements (i.e. to the information contained in hydrodynamic fields) then the level 1 model is sufficient. Predictions based on the level 1 model agree well with results of observations. The level 2 model is therefore superfluous. We shall however ask the following question: can the new information and the new insight gained on the level 2 be used to contribute the level 1 model? This question can be rephrased also as follows: Level 1 models are often introduced as certain approximations of more detailed level 2 models (we can think for example about the way the Lorenz dynamical system [3] was introduced). Can a more complete analysis of the level 2 model contribute to the analysis of solutions of the governing equations of the level 1 model?

In this note we shall introduce an approach to the analysis of this question. The approach is based on the following observation. If the experience shows that the level 1 model is sufficient then it means that the level 2 dynamics approaches the level 1 dynamics. By this we mean that there exists an immersion I: level 1 state space → level 2 state space such that the asymptotic part of the level 2 trajectories can be well approximated by the level 1 trajectories lifted to the level 2 state space by the immersion I. We shall call the image of I a level 2 asymptotic subspace. The new information that arises in this analysis of the level 2 dynamics is the structure (in particular the geometrical structure) of the level 2 asymptotic subspace. We conjecture that this structure is the natural structure for the level 1 state space and that some important properties of the level 1 trajectories can be read already from the structure. In other words, we conjecture that important changes in the behaviour of level 1 trajectories are accompanied by important changes of the level 2 asymptotic subspace. We show (section 2) that in the special case when the level 1 model is equilibrium thermodynamics (i.e. level 1 model is a static not a dynamical model), the geometrical structure of the level 2 asymptotic subspace is the structure of equilibrium thermodynamics. We then suggest to regard the geometrical structure of the level 2 asympotic subspace as the thermodynamical structure of the level 1 model also in the case when the level 1 model is a dynamical model. In section 3 we take hydrodynamics as the level 1 model and Boltzmann kinetic theory as the level 2 model. Well known results about solutions of the Boltzmann kinetic equation are then used to illustrate the point of view taken in this note.

2. Illustration: Approach to Equilibrium Thermodynamics

In this section we consider only closed physical systems. The level 1 model is chosen to be equilibrium thermodynamics (i.e. in this particular case there is no vector field defined in the level 1 state space). As the level 2 model we choose the Boltzmann kinetic theory (i.e. one particle distribution function serves as the level 1 state variable and the Boltzmann kinetic equation as the governing equation). At the end of this section we shall then generalize the results to a large family of level 2 dynamical theories.

It is a well known experience that closed physical systems reach (as time goes to infinity) states (called equilibrium states) and at the equilibrium states their behaviour is well described by equilibrium thermodynamics. This means, from the point of view of level 2, that the level 2 trajectories reach a level 2 asymptotic subspace (in this case it is a subspace of fixed points of the level 2 dynamics). The geometrical structure of this level 2 asymptotic subspace is then expected to represent the structure of equilibrium thermodynamics. We shall now show that this is indeed so. First, we recall the geometrical formulation of equilibrium thermodynamics (Refs. [4]-[12]), second, we formulate the level 2 vector field in such a way (see Refs. [13]-[30]) that the geometrical structure of the level 2 asymptotic subspace as well as the equilibrium thermodynamic interpretation of the geometrical structure is manifestly revealed.

The state variables chosen in equilibrium thermodynamics are: $n = N/V$ and $e = E/V$, where N is the number of moles, E is the energy and V is the volume (only one component systems are considered). Since we shall later (in section 3) extend some of the concepts introduced in this section to general level 1 models, it will be convenient to introduce already at this stage all concepts and relations in terms of general level 1 state variables. The symbol $\varphi = (\varphi_1,\dots,\varphi_k)$ will be used to represent the level 1 state variable. In the context of equilibrium thermodynamics $k = 2$, $\varphi_1 = n$, $\varphi_2 = e$. The level 1 state space is two dimensional. In general however, some of the symbols φ_i, $i = 1,\dots,k$ may represent functions (fields) and the dimension of the level 1 state space may be thus infinite. We emphasize that in this section the level 1 model is only equilibrium thermodynamics and the general notation is used only to avoid repetitions later in section 3.

The thermodynamical structure is specified on the level 1 by specifying a function

$$\begin{aligned} &\text{level 1 state space} \xrightarrow{\Phi} \mathbb{R} \\ &\varphi \longmapsto \Phi(\varphi) = \Phi \end{aligned} \tag{1}$$

In equilibrium thermodynamics $\Phi = s = S/V$, where S is the entropy. The individual features of the system under consideration are expressed in the function (1). An alternative formulation of the level 1 thermodynamical structure is achieved with the help of a generating function

$$G = G(\varphi\ ;\ \varphi^*) = -\Phi(\varphi) + \sum_{i=1}^{k} (\varphi_i, \varphi_i^*) \quad . \tag{2}$$

The quantities φ_i^* are called conjugate to the state variables φ_i. By $(,)$ we denote L_2 scalar product. In equilibrium thermodynamics $\varphi_1^* = -\mu/T$, $\varphi_2^* = 1/T$, where T is, the temperature and μ is the chemical potential. The conjugate state variables are introduced by

$$\varphi_i^* \overset{\text{def}}{=} \frac{\delta\Phi}{\delta\varphi_i} \tag{3}$$

(in equilibrium thermodynamics $1/T = \partial s/\partial e$, $-\mu/T = \partial s/\partial n$). In order that the relation $\varphi \mapsto \varphi^*$ defined by (3) is one-to-one the function Φ has to be convex (or concave). Solutions of

$$\frac{\delta G}{\delta\varphi} = 0 \tag{4}$$

(in equilibrium thermodynamics solutions of $\partial G/\partial n = 0$, $\partial G/\partial e = 0$), where $\delta/\delta\varphi$ denotes the Volterra functional derivative if φ is a function and the ordinary partial derivative of φ is an element of a finite dimensional space, are denoted

$$\varphi_{eq}(\varphi^*) \tag{5}$$

and called level 1 equilibrium states. We introduce

$$G(\varphi_{eq}(\varphi^*)\ ;\ \varphi^*) = G^* \tag{6}$$

In equilibrium thermodynamics $G^* = - P/T$, where P is the pressure. The relation (6) between G^* and φ^* is an alternative formulation of the relation (1) between Φ and φ. We note that the inverse of the conjugation mapping (3) is:

$$\varphi_i = \frac{\delta G^*}{\delta \varphi_i^*} \tag{3'}$$

(in equilibrium thermodynamics $e = - \partial(^{\partial(P/T)}/_{\partial(1/T)}$, $n = {}^{\partial(P/T)}/_{\partial(\mu/T)}$. The geometrical formulation of the level 1 thermodynamical structure is introduced as follows. We introduce a new space, called level 1 & 1^* extended state space, as a cartesian product of level 1 state space, conjugate level 1 state space and a one dimensional space, called a G - state space. Elements of the level 1 & 1^* extended state space are thus

$$(\varphi\ ,\ \varphi^*\ ,\ G)\quad . \tag{7}$$

In equilibrium thermodynamics the level 1 & 1^* extended state space is a five dimensional space with coordinates (n, e, μ/T, 1/T, G). The level 1 equilibrium states (5) form a subspace of the level 1 & 1^* extended state space. The geometrical structure of this subspace is the geometrical formulation of the level 1 thermodynamical structure (Refs. [4]-[10]).

Now we turn our attention to the level 2 description. The level 2 state variables will be denoted formally by f (e.g. f = f($\underline{r}$,$\underline{v}$) = one partical distribution function, $\underline{r}$ is the position vector, $\underline{v}$ denotes the velocity). Typically, the level 2 state space is an infinite dimensional state space. We recall that, according to the terminology introduced in section 1, the level 2 description is more detailed (microscopic) than the level 1 description. How can we formulate level 1 thermodynamics from the point of view of level 2? We proceed as follows (Refs. [10]-[14]). We introduce projection mappings P and Q by:

$$\begin{aligned} &\textbf{level 2 state space} \xrightarrow{P} \textbf{level 1 state space}\\ &f \xmapsto{P} \varphi(f)\\ &\textbf{level 2 state space} \xrightarrow{Q} R\\ &f \xmapsto{Q} \Phi(f)\quad . \end{aligned} \tag{8}$$

The individual features of the system under consideration are expressed in the projection mappings P and Q. In the case when the level 1 model is equilibrium thermodynamics $f \overset{P}{\mapsto} (n(f),e(f))$ and $f \overset{Q}{\mapsto} s(f)$. In order to obtain the fundamental relation (6) of the level 1 thermodynamic we proceed similarly as before. We introduce a generating functional

$$G = G(f\ ;\ \varphi^*) = -\ \Phi(f) + \sum_{i=1}^{k} (\varphi_i(f), \varphi_i^*) \tag{9}$$

Solutions of

$$\frac{\delta G}{\delta f} = 0 \tag{10}$$

denoted

$$f_{eq}(\varphi^*) \tag{11}$$

are called level 2 equilibrium states. As in (6) we introduce

$$G(f_{eq}(\varphi^*)\ ;\ \varphi^*) = G^* \quad . \tag{12}$$

The relation (12) is now the fundamental relation of the level 1 thermodynamics that corresponds to the projection mappings P,Q introduced in (8). The geometrical formulation is introduced as follows. We introduce a new space, called level 2 & 1^* extended state space, as a cartesian product of level 2 state space, conjugate level 1 state space and a one dimensional space called G - state space. Elements of the level 2 & 1^* extended state space are thus

$$(f\ ,\ \varphi^*\ ,\ G) \quad . \tag{13}$$

We now regard (9) as a relation in the level 2 & 1^* extended state space. The level 2 equilibrium states (11) form a subspace of the level 2 & 1^* extended state space. The geometrical structure of this subspace is the geometrical formulation of the level 1 thermodynamical structure.

Now we introduce a time evolution into the level 2 state space. We shall write the governing equation formally as

$$\frac{\partial f}{\partial f} = R^{(2)}(f) \quad . \tag{14}$$

To begin with, we take $f \equiv f(\underline{r},\underline{v}) \equiv$ one particle distribution function and the Boltzmann kinetic equation [15],

$$\frac{\partial f}{\partial t} = -\ v_\alpha \frac{\partial f}{\partial r_\alpha} + \textbf{collision term} \tag{15}$$

as the level 2 governing equation. The summation convention is used in Eq. (15), α = 1,2,3. The explicit form of the collision term will not be needed.

Our aim now is to extend the time evolution (15) (defined in the level 2 state space) to the level 2 & 1^* extended state space. The extension is expected to be such that the time evolution in the conjugate level 1 state space and in the G - state space will reflect important qualitative properties of the time evolution in the level 2 state space. In the case discussed in this section (i.e.

level 1 model is equilibrium thermodynamics) there is clearly no time evolution in the conjugate level 1 state space. The problem is thus to formulate the level 2 time evolution in such a way that the corresponding generating function G is revealed. On the basis of the results concerning properties of solutions of the Boltzmann equation (15) that have already been obtained by BOLTZMANN [15] we can expect that the generating function corresponding to (15) is the generating functional (9) with

$$
\begin{aligned}
n(f) &= \frac{1}{V} \int d\underline{r} \int d\underline{v}\; f(\underline{r},\underline{v},t) \\
e(f) &= \frac{1}{V} \int d\underline{r} \int d\underline{v}\; \frac{1}{2} v^2\, f(\underline{r},\underline{v},t) \\
s(f) &= -\frac{1}{V} \int d\underline{r} \int d\underline{v}\; f(\underline{r},\underline{v},t)\, \ln f(\underline{r},\underline{v},t)
\end{aligned}
\tag{16}
$$

(It can be easily verified that the thermodynamic relation (12) that corresponds to (16) is the thermodynamic relation of an ideal gas). Can we formulate the Boltzmann equation (15) in such a way that the generating functional (9), (16) appears manifestly? The answer to this question has been found in Refs. [16], [17].

By using results obtained in these references we suggest the following time evolution equations in the level 2 & 1^* extended state space:

$$
\frac{\partial}{\partial t} \begin{bmatrix} f \\ \varphi^* \\ G \end{bmatrix} = \begin{bmatrix} \{f,G\} \\ 0 \\ \{G,G\} \end{bmatrix} \tag{17}
$$

where G is the generating functional (9), (16) and {,} is a bracket operation. The explicit specification of the bracket operation can be found in GRMELA [16], [17]. Here, we shall only list the following properties:

$$
\{A,B\} = \{A,B\}_{rev} + \{A,B\}_{irr} \tag{18}
$$

$$
\begin{aligned}
\{A,B\}_{rev} &= -\{B,A\}_{rev} \\
\{A,B\}_{rev} &= A\{B,C\}_{rev} + \{A,C\}_{rev}\, B \\
\{A,\{B,C\}_{rev}\}_{rev} &+ \{B,\{C,A\}_{rev}\}_{rev} + \{C,\{A,B\}_{rev}\}_{rev} = 0
\end{aligned}
\tag{19}
$$

$$
\{A,s\}_{rev} = 0, \ \{A,n\}_{rev} = 0 \qquad \text{for all } A \tag{20}
$$

$$
\{A,B\}_{irr} = \{B,A\}_{irr}
$$

$\{A,B\}_{irr}$ is a linear function of $\frac{\delta A}{\delta f}$; and a convex function of $\frac{\delta B}{\delta f}$

$$\{A,A\}_{irr} \leq 0 \tag{21}$$

$$\{A,e\}_{irr} = 0, \; [A,n]_{irr} = 0 \quad \text{for all } A \; .$$

By A,B,... we denote sufficiently regular functionals of f; s,e,n are functionals introduced in (8) and, in the case of the Boltzmann kinetic theory, specified in (16). We note that the properties (19) tell us that the bracket $\{,\}_{rev}$ is the Poisson bracket. The first line in Eq (17) is completely equivalent to the Boltzmann equation (15) (see Refs. [16],[17]). This can be verified by straighforward calculations. We see immediately from (17) that the generating functional G generates manifestly both dynamics in the level 2 state space and the corresponding equilibrium thermodynamics structure. It follows from (17)-(21) that elements of the level 2 state space approach equilibrium states introduced in (11).

This means that the level 2 asymptotic state space is the subspace formed by level 2 equilibrium states. We recall that the geometrical structure of this subspace has been identified with the structure of equilibrium thermodynamics. The first line in (17) is just a recasting of the Boltzmann equation. But it is the recasting that allows to see the properties of its solutions that interest us. We recall that in the context of dynamical systems recosting means sometimes solving. For example if we change coordinates (i.e. if we recast the vector field) so that the vector field becomes particularly simple, we solve the governing equations of the dynamical system. In the case of the recasting (17) we have not solved the Boltzmann equation but we have extracted a property of its solution that interests us. Summing up, the time evolution equations (17) illustrate our conjecture formulated in section 1. We have considered however only a particular case in which level 2 is the Boltzmann kinetic theory and level 1 is equilibrium thermodynamics. The immersion mentioned in section 1 is:

$$\left[\frac{\mu}{T}, \frac{1}{T}\right] \mapsto f_{eq}(\underline{r}, \underline{v}; \frac{\mu}{T}, \frac{1}{T})$$

The results obtained in Refs. [16]-[30] allow to generalize the illustration of the conjecture to the case in which level 1 remains only equilibrium thermodynamics but level 2 dynamical theories encompass a large class of dynamical theories. The recasting (17) has been explicitely identified for a large class of dynamical equations. In every case the functionals n,e,s as well as the brackets $\{,\}_{rev}$, $\{,\}_{irr}$ are different. They all share however the properties (18)-(21) and thus the thermodynamic interpretation (17) can be applied to all of them.

3. Illustration: Approach to Hydrodynamics

In this section we shall consider a general case. The physical systems under consideration may be open and both the level 1 and the level 2 models are dynamical models. We recall that, according to the terminology introduced in section 1, the level 2 description is more detailed (microscopic) than the level 1 description. In the general case we have, in addition to the time evolution in the level 2 state space (governed by Eq. (14), also a time evolution in the level 1 state space. Its governing equations will be written formally as

$$\frac{\partial \varphi}{\partial t} = R^{(1)}(\varphi) \quad . \tag{22}$$

Similarly as in the previous section we now look for an extension of the time evolution in the level 2 state space (Eq. (14)) and the time evolution in the level 1 state space (Eq. (22)) into the level 2 & 1^* extended state space. We shall write the time evolution equation in the level 2 & 1^* extended state space formally as

$$\frac{\partial F}{\partial t} = R^{(1,2)}(F) \quad , \tag{23}$$

where $F \equiv (f, \varphi^*, G)$. The problem that we want to discuss is the following: given (14) and (22) found (23) that satisfies the four requirements introduced below:

(i) Trajectories generated by (23) and projected on the first two components of the level 2 & 1^* extended state space are identical or at least close to the trajectories generated by (14) and by

$$\frac{\partial \varphi^*}{\partial t} = R^{(1)*}(\varphi^*) \tag{24}$$

that is (22) transformed into the conjugate level 1 state space by the conjugation mapping (3).

(ii) Trajectories generated by (23) and projected on the last two components of the level 2 & 1^* extented state space characterize the "gross" (macroscopic) behaviour of the time evolution in the level 2 state space generated by Eq. (14). In other words, (23) manifestely expresses the separation into the slow motion (projection on the last two components) and the fast motion that manifestly approaches the slow motion.

(iii) The level 2 asymptotic subspace as well as its geometrical structure is revealed. We expect that this will be done as in the previous section by an explicit appearence of the generating functional.

(iv) If the level 1 model is equilibrium thermodynamics then (23) coincides with (17).

This problem has to be of course studied separately for every particular choice of Eqs. (14) and (22). Encouraged by the universality of (17) we make however a bold suggestion. We expect that (23) assumes the form

$$\frac{\partial}{\partial t}\begin{bmatrix} f \\ \varphi^{*} \\ G \end{bmatrix} = \begin{bmatrix} \{f,G\} \\ R^{(2)*}(\varphi^{*}) \\ \{G,G\} + (\varphi(f)\ ,\ R^{(1)*}(\varphi^{*})) \end{bmatrix} \tag{25}$$

The generating functional G (see Eq. (9)) and the bracket operation {,} will be different for different equations (14) and (22) but the properties (18)-(22) will be shared by all of them. If we succeed to recast Eqs. (14) and (22) into the form (25) then the requirements (i)-(iv) introduced above are clearly satisfied.

In the rest of this paper, we shall illustrate Eq. (25) on an example: In the example we take the Boltzmann kinetic equation (15) as representing Eq. (14) and the Euler or the Navier-Stokes-Fourier hydrodynamic equations as representing Eq. (22). It means that

$$\varphi \equiv (n(\underline{r},t),\ e(\underline{r},t),\ \underline{u}(\underline{r},t)) \tag{26}$$

are the hydrodynamic fields (mass density, energy and velocity). Solutions of the Boltzmann kinetic equation as well as hydrodynamic euqations have been studied extensively. We recall in particular the following three results:

(1) CHAPMAN [31] and ENSKOG [32] have suggested a method for finding the immersion I introduced in section 1. The Chapman-Enskog method is a perturbation method, the collision term in the Botzmann equation (15) is considered as a dominant term. The zero approximation of I is the mapping $(n(\underline{r},t),\ e(\underline{r},t),\ \underline{u}(\underline{r},t))$ $\rightarrow$ local equilibrium distribution function. Application of the perturbation method leads then to corrections.

(2) An integral part of hydrodynamics is the so called local equilibrium assumption. According to this assumption: a) Pointwise for all $\underline{r}$ and t the equilibrium thermodynamics structure (see section 2) is attached to the state variables $n(\underline{r},t)$, $e(\underline{r},t)$. b) The local pressure $p(\underline{r},t)$ that appears in the hydrodynamic equation is pointwise the equilibrium pressure introduced in Eq. (6).

(3) There is large experimental evidence for the approach of the kinetic theory dynamics to hydrodynamics. Predictions calculated from hydrodynamic equations have been found in a good agreement with results of measurements of hydrodynamic fields in a large class of situations.

Now we shall introduce the generating functional G and the bracket operation {,} such that (25) will reflect manifestely the results (1)-(3) mentioned above.

The results (1) and (2) suggest the following generating functional. The projection operators P and Q introduced formally in (8) are

$$f \overset{P}{\longmapsto} (n(f;\ \underline{r},t),\ \underline{u}(f;\ \underline{r},t),\ e(f;\ \underline{r},t)) \tag{8'}$$

where

$$n(f;\ \underline{r},t) \qquad = \int d\underline{v}\ \ f(\underline{r},\underline{v},t)$$

$$n(f;\ \underline{r},t)\ \underline{u}(f;\ \underline{r},t) = \int d\underline{v}\ \ \underline{v}\ f(\underline{r},\underline{v},t)$$

$$n(f;\ \underline{r},t)\ e(f;\ \underline{r},t) = \int d\underline{v}\ \ \frac{1}{2}\, v^2\ f(\underline{r},\underline{v},t)$$

and

$$f \overset{Q}{\longmapsto} (s(f;\ \underline{r},t)),$$

where

$$n(f;\underline{r},t)\ s(f;\underline{r},t) = -\int d\underline{v}\ \ f(\underline{r},\underline{v},t)\ \ln f(\underline{r},\underline{v},t)$$

We can verify easily that the equilibrium state f_{eq} corresponding to this generation functional (see Eq. (1)) is the local equilibrium distribution and the thermodynamical structure implied by the generating functional is the structure introduced by the local equilibrium assumption (more specifically, the fundamental thermodynamic relation (12) in the ideal gas thermodynamic relation).

The bracket {,} that we choose is the some bracket as the bracket introduced in Refs. [16] and [17] in the context of the formulation (17). This means that the first line (25) is:

$$\frac{\partial f}{\partial t} = -\frac{\partial}{\partial r_\gamma}\left[(v_\gamma + \frac{u^*_\varphi}{e^*})\ f\right] + \frac{\partial}{\partial v_\gamma}\left[(\frac{1}{2}\, v^2\, \frac{\partial e^*}{\partial r_\gamma} + v_\beta\, \frac{\partial u^*_\beta}{\partial r_\gamma} + \frac{\partial u^*}{\partial r_\gamma})\ \frac{1}{e^*}\ f\right] + \text{collision term} \tag{27}$$

We shall make three remarks concerning Eq. (27).

Remark 1. We note that if we apply on Eq. (27) the projection P introduced in (8'), use $\int d\underline{v}\ v_\alpha\ v_\beta\ f(\underline{r},\underline{v},\underline{t}) = p(\underline{r},t)\delta_{\alpha\beta}$, where p is the pressure arising in the thermodynamical structure corresponding to the generation functional above, and use the thermodynamic structure, then we find that the hydrodynamic fields n, $\underline{u}$, and e do not evolue in time. This means that indeed (27) describes approach to the time evolution governed of the Euler hydrodynamic equations. We can regard Eq. (27) also as the Boltzmann equation modulo the Euler hydrodynamic equations. We note that Eq. (27) is a generalization of the equation (see e.g. Ref. [31])

$$\frac{\partial f}{\partial t} = -\frac{\partial}{\partial r_\gamma}\left[\ (v_\varphi - \underline{u}_\gamma) f\ \right] + \text{collision term} \tag{28}$$

that is the Boltzmann kinetic euqation modulo the first Euler hydrodynamic equation (i.e. the equation $\partial n/\partial t = -\partial/\partial r_\gamma (n\ u_\gamma)$). Since $\underline{u}^*/e^* = -\underline{u}$ (this follows from the conjugation mapping (3')) the first term on the right hand since of Eq. (27) is the same as the first term on the right hand side of Eq. (28).

Remark 2. Let us consider Eq. (27) as nonautonomous equation, the quantities n^*, $\underline{n}^*$, e^* are assumed to evolve in time according to the Euler hydrodynamic equations on which the conjugation mapping (3') has been applied. The following question arises. How close are the trajectories calculated from (27) to the trajectories calculated form (15)?. This is a difficult question that we shall not discuss here. We only note that the zero Chapman-Enskog approximation is the exact asymptotic solution of Eq. (27).

Remark 3. Let us consider the example in which Eq. (14) represents still the Boltzmann equation (15) but Eq. (22) represents the Navier-Stokes-Fourier hydrodynamic equations rather than the Euler hydrodynamic equations. In this case the previous analysis remains valid except that the operator Q given by (8') is replaced by,

$$
\begin{aligned}
n(f;\, \underline{r},t)\; s(f;\, \underline{r},t) = & \int d\underline{v}\; f(\underline{r},\underline{v},t)\; \ln f(\underline{r},\underline{v},t) \\
& + \int d\underline{r}' \int d\underline{v}' \int d\underline{v}\, \Big[B(\underline{r},\, \underline{r}',\underline{v},\underline{v}')\; v_\alpha\, v_\beta\, v'_\beta \\
& + C(\underline{r},\, \underline{r}',\, \underline{v},\, \underline{v}')\, \frac{1}{2} v^2\, \frac{1}{2} v'^2\, v_\alpha\Big] \\
& \frac{\partial}{\partial r'_\alpha} (f(\underline{r}',\underline{v}',t))\; f(\underline{r},\underline{v},t),
\end{aligned}
\qquad (29)
$$

Where B and C are determined by requiring that the equilibrium states (i.e. solutions of Eq. (11)) are the local equilibrium distribution functions corrected by the first Chapman-Enskog correction. We can now readily write equation that corresponds to Eq. (27). This equation is then the Boltzmann equation modulo the Navier-Stokes-Fourier hydrodynamic equations.

Finally, we would like to emphasize that the discussion presented in this section is very preliminary. The observations that we have made shows however a route that can be taken.

4. Concluding remarks

We have argued that the thermodynamical structure that can be associated with a model of a physical system (static or dynamic) arises from an analysis of solutions of a dynamical model of the same physical system but formulated on a more microscopic level of description. Let us call the original level of description level 1 and the more microscopic level of description level 2. The model formulated in the level 1 is called level 1 model, similarly, the model formulated on the level 2 will be called level 2 model. We have seen that the thermodynamical structure associated with the level 1 model is the geometrical structure of the state space used to describe states on the level 1. If we restrict ourselves only to one level of description (say, level 1) we do not expect, in general, any direct relation between the level 1 dynamics and the level 1 thermodynamics. An undirect relation however exists since the level 1 thermodynamical structure is a consequence of the level 2 dynamics and both the level 1 and level 2 models are models of the same physical system. We expect that some radical changes in the behaviour of the modeled physical system will reflect themselves on both level 1 and level 2 (recall for example manifestation of phase transitions in equilibrium thermodynamics and in terms of, say, pair correlation function). Since the level 2 model presents itself on the level 1 in the thermodynamic structure, we expect that some radical changes in the behaviour (instabilities, transition to chaos) can be read also from the thermodynamic structure. This application of the observations made in this paper remains however at this point only a conjecture.

References

1. S. Chandrasekhar, Hydrodynamic and Hydromagnetic Stability, Oxford (1961).
2. I. Müller, Thermodynamics, London, Pitman (1985).
3. E.N. Lorenz, Tellus, 16, 1 (1964).
4. J.W. Gibbs, Collected Works, Longmans, Green and Co., New York (1928).
5. L. Tisza, Generalized Thermodynamics, MIT, Cambridge, MA (1966).
6. H.B. Callen, Thermodynamics, Wiley, New York 1960, 2nd edition (1985).
7. R. Hermann, Geometry, Physics, and Systems, Marcel Dekker, New York (1973).
8. F. Weinhold, J. Chem. Phys. 63, 2479 (1975); 65, 559 (1976).
9. P. Salamon, E. Ihring, R.S. Berry, J. Math. Phys. 24, 2515 (1983).
10 M. Grmela, Methods of Thermodynamics in Dynamical Systems, Preprint, Ecole Polytechnique de Montréal (1987).
11 C. de Dominics and P. Martin, J. Math. Phys. 15, 14 (1964).
12. M.S. Green, J. Math. Phys. 9, 875 (1968).
13. J.D. van der Waals, Ph.D. Disertation, Leiden (1873).
14. N.G. van Kampen, Phys. Rev. 135, A362 (1964).
15. L. Boltzmann, in Wissenschaftlichen Abhandlungen von Ludwig Boltzmann, Chelsea, N.Y., vol. 2 (1968).
16. M. Grmela, Contemp. Math. 28, 125 (1984).
17. M. Grmela, Physica D, 21D, 179 (1986).
18. G. Baumann and T.F. Nonnemacher, Phys. Lett. A 122, 149 (1987).
19. J.E. Marsden and A. Weinstein, Physica D, 4D, 394 (1982).
20. P.J. Morrison, Phys. Lett. A 80, 383 (1980).
21. P.J. Morrison and J.M. Greene, Phys. Rev. Lett. 45, 790 (1980).
22. I.E. Dzyaloshinskii and G.E. Volovick, Ann. Phys. 125, 67 (1980).
23. Z.R. Iwinski and L.A. Turski, Lett. Appl. Sci. Eng. 10, 174 (1976).
24. W. van Saarloos, D. Bedeaux and P. Mazur, Physica 107A, 109 (1981).
25. A.N. Kaufman, Phys. Lett. A 100, 419 (1984).
26. P.J. Morrison, Phys. Lett. A 100, 423 (1985).
27. P.J. Morrison, Physica D, 180D, 410 (1986).
28. L.A. Turski, Phys. Lett. A 125, 461 (1987).
29. D.D. Holm, J.E. Marsden, T. Ratiu and A. Weinstein, Phys. Rep. 123, 1 (1985).
30. M. Grmela, Phys. Lett. A, in press (1988).

31. S. Chapman and T.G. Cowling, "The Mathematical Theory of Non-Uniform Gases", Cambridge (1952).
32. D. Enskog, in Kinetic Theory, vol. 3, ed. S. Brush, Pergamon Press, N.Y. (1972).

A Functional Analytic Approach to an Optimization Problem of Hydrodynamic Propulsion*

H.P. Urbach

Philips Research Laboratories, Postbox 80 000,
NL-5600 JA Eindhoven, The Netherlands

1. Introduction

Several problems in the optimization theory of hydrodynamic propulsion are also of interest to the mathematician. The aim of the research is to determine for a given propulsive device such as a propeller or a wing, that periodic motion by which a prescribed mean value of the thrust is generated at minimum loss of energy. In addition to the demand that a prescribed mean value of thrust is to be generated, the periodic motions among which an optimum is sought are often also subjected to other constraints, for example on their amplitude.

It is desirable to consider large sets of periodic motions and often the sets are so large that they contain infinitely many linear independent motions. The corresponding optimization problem is then infinitely dimensional and functional analysis becomes an indispensable tool in the study.

In SPARENBERG [4] several models are described which are used in the study of the working of several propulsive devices such as the screw propeller and the flexible and rigid wing, and results obtained in optimizing these devices are given.

The problem which we shall discuss here concerns the optimization of small amplitude motions of a rigid profile moving in an unbounded incompressible and inviscid fluid. The flows induced by the motions are calculated using the linear theory. We shall sketch the mathematical difficulties which are involved in optimizing these motions and describe the results obtained. Mathematical proofs will appear in forthcoming papers.

LIGHTHILL [3] and WU [9] have studied optimum thrust generation by a rigid profile performing pure harmonic motions with, in general, both a heaving and a pitching component. SPARENBERG & THOMAS [5] have proved the existence of an optimum motion in the class of all arbitrarily periodic heaving motions which satisfy a constraint on their amplitudes. Since these motions are arbitrarily periodic and not pure harmonic in general, the optimization problem is infinitely dimensional and the existence of an optimum is not trivial.

---

* This research was sponsored by the Netherlands Organization for the Advancement of Pure Scientific Research (Z.W.O.), Grant 00-63-173.

The problem which we shall study in the following is a generalization of the problem in [5]. We also consider arbitrarily periodic motions, but these consist in general of both a heaving and a pitching component. Due to the presence of the pitching component, the existence proof given in [5] cannot be carried over. By introducing an additional constraint on the part of the thrust generated by the suction at the leading edge of the profile, existence of an optimum can be proved. In Section 2 we briefly discuss the problem of existence. In Section 3 a number of properties of the optimum motions are described which can be derived a priori and which therefore are useful in order to check the results obtained by numerical calculations. Furthermore, the character of the optimum motions as a function of the required mean thrust is discussed. Finally, in Section 4 the mathematics used in deriving the properties is sketched briefly.

2. On the Existence of Optimum Motions

We consider the 2D flow problem of a rigid profile of length 2ℓ moving at constant speed U through an unbounded incompressible inviscid fluid while making small periodic motions with period τ_0. The fluid is assumed to be at rest far ahead of the profile. Let (x,y) be a Cartesian coordinate system with respect to which there is a homogeneous incoming flow in the negative x-direction with velocity U at $x = +\infty$ and the position of the profile at time t is given by

$$y = h(x,t) = \frac{\ell}{2}\, a(t) + \alpha(t)x\ , \quad -\ell \leq x \leq \ell\ , \tag{2.1}$$

where a and α are τ_0-periodic dimensionless functions (see Fig. 1).

It is known ([1], [3], [8]) that by using the linear theory and the Kutta condition at the trailing edge, all flow quantities can be expressed in terms of ℓ, U, τ_0 and the functions a and α. The momentary thrust consists of two parts, namely the x-component of the pressure jump integrated across the profile, and the suction force at the leading edge. The latter is always nonnegative, but the former can be negative, even to the extent that the total thrust is negative.

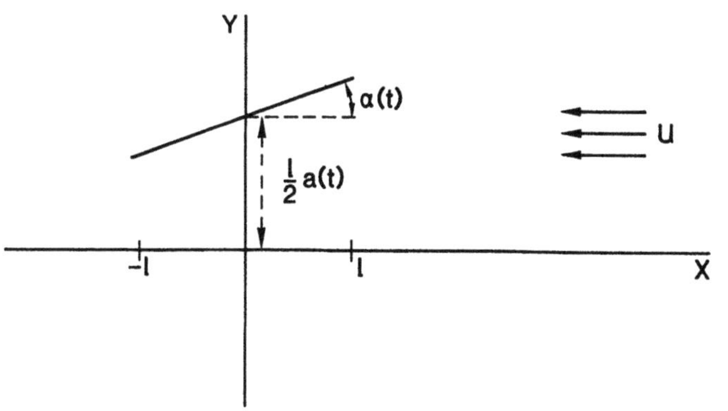

Figure 1. The rigid flat profile of length 2ℓ in uniform parallel flow

The mean increase of kinetic energy in the fluid during a period is the mean loss of energy. Let $C_E(a,\alpha)$, $C_T(a,\alpha)$ and $C_{T^S}(a,\alpha)$ be the mean loss of energy, the mean generated total thrust and the mean suction force, respectively. If we write

$$a(t) = \sum_{m=-\infty}^{\infty} \hat{a}(m)e^{im\omega_0 t}, \quad \alpha(t) = \sum_{m=-\infty}^{\infty} \hat{\alpha}(m)e^{im\omega_0 t},$$

where $\omega_0 = \frac{2\pi}{\tau_0}$, then we have ([6])

$$C_E(a,\alpha) = \sum_{m=-\infty}^{\infty} (\overline{\hat{a}(m)},\overline{\hat{\alpha}(m)})\, E(m\sigma_0) \begin{pmatrix} \hat{a}(m) \\ \hat{\alpha}(m) \end{pmatrix},$$

$$C_T(a,\alpha) = \sum_{m=-\infty}^{\infty} (\overline{\hat{a}(m)},\overline{\hat{\alpha}(m)})\, T(m\sigma_0) \begin{pmatrix} \hat{a}(m) \\ \hat{\alpha}(m) \end{pmatrix},$$

$$C_{T^S}(a,\alpha) = \sum_{m=-\infty}^{\infty} (\overline{\hat{a}(m)},\overline{\hat{\alpha}(m)})\, T^S(m\sigma_0) \begin{pmatrix} \hat{a}(m) \\ \hat{\alpha}(m) \end{pmatrix},$$

where $\sigma_0 = \omega_0 \ell/U$ and $E(\sigma)$, $T(\sigma)$ and $T^S(\sigma)$ are (2,2) complex hermitian matrices of which the elements are $O(\sigma^2)$ for $|\sigma| \to \infty$. The quantities C_E, C_T and C_{TS} are dimensionless. For brevity, some factors appearing in the formulae given in [6] have been absorbed into the matrices.

We introduce the Hilbert space

$$H^1_{\tau_0} = \{f \in C(\mathbb{R});\ f(t+\tau_0) = f(t)\ \forall t,\ f' \in L^2_{loc}(\mathbb{R})\}$$

with scalar product

$$(f,g)_1 = \sum_{m=-\infty}^{\infty} (1+m^2)\hat{f}(m)\overline{\hat{g}(m)} .$$

The motions which we consider will always satisfy $a,\alpha \in H^1_{\tau_0}$. It is clear that C_E, C_T and C_{TS} are continuous on $H^1_{\tau_0} \times H^1_{\tau_0}$.

Let the external force exerted on the profile in order to sustain its motion be applied at point $P = (x_p,y_p)$ which is situated either on the profile itself or on its extension. P has fixed x-coordinate and is rigidly connected to the profile so that at every time t we have

$$y_p(t) = h(x_p,t) = \frac{\ell}{2}\, a(t) + \alpha(t)x_p . \tag{2.2}$$

We shall now state the optimization problem. For given $\ell, U, \tau_0 > 0$ and given $0 < T_0 < \infty$, $0 \leq r < \infty$, $0 < C_\infty \leq \infty$ and $x_p \in \mathbb{R}$ we define the problem:

(P) minimize $C_E(a,\alpha)$ subject to $(a,\alpha) \in A$,

where

$$A = \{(a,\alpha) \in H^1_{\tau_0} \times H^1_{\tau_0};\ C_T(a,\alpha) = T_0,\ C_{T^S}(a,\alpha) \leq rT_0,\ |h(x_p,t)| \leq C_\infty\ \forall t\} .$$

Hence the set of admissible motions all generate mean total thrust T_0, while the part of this thrust generated by the suction does not exceed rT_0 and the amplitude of the motion of point P does not exceed C_∞. $C_\infty = \infty$ corresponds, of course, to the case that the motion of P is not constrained.

For the study of problem (P) the following fact is of fundamental importance (for details see [6]). There exists a bijective linear change of variables $(a,\alpha) \to (z,\zeta)$ of the form:

$$\begin{pmatrix} \hat{z}(m) \\ \hat{\zeta}(m) \end{pmatrix} = Q(m\sigma_0) \begin{pmatrix} \hat{a}(m) \\ \hat{\alpha}(m) \end{pmatrix} \quad \forall m \in \mathbb{Z} , \tag{2.3}$$

where $Q(\sigma)$ is a (2,2) complex matrix which is $O(1)$ for $|\sigma| \to \infty$, such that C_E, C_T and C_{T^S} expressed in (z,ζ) become

$$C_E(z) = \sum_{m=-\infty}^{\infty} 2m^2 \sigma_0^2 |\hat{z}(m)|^2 , \tag{2.4}$$

$$C_T(z,\zeta) = C_E(z) + \sum_{m=-\infty}^{\infty} 4m^2 \sigma_0^2 w(m\sigma_0)\hat{z}(m)\overline{\hat{\zeta}(m)} \equiv C_E(z) + R_T(z,\zeta) , \tag{2.5}$$

$$C_{T^S}(\zeta) = 8|\hat{\zeta}(0)|^2 + \sum_{m=-\infty}^{\infty} 2m^2 \sigma_0^2 |\hat{\zeta}(m)|^2 , \tag{2.6}$$

where $w(\sigma)$ is a complex function which is $o(1)$ for $|\sigma| \to \infty$ and where we introduced the bilinear map R_T. Hence C_E does not depend on ζ and C_{T^S} does not depend on z.

Using the fact that $(a,\alpha) \to (z,\zeta)$ is an isomorphism $H^1_{\tau_0} \times H^1_{\tau_0} \to H^1_{\tau_0} \times H^1_{\tau_0}$ and applying the theory of weak topologies on Hilbert spaces, one can deduce from (2.4), (2.5), (2.6):

(i) C_E is convex and hence weakly under semi-continuous (u.s.c.) on $H^1_{\tau_0}$;
(ii) R_T is weakly sequentially continuous on $H^1_{\tau_0} \times H^1_{\tau_0}$. Hence, C_T is weakly sequentially u.s.c. on $H^1_{\tau_0} \times H^1_{\tau_0}$;
(iii) C_{T^S} is convex, hence weakly u.s.c. on $H^1_{\tau_0}$.

Due to the presence of the equality constraint $C_T(a,\alpha) = T_0$, the set A is not weakly closed in $H^1_{\tau_0} \times H^1_{\tau_0}$. Hence, although C_E is weakly u.s.c. we cannot apply the general theorem which says that a functional which is u.s.c. assumes its infimum on a compact set. Using the transformation (2.3) one can nevertheless show:

Theorem: For all values of the parameters as stated above, problem (P) has at least one solution.

The proof will be given in [6].

Given a solution, every other motion obtained by a phase shift is of course also a solution. It is not true in general that, apart from phase shifts, the optimum motion is unique. In fact, for some values of the parameters two essentially different solutions have been calculated numerically.

Finally, some comments on the constraint on the suction force. When $C_\infty = \infty$ and $r = \infty$ (no constraint on the suction), problem (P) has no solution. In fact, a sequence $\{(z_n,\zeta_n)\}$ such that $\lim_{n\to\infty} C_E(z_n) = 0$ and $C_T(z_n,\zeta_n) = T_0 \ \forall n$, is readily constructed and hence, if (z,ζ) is a solution, then we must have $C_E(z) = 0$ and $C_T(z,\zeta) = T_0$ (> 0). But (2.4) and (2.5) immediately show that this is impossible. When $C_\infty < \infty$ it has not been proved that the constraint on the suction force is necessary for the existence of an optimum, but in any case this constraint is sufficient. Furthermore, the introduction of a bound on the suction at the leading edge is not merely mathematically convenient, but also useful from the mechanical point of view, because motions which generate large suction forces are likely to cause stalling.

3. Some Properties of the Optimum Motions

The ratio of the mean useful power and the mean total power which is to be supplied to the system in order to sustain a given motion (a,α) is the efficiency $\eta(a,\alpha)$ of that motion. In terms of the dimensionless quantities introduced above there holds

$$\eta(a,\alpha) = \frac{C_T(a,\alpha)}{C_E(a,\alpha) + C_T(a,\alpha)}$$

It follows from (2.6) that the mean suction force vanishes if and only if $\hat{\zeta}(m) = 0$ $\forall m \in \mathbb{Z}$. Then, by (2.5): $C_E = C_T$. We conclude therefore, that all motions for which the mean suction force vanishes have efficiency 0.5. Therefore, problem (P) is trivial when $r = 0$.

Now, let $C_\infty = \infty$ and T_0, r be fixed. For every $m \in \mathbb{N}$ one can calculate the optimum m^{th} harmonic motion. Let it be denoted by h_m and let η_m be its efficiency. Motion h_m is unique modulo phase shifts and η_m can be shown to be independent of T_0 and to be a monotonically decreasing function of m with $\lim_{n\to\infty} \eta_m = 0.5$. Since h_1 is more efficient that all other h_m, it follows that in fact h_1 is more efficient than any other τ_0-periodic motion generating the same mean thrust T_0 while the mean suction force does not exceed rT_0.

It is clear that when the optimum pure harmonic motions are scaled such that for each of them the motion of point P has the same amplitude, say 1, the lower harmonics generate smaller mean thrust. Hence, when $C_\infty = 1$, there exists a threshold value $\tilde{T}_0 = \tilde{T}_0(\sigma_0,r,\bar{x}_p)$ with $\bar{x}_p \equiv x_p/\ell$, such that when $T_0 \leq \tilde{T}_0$, the optimum lowest harmonic is the solution of (P) whereas, for $T_0 > \tilde{T}_0$, the optimum lowest harmonic for which the amplitude of the motion of P does not exceed 1 cannot generate the required thrust. So, the question rises: what kind of motion is optimum when $T_0 > \tilde{T}_0$?

The following properties of the optimum motions can be derived a priori (details will be published elsewhere):

a) For every $T_0 > \tilde{T}_0(\sigma_0,r,\bar{x}_p)$ infinitely many Fourier coefficients of both the heaving and the pitching part of the solution are nonzero, except perhaps when for some $m \in \mathbb{N}$: $T_0 = \tilde{T}(m\sigma_0,r,\bar{x}_p)$ in which case it is not excluded that h_m is a solution.

b) Let (a_0,α_0) be a solution of (P) with $C_\infty = 1$ and define

$$I^+ = \{t \in [0,\tau_0];\ \frac{\ell}{2} a_0(t) + \alpha_0(t)x_p = C_\infty\} ,$$

$$I^- = \{t \in [0,\tau_0];\ \frac{\ell}{2} a_0(t) + \alpha_0(t)x_p = - C_\infty\} .$$

Then, when $T_0 \geq \tilde{T}_0$ we have: $I^+ \neq \emptyset$ and $I^- \neq \emptyset$. Hence, there exist at least two times t_1, t_2 per period such that point P achieves the upper bound at time t_1 and the lower bound at t_2.

c) Both heaving and pitching part of every optimum motion are infinitely differentiable at all times $t \notin \partial I^+ \cup \partial I^-$.

Since the optimum lowest harmonic has the highest efficiency, it it to be expected that when $T_0 > \tilde{T}_0(\sigma_0,r,x_p)$ and T_0 is not very large, the lowest mode will be the dominant one in the Fourier expansion of the solution. Numerical calculations carried out for the case

$\tau_0 = 1$ sec, $\ell = 15$ cm, $U = 0.9$ m/sec (hence $\sigma_0 = \frac{\pi}{3}$), $r = 0.4$,

$C_\infty = 0.35\ell = 5.25$ cm, $x_p = 7.5$ cm (quarter chord point)

confirm this. Let b(t) be the function such that

$$\frac{\ell}{2} b(t) = \frac{\ell}{2} a(t) + \frac{\ell}{2} \alpha(t) ,$$

i.e., $\frac{1}{2}b(t)$ is the motion of the quarter chord point made dimensionless by ℓ. For all $T_0 \leq \tilde{T}_0(\frac{\pi}{3},0.4,0.5) = 0.0665$ the solution of (P) is the optimum lowest harmonic with efficiency $\eta_1 = 0.745$ independent of T_0. When $T_0 = 0.0665$, the amplitude of the motion of P equals C_∞. The function $\frac{1}{2}b$ corresponding to this optimum is shown in Fig. 2. Furthermore, the function $\frac{1}{2}b$ of two numerically calculated solutions corresponding to two values of T_0 larger than $\tilde{T}_0$, namely for $T_0 = 0.2$ and

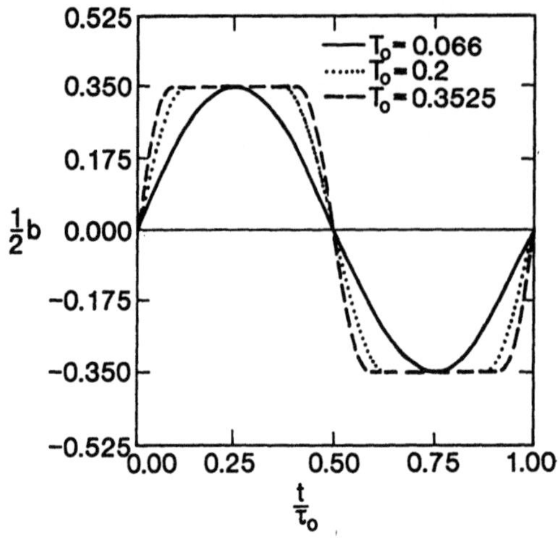

Figure 2. The function $\frac{1}{2}b$ corresponding to the optimum motions for $T_0 = 0.0665$, $T_0 = 0.2$ and $T_0 = 0.3525$

$T_0 = 0.3525$, are also shown in Fig. 2. It is seen that I^+ and I^- are intervals with positive lengths which increase when T_0 is increased.

For all T_0 with $0.0665 < T_0 < 0.3525$, the calculated solutions have dominant lowest mode and the numerical study indicates that the solutions are unique apart from phase shifts. However, when $T_0 = 0.3525$, a second solution was found which has dominant *second* mode. The function $\frac{1}{2}b$ corresponding to this second solution is shown in Fig. 3. When $0.3525 < T_0 < 1.244$, the solutions all have dominant second mode and are unique apart from phase shifts. The function $\frac{1}{2}b$ corresponding to $T_0 = 0.6$ is shown in Fig. 3. At $T_0 = 1.244$ a second solution is found having dominant third mode and for $T_0 > 1.244$ the calculated solutions all have dominant third mode.

In Fig. 4 the efficiency of the optimum motions is drawn as function of the required mean thrust. For $T_0 \leq 0.0665$ the efficiency is $\eta_1 = 0.745$ whereas for $T_0 > 0.0665$ the efficiency decreases rapidly with T_0. At the transitional values 0.3525 and 1.244 the slope of the efficiency curve suddenly changes. $T_0(\frac{2\pi}{3},0.4,0.5)$ happens to be identical to 1.244 and the solution having dominant second mode is

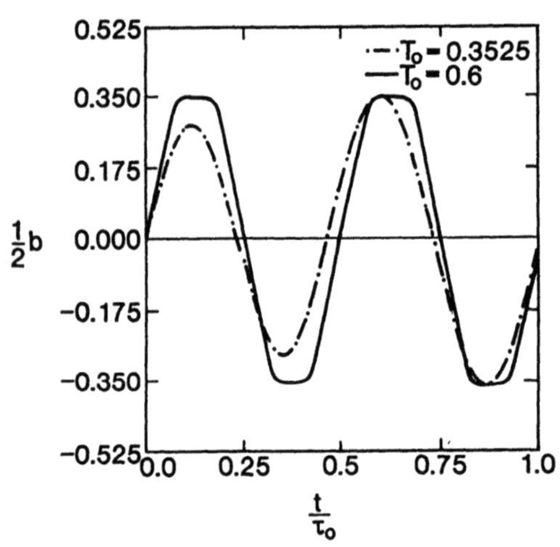

Figure 3. The function $\frac{1}{2}b$ corresponding to the optimum motions for $T_0 = 0.3525$ and $T_0 = 0.6$

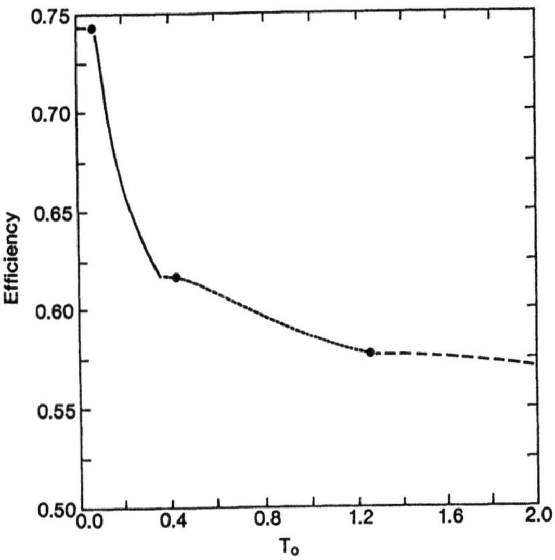

Figure 4. The efficiency of optimum motions as a function of T_0. The continuous part and the left and right dashed parts of the curve correspond to solutions having dominant first, second and third mode, respectively. The dots indicate the efficiency of the optimum first, second and third harmonic motion.

in fact the optimum second harmonic. Finally, we remark that all calculated solutions were observed to generate the maximum allowed mean suction force rT_0. Hence, the constraint on the suction force was always found to be active.

4. On the Lagrange Multiplier Rule

The equation satisfied by all solutions of the optimization problem is the Lagrange multiplier rule. However, the classical Kuhn-Tucker theorem cannot be applied in case $C_\infty < \infty$ because the constraint on the amplitude of the motions contains the maximum norm and this functional is not Gateaux differentiable.

It is possible to derive a generalized Lagrange multiplier rule. An attractive method to do this is using the Dubovitskii-Milyutin lemma which is a direct corollary of the Hahn-Banach theorem (see [2]). The advantage of using this lemma instead of other theorems of nonsmooth analysis is that it has a geometrical interpretation which makes it relatively easy to use. Generally speaking, the Dubovitskii-Milyutin lemma is applicable when constraints which are nonsmooth are convex.

From the thus obtained generalized Lagrange multiplier rule, the properties a), b) and c) stated in Section 3, can be deduced. The details will appear elsewhere.

References

1. R.F. Bisplinghoff, H. Ashley, R.L. Halfman: *Aeroelasticity* (Addison-Wesley, Reading)
2. H. Halkin: JOTA. 6 (1970), 138-148
3. M.J. Lighthill: J. Fluid Mech. 44 (1970), 265-303

4. J.A. Sparenberg: *Principles of Hydrodynamic Propulsion* (Martinus Nijhoff, The Hague, Boston, Lancaster 1984)
5. J.A. Sparenberg, G.E.F. Thomas: Math. Meth. Appl. Sci. 3 (1981), 424-434
6. H.P. Urbach, to appear in J. Appl. Math. Opt.
7. T.Y. Wu: J. Fluid Mech. 46 (1971), 337-355
8. T.Y. Wu: J. Fluid Mech. 46 (1971), 521-544

Nonlinear Problems in the Kinetic Theory of Gases

C. Cercignani
Dipartimento di Matematica, Politecnico di Milano
Piazza Leonardo da Vinci, 32, I-20133 Milano, Italy

1-Introduction.

In the previous meeting of this series, held in Bad Honnef, I presented a survey lecture on "Half-space Problems in the Kinetic Theory of Gases" [1]. In the concluding remarks of that paper, I pointed out that it was not hopeless to try to obtain new results for nonlinear boundary value problems in kinetic theory. I am glad to report that there has been progress in this direction. It is for this reason that, although there are interesting new results on the initial value problem for nonlinear kinetic equations [2-4], I shall be dealing here with the recent progress on boundary value problems for nonlinear kinetic equations. These results refer to discrete velocity models and are the outcome of my joint work with R. Illner and the late Marvin Shinbrot [5,6]. A more recent result, obtained in collaboration with these authors and M. Pulvirenti [7] deals with half-space problems and is thus directly related to my survey [1].

The plan of the paper is as follows. In Sect.2 the basic notations about the discrete velocity models will be recalled. Sect.2 is devoted to the proof of an existence theorem in a slab. Sect. 3 deals with a particular but interesting two-dimensional problem. Finally the half-space problem is dealt with in Sect. 4.

2- Discrete Velocity Models.

A discrete velocity model [8] is a set of partial differential equations

$$\frac{\partial f^i}{\partial t} + \underline{u}_i . \nabla f^i = C^i(f) \quad , \quad (i=1,\ldots,n) \quad , \tag{2.1}$$

where $f = (f^1,\ldots,f^n)$ are the particle densities associated with the n admissible velocities $\underline{u}_i \in \mathbb{R}^3$, and where each C^i has the form

$$C^i(f) = \sum_{j,k,l} A^{ij}_{kl} (f^k f^l - f^i f^j) \quad . \tag{2.2}$$

What is actually needed is the following. Let $\mathscr{C}^o$ denote the continuous functions on a set with the usual topology of uniform convergence. Denote the non-negative functions in $\mathscr{C}^o$ by $\mathscr{C}^o_+$, and denote the Cartesian product of n copies of $\mathscr{C}^o_+$ by $(\mathscr{C}^o_+)^n$. Let

$$\rho(f) = \sum_i f^i \tag{2.3}$$

and define

$$C^i_\gamma(f) = C^i(f) + \gamma f^i \rho(f) \quad . \tag{2.4}$$

My first assumption will be that for $\gamma > 0$ large enough C^i_γ maps $(\mathscr{C}^o_+)^n$ continuously into itself. I also need the three conservation laws of mass, momentum and energy:

$$\sum_i C^i(f) = 0 \quad , \tag{2.5}$$

$$\sum_i \underline{u}^i C^i(f) = 0 \quad , \tag{2.6}$$

$$\sum_i |\underline{u}_i|^2 C^i(f) = 0 \quad . \tag{2.7}$$

There is also an analogue of the Boltzmann lemma [9]:

$$\sum_i \log f^i C^i(f) \leq 0 \quad . \tag{2.8}$$

In Sects. 3 and 4 I shall only need Eq. (2.5) and the first component of Eq. (2.6).

3. <u>The Slab Problem.</u>

In this section I shall be concerned with *steady*, one-dimensional models for which f is independent of t and independent of y and z, so that Eq. (2.1) takes the form

$$\xi_i \cdot \frac{df^i}{dx} = C^i(f) \ , \ (i=1,\ldots,n) \qquad (2.9)$$

where ξ_i are the x-components of the vectors $\underline{u}_i$. Eq. (2.9) is to be solved in the slab $0<x<d$ under the additional assumption that

$$\xi_i \neq 0 \ , \ i=1,\ldots,n \ . \qquad (2.10)$$

Let me state now the boundary conditions under which Eq. (2.9) is solved. The simplest set is (problem A):

$$\text{if } \xi_i>0, \text{ then } f^i(0)=\alpha_i ; \quad \text{if } \xi_i<0, \text{ then } f^i(d)=\beta^i . \qquad (2.11)$$

where α_i and β_i are given non-negative constants.

The second set of boundary conditions is the following (Problem B):

$$\text{if } \xi_i>0, \text{ then } f^i(0)=\alpha_i ; \quad \text{if } \xi_i<0, \text{ then } f^i(d)=\beta^i \sum_j{}^{+} \xi_j f^j(d). \qquad (2.12)$$

Here by Σ^+ I mean that the sum is over the *positive* velocities ξ_i. and I assume not only that the α_i and β_i are non-negative, but also that the β_i are normalized by the condition:

$$\Sigma^- \xi_i \beta^i = -1 \qquad (2.13)$$

where, as the notation indicates, the sum is over the negative values of the velocity component. Notice that

$$\sum_i \xi_i f^i(d) = 0 \qquad (2.14)$$

i.e. the net flow at the upper wall is zero. This means that there is no flow through the other wall as well, because of the equation of mass conservation implied by Eq. (2.5).

In order to prove existence for Problem B, I define C_γ^i as in Eq. (2.4), with γ large enough and solve the equation

$$\xi_i \frac{df^i}{dx} + \gamma f^i \rho(f) = C_\gamma^i(f) , \qquad (2.15)$$

with the boundary condition (2.12). This problem B_γ is clearly equivalent to problem B. The first step is to prove

Lemma 1- Let $g \in (\mathscr{C}^o)^n$. For γ large enough, the linear equation

$$\xi_i \frac{df^i}{dx} + \gamma f^i \rho(|g|) = C_\gamma^i(|g|) , \qquad (2.16)$$

with the boundary conditions (2.12) has a unique solution $f \in (\mathscr{C}_+^o)^n$.

The proof is trivial, being based on the solution of two initial value problems (to be solved in sequence) for a system of ordinary differential equations.

It is now possible to prove the main theorem on Problem B, *i. e.*:

Theorem 2. Problem B has a solution in $(\mathscr{C}_+^o)^n$.

Proof. It is sufficient to show that Problem B_γ has a solution. For γ large enough, define the mapping $T:(\mathscr{C}_+^o)^n \to (\mathscr{C}_+^o)^n$ by the formula Tg=f, where f is the solution of Eq. (2.16) with the boundary conditions (2.12). It is easy to conclude that T is continuous and its restriction to bounded sets is compact . Now, let f be any solution of the equation $f=\lambda Tf$, with $0<\lambda<1$. Then, f is non-negative, differentiable and satisfies Eq. (2.15) with λC_γ^i in place of C_γ^i with the boundary conditions (2.12) with $\lambda\alpha^i$ in place of α^i. Let $c=\max|\xi_i|$ and $p(f)=\sum_i \xi_i^2 f_i$. Using the properties of C^i,expressed by Eqs. (2.5) and (2.6) and Eqs. (2.12-13) it is easy to find the *a priori* inequality:

$$p(x) \le 2c\textstyle\sum^+ \xi_i \alpha^i \qquad (2.17)$$

This shows that any solution of the equation $f=\lambda Tf$ is bounded.

The result now follows from the version of the Leray-Schauder theorem proposed by Schaefer [10,11].

A similar argument can be used [5] to prove

Theorem 3. Problem A has a solution in $(\mathscr{C}^{o}_{+})^{n}$.

It is to be remarked that the solutions of Problems A and B may not be unique . However it is easy to prove uniqueness for d small enough because then T is a contraction.

3. Existence for a Two-dimensional Model.

In this section I describe the solution of a boundary value problem associated with a 4-velocity model in a rectangle R=[0,a]x[0,b] [6]. The problem to be solved reads as follows:

$$\frac{\partial f^1}{\partial x} + f^1 f^2 = f^3 f^4, \qquad f^1(0,y) = \phi^1(y) \quad , \tag{3.1}$$

$$-\frac{\partial f^2}{\partial x} + f^1 f^2 = f^3 f^4, \qquad f^2(a,y) = \phi^2(y) \quad , \tag{3.2}$$

$$\frac{\partial f^3}{\partial x} + f^3 f^4 = f^1 f^2, \qquad f^3(x,0) = \phi^3(x) \quad , \tag{3.3}$$

$$-\frac{\partial f^4}{\partial x} + f^3 f^4 = f^1 f^2, \qquad f^4(x,b) = \phi^4(x) \quad . \tag{3.4}$$

In order to use a Leray-Schauder technique as in the case of the slab , one first studies the following boundary value problem involving only derivatives with respect to x:

$$\frac{\partial f^1}{\partial x} + f^1 f^2 = h^1 \quad , \qquad f^1(0) = \phi^1 \quad , \tag{3.5}$$

$$-\frac{\partial f^2}{\partial x} + f^1 f^2 = h^2 \quad , \qquad f^2(a) = \phi^2 \quad , \tag{3.6}$$

where the functions h^i and the boundary data ϕ^i are given.

The first result is contained in

Lemma 4. Let $\phi^i \in \mathbb{R}_+$, $h^i \in \mathscr{C}^o_+[0,a]$. Then the problem (3-5)-(3.6) has a unique solution $f^i \in \mathscr{C}^o_+[0,a]$.

The proof uses an easy monotonicity argument [6].

Continuous dependence upon the data is proved in

Lemma 5. Let (f^1,f^2) denote the solution of the problem (3.5)--(3.6) and (F^1,F^2) the solution of the same problem with (H^1,H^2) in place of (h^1,h^2) and (Φ^1,Φ^2) in place of (ϕ^1,ϕ^2). Take $\varepsilon>0$. Then, there exists a $\delta>0$ such that

$$|F^1-f^1| + |F^2-f^2| < \varepsilon \tag{3.7}$$

if

$$|H^1-h^1|+|H^2-h^2|+|\Phi^1-\phi^1|+|\Phi^2-\phi^2| < \delta \quad . \tag{3.8}$$

Here $|.|$ is the usual sup norm in $\mathscr{C}^o_+$ and ε can be chosen to depend only on a,δ and $M=\max(|H^1|,|h^1|,|H^2|,|h^2|,|\Phi^1|,|\Phi^2|,|\phi^1|,|\phi^2|)$.

The proof depends upon the circumstance that the equations for $g^i=F^i-f^i$ can be solved in terms of the other quantities bounded by M and the differences bounded by δ.

It is now possible to prove

Lemma 6- Let $\phi^i \in \mathscr{C}^o_+$, $h^i\in \mathscr{C}^o_+$. Then Eqs.(3.5-6) have a unique solution $f^i\in\mathscr{C}^o_+$. With the boundary data ϕ^i fixed, let S_ρ denote the set of all solutions of (3-5)-(3.6) with $(h^1,h^2) \in {}_+B^{01}_\rho$ (the cone of positive functions in the ball of radius ρ in the set of all functions continuous in x and differentiable in y). Then the closure of S_ρ is a compact subset of $(\mathscr{C}^o_+)^2$.

The set S_ρ is equicontinuous because f^i and $\partial f^i/\partial x$ are bounded and the moduli of continuity of h^1 and h^2 are controlled by their membership in B^{01}_ρ. Lemma 5 immediately leads to the conclusion.

One has merely to make the lexicographic change of replacing x by y and the indices 1 and 2 by 3 and 4 to prove similar results for the system

$$\frac{\partial f^3}{\partial y} + f^3 f^4 = h^3 , \qquad f^3(0) = \phi^3 \tag{3.9}$$

$$-\frac{\partial f^4}{\partial y} + f^3 f^4 = h^4 , \qquad f^4(a) = \phi^4 \tag{3.10}$$

One can state as an immediate corollary :

Lemma 7 Let $(g^1, g^2, g^3, g^4) \in (\mathcal{E}^0_+(R))^4$. Then the equations

$$\frac{\partial f^1}{\partial x} + f^1 f^2 = g^3 g^4 , \qquad f^1(0,y) = \phi^1(y) , \tag{3.11}$$

$$-\frac{\partial f^2}{\partial x} + f^1 f^2 = g^3 g^4 , \qquad f^2(a,y) = \phi^2(y) , \tag{3.12}$$

$$\frac{\partial f^3}{\partial y} + f^3 f^4 = g^1 g^2 , \qquad f^3(x,0) = \phi^3(x) , \tag{3.13}$$

$$-\frac{\partial f^4}{\partial y} + f^3 f^4 = g^1 g^2 , \qquad f^4(x,b) = \phi^4(x) , \tag{3.14}$$

have a solution $f=(f^1,f^2,f^3,f^4) \in (\mathcal{E}^0_+(R))^4$.

In proving this lemma it is clear that solving the above system means defining two mappings: one, to be denoted by T_x, of the pair (g^3,g^4) into (f^1,f^2); the other, T_y, of (g^1,g^2) into (f^3,f^4). Finally the solution of the system defines a further mapping, T, of (g^1,g^2,g^3,g^4) into (f^1,f^2,f^3,f^4). It is clear that any solution of (3.1)-(3.4) is a fixed point of T.

One can prove now

Lemma 8 The operator T^2 is compact.

In fact this amounts to showing that $T_x T_y$ and $T_y T_x$ are compact. This easily follows from Lemma 6 [6].

Next, by a suitable use of boundary conditions and application of Schaefer's theorem [10,11] one proves [6]:

Lemma 9. The operator T^2 has a fixed point in $(\mathcal{E}^0_+)^4$.

The desired results, *i.e.* the existence of a fixed point of T rather than of T^2, is obtained by looking at the structure of T^2. In fact, this operator is such that if (h^1,h^2,h^3,h^4) is a fixed point of T^2 then $T_x T_y$ carries the pair (h^1, h^2) into itself. If one calls (g^3, g^4) the pair into which T_y carries (h^1, h^2), then $f \equiv (h^1, h^2, g^3, g^4)$ is a fixed point of T and hence a solution of the problem.

Theorem 10. Let $(\phi^1, \phi^2) \in (\mathscr{C}_+^0(0,b))^2$, $(\phi^3, \phi^4) \in (\mathscr{C}_+^0(0,a))^2$. Then the problem (3.1-4) has a solution $f=(f^1, f^2, f^3, f^4) \in (\mathscr{C}_+^0(R))^4$.

4. The half space problem.

The half space problem to be treated in this section is a limiting case of the slab problem treated in Sect. 2. One drops the boundary condition at x=d and solves the equation in $(0,\infty)$ with the condition that the solution remains bounded. It is possible to show [7] that there are, for given data at x = 0, solutions which, when $x \to \infty$, tend to an equilibrium solution m_i (a discretized version of a Maxwellian [8]). To this end one approximates the half-space problem with problems in a slab, where the boundary condition at x=d is chosen to be of a suitable form (e.g. $f^i(d)=0$ for $\xi_i<0$). Then one proves

Theorem 11. There are boundary conditions at x=0 leading to nonnegative, continuous solutions of the slab problem possessing an upper bound independent of d.

The proof proceeds as in Sect. 2; a look at the estimates indicates that they do not depend on d.

In order to pass to the limit one chooses a sequence of values of d increasing to ∞. The family of corresponding solutions $\{f_n\}$ is bounded and equicontinuous. Then one can extract a subsequence converging uniformly on every bounded interval and pointwise in $[0,\infty)$. The limit f solves Eq. (2.1), satisfies the boundary condition at x=0 and is non-negative. In order to show that it approaches a Maxwellian distribution one first shows:

Theorem 12. There are bounded, non-negative solutions f of Eq. (2.1) satisfying the boundary condition $f^i=\alpha^i$ for $\xi^i>0$ at x=0 and such that, when $x \to \infty$, $\text{dist}(f(x),M) \to 0$, where M is the Maxwellian manifold associated with the invariants of f.

Only the last part of the statement has not been proved. In order to prove it, we consider the steady version of the H-theorem, consequence of Eq.(2.8). According to this $J(f) = \sum_i \xi_i f_i \log f_i$ is nonincreasing and has a derivative in L^1; hence there is a sequence of x_k tending to infinity such that $dJ/dx(x_k)$ goes to zero when $k \to \infty$. From this sequence we can extract a subsequence (to be again denoted by $\{x_k\}$) such that $f(x_k)$ tends to a limit g. Using Eq. (2.8) we conclude that g is Maxwellian. Clearly g has the invariant fluxes of f. By contradiction one shows [7] that any other sequence $\{f(x_k)\}$ must approach the manifold M of Maxwellians having the invariant fluxes of g and the theorem is proved.

In the case $j(f)>0$, the Maxwellian manifold consists of at most finitely many points. But then $\text{dist}(f(x),M)$ can only converge to zero as $x \to \infty$ if f actually converges to a Maxwellian. Hence the following result holds:

Theorem 13. If $j(f)>0$ (where j was defined in Sect.2), there is a Maxwellian m with invariants of f such that f(x) tends to m.

A well known result on steady one-dimensional flows of an Euler gas indicates that if transversal bulk motion is not allowed there are exactly two Maxwellians having given invariant fluxes; one of them is subsonic, the other supersonic. The asymptotic analyses performed with linear tools [1] lead to the conjecture that the limit actually reached is the subsonic Maxwellian.

REFERENCES

1. C. Cercignani: in Trends in Applications of Pure Mathematics to Mechanics, ed. by E. Kröner and K. Kirchgässner, LNP **249** (Springer, Berlin, 1986) p. 35.

2. C. Cercignani and M. Shinbrot: Comm. Part Diff. Eq., 12, 307 (1987).

3. C. Cercignani: J. Stat. Phys. (to appear, 1987).

4. C. Cercignani: J. Stat. Phys. (to appear, 1988).

5. C. Cercignani, R. Illner and M. Shinbrot: Duke Math. J., 55, 889 (1987).

6. C. Cercignani, R. Illner and M. Shinbrot: Comm. Math. Phys. (to appear, 1988).

7. C. Cercignani, R. Illner, M. Pulvirenti and M. Shinbrot: submitted to J. Stat. Phys.

8. R. Gatignol: Théorie cinétique des gaz à répartition discrète de vitesses LNP **36**, (Springer, Berlin, 1975).

9. C. Cercignani: The Boltzmann Equation and its Applications (Springer, New York, 1987).

10. H. Schaefer: Math. Annalen 129, 415 (1955).

11. D. R. Smart, Fixed Point Theorems (Cambridge University Press, London, 1974).

Index of Contributors

MIX
Papier aus verantwortungsvollen Quellen
Paper from responsible sources
FSC® C105338

If you have any concerns about our products,
you can contact us on
ProductSafety@springernature.com

In case Publisher is established outside the EU,
the EU authorized representative is:
Springer Nature Customer Service Center GmbH
Europaplatz 3, 69115 Heidelberg, Germany

Printed by Libri Plureos GmbH
in Hamburg, Germany